Modern
Textile
Characterization
Methods

INTERNATIONAL FIBER SCIENCE
AND TECHNOLOGY SERIES

Series Editor

MENACHEM LEWIN

Hebrew University of Jerusalem
Jerusalem, Israel

Herman F. Mark Polymer Research Institute
Polytechnic University
Brooklyn, New York

Editorial Advisory Board

Modern Textile Characterization Methods

edited by

Mastura Raheel
University of Illinois at Urbana-Champaign
Urbana, Illinois

CRC Press
Taylor & Francis Group
Boca Raton London New York

CRC Press is an imprint of the
Taylor & Francis Group, an **informa** business

CRC Press
Taylor & Francis Group
6000 Broken Sound Parkway NW, Suite 300
Boca Raton, FL 33487-2742

First issued in paperback 2019

ISBN-13: 978-0-8247-9473-6 (hbk)
ISBN-13: 978-0-367-40143-6 (pbk)

Library of Congress Cataloging-in-Publication Data

Modern textile characterization methods / edited by Mastura Raheel.
     p. cm. —(International fiber science and technology series ;13)
    Includes index.
    ISBN 0–8247–9473–7 (hardcover : alk. paper)
    1. Textile fabrics—Testing. 2. Textile fibers—Testing. 3. Non-destructive testing. I. Raheel, Mastura. II. Series.
TS1449.M5763 1995
677′.0287—dc20
                                             95–44165
                                               CIP

Visit the Taylor & Francis Web site at
http://www.taylorandfrancis.com

and the CRC Press Web site at
http://www.crcpress.com

Preface

New developments in fiber science and technology have resulted in fibers with tailored properties, thus expanding their uses beyond the domain of conventional textiles. The classical as well as nonclassical applications of fiber assemblies have placed stringent standards of performance that require precise monitoring of structure–property relationships in fibrous systems. These monitoring techniques must result in objective measurements that are based on sound scientific principles. A large body of knowledge exists on the physical, mechanical, and chemical properties of textiles/fiber assemblies. Also, standard methods have been developed by several national and international organizations such as the American Society for Testing and Materials (ASTM), the American Association of Textile Chemists and Colorists (AATCC), the European Standardisation Committee (CEN), the International Standards Organization (ISO), and others to assess fiber/textile physical, mechanical, chemical, and selected aesthetic properties.

Recently major strides have been made in the development and use of state-of-the-art engineering methods to characterize and assess the properties of polymers, single fibers, and textile assemblies at various stages of development, processing, manufacture, and end use. These methods are neither routinely used by the textile industry nor are all included in books dealing with standard test methods for fibers and textiles.

This volume attempts to bring together selected state-of-the-art methods, along with the scientific basis of these methods and their applications in the vastly diversified field of polymers, fibers, and textiles. Included in this volume are contributions by renowned researchers on polymer characterization methods such as scanning and transmission electron microscopy (SEM and TEM), x-ray diffraction, differential scanning calorimetry (DSC), and nuclear magnetic resonance (NMR). This book also examines surface characterization of fibers using SEM; chromatographic techniques to identify fibers and evaluate internal pore volume in fibers and pore structure patterns in textiles with emphasis on their applications in dyeing, finishing, and composite-making technologies; micromeasurement of single-fiber mechanical properties; objective measurement of fabric hand and its applications; color measurement and control; and methods for evaluating chemical and microbiological barrier properties of textiles.

It is hoped that this volume will fill the gap that exists between the currently employed standard methods for textile testing and the recent advances that have been made in methodology development to assess the characteristics of polymers, single fibers, fibrous systems, and associated processes. It is assumed that the readers are familiar with the fundamentals of fiber science and textile processes. The book should be very useful to those individuals and organizations involved with research and development, process control, and product analysis in the polymer, textile, and related industries. It is hoped this will serve as a valuable reference book for education and research in areas of polymers, textiles, and related sciences.

Mastura Raheel

Contents

Contributors

Keith R. Beck, Ph.D.   Professor, Department of Textile Engineering, Chemistry, and Science, College of Textiles, North Carolina State University, Raleigh, North Carolina

Noelie R. Bertoniere, Ph.D   Research Leader, Textile Finishing Chemistry Research Unit, Southern Regional Research Center, Agricultural Research Service, United States Department of Agriculture, New Orleans, Louisiana

Peter L. Brown   Associate, Fabrics Division, W. L. Gore & Associates, Inc., Elkton, Maryland

Patrick Tak Fu Chong, Ph.D., F.S.D.C.   Color Scientist, Research and Development, Spartan Mills, Spartanburg, South Carolina

Phillip H. Geil, Ph.D   Professor, Department of Materials Science and Engineering, University of Illinois at Urbana-Champaign, Urbana, Illinois

Wilton R. Goynes   Research Chemist, Department of Fiber Physics and Biochemisty, Southern Regional Research Center, Agricultural Research Service, United States Department of Agriculture, New Orleans, Louisiana

Ian R. Hardin, Ph.D.   Professor and Head, Department of Textiles, Merchandising, and Interiors, University of Georgia, Athens, Georgia

Sueo Kawabata, Dr.Eng. Professor, Department of Materials Science, The University of Shiga Prefecture, Hikone City, and Professor Emeritus, Kyoto University, Kyoto, Japan

Bernard Miller, Ph.D Vice President, Research, TRI/Princeton, Princeton, New Jersey

Masako Niwa Professor, Department of Textile and Apparel Science, Nara Women's University, Nara, Japan

Mastura Raheel, Ph.D. Professor of Textile Science, Department of Natural Resources and Environmental Sciences, University of Illinois at Urbana-Champaign, Urbana, Illinois

Ludwig Rebenfeld, Ph.D. President Emeritus and Research Associate, TRI/Princeton, Princeton, New Jersey

Jeffrey O. Stull, M.S. Ch.E. President, International Personnel Protection, Inc., Austin, Texas

Ilya Tyomkin, Ph.D. Senior Scientist, TRI/Princeton, Princeton, New Jersey

Yiqi Yang, Ph.D. Professor, Department of Chemical, Energy and Environmental Research, The Institute of Textile Technology, Charlottesville, Virginia

Modern
Textile
Characterization
Methods

1

Introduction: Developments in Textile Characterization Methods

MASTURA RAHEEL   University of Illinois at Urbana-Champaign, Urbana, Illinois

Textile characterization must take into consideration an in-depth understanding of the nature of fiber-forming materials (polymers), fiber structure, its physical, mechanical, and chemical properties, and how these properties relate to further engineering operations that result in fabrics/textiles and finished products. The end-use performance of finished products will depend upon all these factors, and can be predicted on the basis of fundamental theories of fiber science and sound characterization methods.

Fundamental theories of fiber science have evolved from the classical theories of physics, chemistry, polymer science, and engineering. The greatest advances in textile materials have been where linear laws of classical physics or physical chemistry can be applied. The difficulties increase when it becomes necessary to take account of quantum and relativistic effects and chemical interactions. Textile systems generally are extraordinarily complex, and the effects of treatments almost invariably go beyond the bound of linearity. Thus predictive mathematical models may very well be nonlinear or only yield empirical statistical correlations. Major strides have been made in the last decade or so in the use of sophisticated methods and mathematical models to characterize textile materials and predict end-use performance. Textile characterization is important at all stages of textile production and processing in order to achieve a product that meets perceived performance needs. The aim of textile characterization is to understand the material structure and behavior as well as the processes sufficiently to be able to predict their consequences, and so to be able to set up control techniques that will lead to products with specified properties.

There are numerous well-known organizations, such as the International Standards Organization (ISO), the American Society for Testing and Materials

(ASTM), the American Association of Textile Chemists and Colorists (AATCC), the European Standardization Committee (CEN), and various others, that develop standard test methods for evaluating and predicting performance of fibrous systems. However, generally, there is a significant time lag between the developments in textile characterization methods and their acceptance as standard methods. The literature is replete with innovative uses of standard methods as well as newer methods and instrumentation for characterizing polymers, fibers, textiles, and their auxiliaries. It is not the intent of this book to include all physical, mechanical, and chemical methods for characterization of fibrous materials, but rather to focus on recent developments in selected characterization methods and their applications to fibrous systems, based on evolving theories of physical, chemical, and engineering sciences.

The book begins with polymer characterization methods. Polymers, the fiber-forming materials, have (or can be manipulated to have) characteristic structures and physicochemical properties. These features have profound impact on fiber and textile properties. In Chapter 2 P. H. Geil, a renowned polymer scientist, discusses in great detail polymer characterization methods. The specific areas of polymer characterization covered in Chapter 2 include (1) chemical structure, including composition and configuration, (2) physical structure, including crystallinity and morphology-related aspects, and (3) physiochemical properties.

Geil mentions the use of traditional methods of characterizating various aspects of polymers but focuses mainly on recent advances in polymer characterization methods. For example, polymer chemical composition and configuration analysis begins with the traditional analytical chemistry techniques of elemental analysis by atomic absorption spectroscopy, x-ray dispersive analysis, and reaction of specific groups in a polymer with specific reagents, but the thrust of his discussion is on Fourier-transform infrared spectroscopy (FTIR) and FT nuclear magnetic resonance (FT-NMR) methods. He explains the theoretical basis of these analytical techniques and provides practical guidance about sample preparation, the analytical technique, and interpretation of results. Also, he describes the usefulness of these techniques in studying textile fibers. Molecular weight determination is described using chromatography processes and also by simpler techniques such as solution and melt viscosity methods. The significance of molecular weight characterization on solution spinning and melt spinning of fibers is described.

The physical structure of polymers and fibers requires a range of techniques for characterization because of the range of size scales, particularly in fibers. The structures of interest fall into the size scale of the individual molecular segment; the relative number of regular and random conformations and their arrangement in space, that is, the degree of crystallinity and orientation; the size and shape of the crystalline and amorphous regions; and the organization and interaction of these crystals in larger structures. Characterization of all these aspects is discussed in great detail with illustrations and examples of polymers and fibers by using a range of techniques. The techniques described include FTIR, electron diffraction (ED), x-ray diffraction [both wide-angle (WAXD) and small-angle scattering (SAXS)], and electron microscopy (EM) [both scanning (SEM) and transmission (TEM)]; also many probe microscopes are described. Geil cautions about the problems in utilizing several techniques (especially electron diffraction) that primarily depend upon appropriate sample preparation. He suggests sample preparation methods and describes their representative results and potential difficulties.

In Chapter 3, W. R. Goynes discusses the importance of structural characterization of fibers and textiles using scanning electron microscopic (SEM) techniques. He focuses on the specifics of sample preparation and microscope operating conditions, bringing to attention the difficulties of obtaining meaningful signals and interpreting those signals. The significance of back-scattered electrons in interpreting changes in elemental composition of fibers/materials is introduced, and the importance of x-rays for elemental analysis is emphasized. Goynes concludes with examples of textile characterization using SEM as a powerful tool. It is well known that surface morphology and characteristic structural features of fibers are dramatically revealed by scanning electron microscope; however, Goynes also presents the effects of physical and chemical treatments on changes in the fibers' characteristic features. This characterization method also provides valuable information regarding process evaluation and product quality control.

Chapter 4 focuses on analytical pyrolysis as a technique to identify and detect small changes in polymers, fibers, and other textile auxiliaries. Analytical pyrolysis (or thermolysis) is a nonoxidative process in which polymers or large molecules break down into characteristic smaller molecules. Instrumental analysis of these pyrolysates, which are structure-specific volatile compounds, provides information about the structure and identity of the parent compound. I. R. Hardin discusses the mechanism of pyrolysis, the types of reactions that occur to give rise to complex mixtures of products, and how these volatile fragments are separated and analyzed using gas chromatography (GC) alone or in conjunction with mass spectrometry or Fourier-transform spectroscopy. He elaborates on these techniques with examples of identifying or detecting small changes in polymers, finishes, and dyes.

Chromatographic and spectroscopic methods are employed for characterization of a wide variety of polymers, fibers, textiles, and textile auxiliaries. In Chapter 5, Y. Yang presents the scientific basis and application of conventional liquid chromatography (LC) for dye identification, separation, and purification. Also, as a powerful tool, LC is employed for analysis of textile finishing processes such as flame retardant, stain resistant, durable press, and others. Packing textile material into the column as a stationary phase is an innovative method for the investigation of pore structure and dyeing and finishing behavior of the specific textile employed as a stationary phase. This technique is useful as well for studying dyeing and finishing mechanisms in textile systems. Yang provides the basic concepts of liquid chromatography as a tool for textile and related materials characterization, and focuses on pore structure and surface area analysis as it relates to textile wet processes. The subjects of color identification, separation and purification, dyeing thermodynamics, sorption isotherms, dye compatibility and dye–fiber interactions are discussed in depth. In a related topic, K. R. Beck, in Chapter 6, focuses on characterization of durable-press finishes for cellulosic textiles using chromatographic and spectrophotometric methods. Beck, a pioneer in the use of chromatographic techniques for analyzing textile finishes, describes analysis of durable press chemicals utilizing thin-layer, gas, and high-performance liquid chromatographic methods as well as spectroscopic methods. The spectroscopic methods included are ultraviolet-visible, near infrared, infrared, nuclear magnetic resonance, and mass spectrometry. Beck illustrates the use of these methods in determining molecular structure, mixture composition, and properties of durable press agents, as well as the mechanism of cross-linking reactions.

In Chapter 7, N. R. Bertoniere describes a technique based on the principles of gel-permeation chromatography. Her focus is on the development of reverse gel permeation column chromatography to assess pore size distribution in cotton cellulose. This method was developed at the Southern Regional Research Center, New Orleans, La. over a period of years by Bertoniere and associates. Bertoniere describes the experimental problems with columns made from cotton cellulose by various methods and proposes meaningful solutions. Reverse gel-permeation chromatography as a tool to elucidate pore structure in different varieties of cotton and jute fibers is described. The effects of caustic mercerization and liquid ammonia treatment on pore size distribution of cotton are explained; the progressive losses in the accessible internal volume of cotton with increasing the degree of cross-linking is used to illustrate increases in resilience accompanied by losses in strength. Of significance is the use of this method in following the differences among conventional cross-linking agents and formaldehyde-free cross-linking agents with respect to the degree to which they alter the pore size distribution in the cross-linked cotton. Bertoniere explains why formaldehyde-free reagents differ in the weight add-on required to impart easy care performance to cotton fabric. Research in this area is ongoing.

Chapter 8, authored by L. Rebenfeld et al., focuses on characterization of pore structure in fibrous networks as it relates to absorbency. They discuss the discontinuous nature of textile materials, their heteroporous nature, and the deceptively high level of porosity in textile materials—which is directly related to absorbency. Nevertheless, the porosity of a textile material is strongly affected by lateral compressive forces to which the material is subjected, hence the pressure dependence of liquid absorption characteristics of textiles. While porosity is an important physical quantity, the dimensions of the pores give a more descriptive way of characterizing the porous nature of a network. Rebenfeld and associates delineate pore volume, which determines liquid absorption capacity, from geometric considerations such as pore throat dimensions that influence liquid flow-through processes, which in turn affect filtration or barrier properties of porous materials. On the basis of the heteroporous nature of fibrous materials, they introduce the concept of pore sizes and their distribution as unimodal, bimodal, and trimodal. To characterize the pore structure in terms of pore volumes and pore throat dimensions they describe the instrumentation of mercury porosimetry used until recently, and the new instrumentation developed at the Textile Research Institute by the authors. These analytical methods are particularly well suited for textiles and other compressible planar materials.

Another topic that has presented much difficulty in the past is that of characterizing single fibers as to their mechanical properties. S. Kawabata, a renowned researcher in the area of polymers, fibers and textiles, presents in Chapter 9 the theoretical basis of direct measurement of the mechanical properties of single fibers. Kawabata describes the advantages of direct "micromeasurement" of single fiber mechanical properties and discusses anisotropy in mechanical properties and the difficulties in measuring very small force and deformation in a single fiber. The mechanical anisotropy of the fiber strictly reflects the microstructure of the fiber and has great implications on the micromechanics of fiber/resin composites. Kawabata also presents the instrumentation developed by the author for this purpose.

The next four chapters focus on new developments in analyzing textile attributes (handle, color, protective qualities) that are not easily measurable as compared to specific textile properties. Chapter 10 deals with objective measurement of fabric hand. S. Kawabata and M. Niwa, the leaders in this area of research, present the significance of fabric hand or handle evaluation on the perception of garment appearance, comfort and tailorability. They analyze and correlate fabric hand judgments by experts with specific fabric properties that express fabric hand characteristic and that can be measured objectively. This is described as objective system for hand evaluation. The nonlinear mechanical properties of a fabric that describe fabric hand, including the weighting system for these properties and the equations that describe these weighting systems, are presented. The mechanical parameters are measured by a set of four instruments known as the Kawabata system or KESF system. Recently, an automated KESF system has been developed.

Color and colorimetry is another elusive but rapidly evolving area of study. P. T. F. Chong, in Chapter 11, provides an extensive background in basic colorimetry and describes the color measuring systems, as well as the developments in color measuring instruments. On the basis of his extensive experience as a color scientist, Chong provides valuable insights into instrument setup, calibration, and verification, as well as sample preparation and color measurement. This is followed by an in-depth presentation of the application of color measuring systems in the textile and textile-related industries. The major applications discussed are color matching, color quality monitoring or screening the color of the products against preset tolerance in color requirement, colorant strength evaluation, and whiteness/yellowness evaluation. Chong also discusses aspects of colorant solution evaluation including colorant strength, dye solubility, solution stability, dye exhaustion characteristics and so on.

Chapter 12 deals with characterization of chemical barrier performance of textile systems. J. O. Stull describes the types of barrier materials, standards pertaining to chemical barrier performance of these materials, and an overview of barrier testing approaches. Three testing approaches are discussed in detail; those pertaining to resistance of material to degradation, chemical penetration resistance, and permeation resistance. The complexities of textile substrate (homogeneous single layer, coated, laminated, microporous, or containing adsorptive components), testing techniques, test conditions, and the impact of multicomponent chemical challenges are brought to focus. For example, using different test methods, or even the same method but different test conditions, can provide different results for the same material and chemical combination. Thus, selection of test method and conditions must be appropriate to the product's application and expected performance.

Degradation resistance testing may show how material/products deteriorate or are otherwise affected, but will not always demonstrate retention of barrier characteristics with respect to specific chemicals. Degradation testing is most useful when retention of specific physical properties is desired or as a screening technique for other chemical barrier testing techniques. Penetration testing should only be used if the wetting or repellency characteristics of materials are to be evaluated. This type of testing is appropriate for the evaluation of material performance against liquid chemicals and can be used for microporous and continuous film-based materials. Vapor transmission test methods are used to measure gross vapor penetration of chemical vapor or gas challenges over relatively short periods of time. This characterization technique is applicable to any film-based material or adsorbent-based material. Chemical permeation testing, however, provides a barrier material's total chemical resistance and can detect very small amounts of permeating chemical. Thus, permeation testing provides the most rigorous of all chemical resistance test methods. Several techniques are presented to provide flexibility in test conditions and applications. Since there are a number of techniques to characterize the barrier performance of materials, careful selection of a test method and its parameters depends on the understanding of the material (textile/product) and its application.

In Chapter 13 P. L. Brown introduces a topic of much interest and concern among health care providers and others, the barrier properties of textiles against microorganisms. Recently, the focus on preventing transmission of infectious microorganisms through barrier materials has grown to include both infection control and personal protection. One major reason for this growth is the risk associated with exposure to blood-borne pathogens as perceived by the health care community. Other potentially hazardous microorganisms (not blood-borne) include Prions, Muerto Canyon virus, and multiple-drug-resistant forms of *Mycobacterium tuberculosis*, staphylococci, and enterococci, to name a few. In addition, biotechnology workers dealing with recombinant DNA, laboratory technicians handling cultures of human pathogens, and veterinary and agricultural workers dealing with zoonotic agents also risk exposure. However, each work environment with potential microbiological hazard may require a different strategy and risk reduction decision.

The basic performance objectives of personal protective clothing products against biohazards are allowing fluid flow, such as air or liquid, while limiting the transfer of potentially pathogenic microbes being transported with them, or else preventing the transfer of fluids and indirectly preventing the transfer of microbes. These two objectives are fundamentally different and require different experimental approaches to the analysis and characterization of the barrier properties of the respective materials to microorganisms. Brown, with his extensive experience as a research scientist and protective product specialist, provides an extensive theoretical background about the types of biohazards, textile substrates, and characterization methods for assessing barrier properties of textiles. He discusses the limitations of laboratory test methods and emphasizes the need for understanding the different microbial, physical, chemical, and thermal stresses imposed on textiles (and finished products) used in personal protection and infection control.

Recognizing the complexities of the different end-use environments for microbial barrier textiles and various stresses that can be imposed on their barrier integrity, Brown discusses developing a realistic strategy related to product evaluation in the laboratory. He suggests developing a feasible testing hierarchy based on combinations of various tests. The degree of hazard associated with exposure to the microbes will dictate how carefully the end-use application for the textile will need to be investigated, how conservative the modeling and experimental approach should be, and the definitions for adequate versus inadequate microbial barrier performance. The ultimate goal is reduction of the risk of product failure during actual use.

In summary, this volume focuses on current and evolving methods of characterizing selected attributes of fibrous materials that are difficult to predict by employing a single standardized test method.

2
Polymer Characterization

PHILLIP H. GEIL   University of Illinois at Urbana-Champaign, Urbana, Illinois

Polymer characterization can be divided into three areas: (1) chemical structure, including composition and configuration; (2) physical structure, including such interactive factors as degree of crystallinity, crystal structure, defects and disorder, conformation, and morphology, and (3) properties, primarily physical but also chemical. In this chapter, I summarize traditional methods of characterization, with references to permit readers to obtain further details of both background and methods, and describe in somewhat greater detail newer methods, all with particular emphasis on techniques applicable to synthetic textiles and textile polymers. For most of the methods I assume an understanding of the terminology and basis for the traditional techniques; such as, for x-ray diffraction, unit cell, crystal structure, Bragg's law, reciprocal lattice, and Ewald's sphere. An excellent recent compilation of polymer characterization techniques is given in Ref. 1.

I. CHEMICAL STRUCTURE

In this section the composition, configuration, and molecular weight (average and distribution) are considered. The basic techniques for all of these can be considered traditional, with improvements primarily in instrumentation. Since there are numerous discussions in general and specific texts, I only summarize them here.

Of first concern is "purification" of the polymer, that is, separation of the polymer from additives, catalyst residue, and impurities, with characterization of all of these often being of interest. This is often difficult in concept and reality for polymers. Methods include [2,3]:

1. Extraction, generally by use of nonsolvents for the polymer, to remove additives, etc. Of concern is the ability to determine completeness and the time required.

2. Solution reprecipitation, to improve purification. Problems include variation in solubility with molecular weight, as well as dependence on branching, cross-linking, tacticity, and crystallinity, separation by which may or may not be desired. For common polymers the *Polymer Handbook* [4] lists potential solvents (see also Ref. 3).

3. Separation, to remove nonpolymer impurities and separate low-molecular-weight fragments produced by analytical, chemical, or thermal degradation. For polymers, separation generally uses liquid/solid (e.g., thin layer, TLC), liquid/liquid, ion exchange, and gel permeation chromatography (GPC) techniques, with the sample inserted following dissolution. TLC, for instance, is simple, rapid, and effective but yields only microgram quantities for subsequent characterization. GPC, especially preparative GPC, yields larger samples. Although usually used for molecular weight fractionation (see later discussion), it is also useful for variations in chemical structure.

A. Composition—Elemental Analysis and Substitutional Groups

Elemental analysis, based on traditional analytical chemistry techniques for low-molecular-weight systems (e.g., combustion), is the usual basis for comparison and calibration of other techniques (see, e.g., Ref. 3). Some journals, such as *Macromolecules,* require elemental analyses for monomers and polymers described in synthesis papers. Atomic absorption spectroscopy, as well as a variety of other techniques, can be used for trace elements, such as from unseparated catalyst residues and stabilizers, with x-ray dispersive analysis, in a scanning (SEM) or transmission electron microscope (TEM), usable for elements (of higher mass than carbon) in particulate additives and impurities [e.g., 5]. In the SEM, x-ray dispersive analysis (see Chapter III), for example, can be applied to particulates on the surface of fibers while in the TEM, sections or specially prepared thin films would be used in the STEM mode.

For determination of substituent groups (e.g., CH_2, $C=O$, $C=C$, etc), standard chemical methods can be used either after breakdown of the polymer (with care to insure against changes in the groups to be tested) or on the polymer itself [2,3]. The methods involve reaction of specific groups with known reagents. More frequently used are infrared spectroscopy techniques (IR) and, increasingly since the 1980s with the development of high-speed computers, Fourier-transform IR (FTIR). As will become obvious, IR is a technique of broad applicability for both chemical and physical characterization of polymers. As discussed in numerous texts (see, in particular, Refs. 6–8 and other references therein as well as Ref. 1), IR absorption is due to excitation of the vibrational motion of groups of nuclei. An IR absorption band, with an intensity proportional to the number of groups in the beam, can be observed for each vibrational degree of freedom (normal mode, type of motion) of a molecule for which the induced or real dipole interacts with the incident light, absorbing energy. Secondary requirements are that the band can be resolved from other bands, now greatly simplified with, for example, deconvolution programs available in FTIR systems, and the intensity is strong enough to detect.

In the complementary Raman spectroscopy technique (for general Raman references related to polymers, see Ref. 1, Chapters 21 and 42), a change in polarizability of the molecule results in inelastic (incoherent, change in wavelength) scattering of the incident beam (usually visible or near IR, laser light). Symmetric vibrations of, for example, a linear CO_2 molecule give rise to Raman scattering, while nonsymmetric vibrations give rise to IR absorption; furthermore, polar bands yield strong IR absorption while nonpolar bands do not. Thus the C-C polymer backbone (or similar bonds in the centrosymmetric O_2 and N_2 molecules) does not absorb in the IR, while substituted groups with C-H, C-F, and C$=$O, because of differences in electronegativity of the atoms, are polar and absorb strongly. The symmetrical vibrations, on the other hand, scatter Raman strongly. In both cases individual groups will absorb or scatter radiation at a number of unique frequencies, with these "characteristic group frequencies" permitting characterization of the samples composition.

Consider the CH_2 group. Its vibrations (Fig. 1), as for other similar types of groups, can be classified as follows, with the frequencies given in the figure:

1. Valency or stretching vibrations (symbol ν) result in a change in one or more (few) bond lengths, and may be symmetrical (ν_s) or asymmetrical (ν_{as}).
2. Planar deformation or bending vibrations (δ) result in a change in one or more bond angles with approximately constant bond lengths.
3. Nonplanar deformation vibrations (γ, ρ, τ) result in a complex change in several angles, with bond lengths nearly constant. Examples include wagging (pendulum vibration perpendicular to the CH_2 plane, symbol γ), rocking (pendulum vibration in the CH_2 plane, symbol ρ), and twisting (rotational vibration about the CH_2 symmetry axis, symbol τ).

The characteristic group frequencies of hydrocarbons are shown in Figure 2. The CH_2 δ vibration (bending, 1460 cm^{-1}, below the range of the figure) is essentially independent of the number or sequence of CH_2 groups or the physical state; it is thus a good internal thickness band for calibrating the amount of sample in the beam if isolatable from neighboring bands. There is some shift, to 1440 cm^{-1}, when the CH_2 is adjacent to an unsaturated C and 1425 cm^{-1} when adjacent to a carbonyl group. On the other hand, the γ, ρ, and τ vibrations are very sensitive to the local environment of the group and thus can be used for determination of configuration (discussed later). Catalogs of characteristic group frequencies are available for both low [11] and high [2,12] molecular weight materials as well as in the computer library systems available for the IR instruments. The latter permits spectral matching.

Methods of polymer sample preparation for FTIR are diagrammed in Figure 3 (see Ref. 8 for details). Of these, the transmission and reflection techniques are

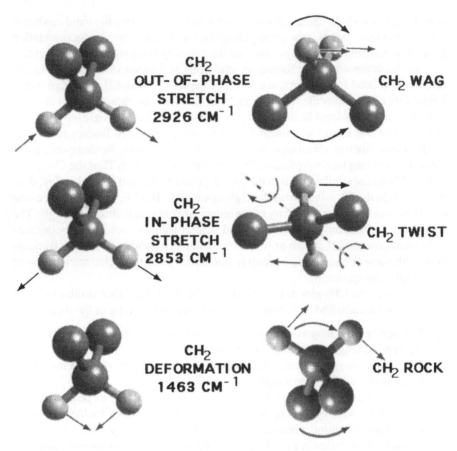

Figure 1 Examples of CH_2 group vibrations (normal modes). The small circles represent H atoms, and the large C atoms with the upper C atoms on the left, for instance, being the next C atoms along the chain. (Modified from Ref. 9.)

of the most use for fibers. Standard transmission methods, using thin molded or drawn films, require uniform-thickness, nonscattering films. Fibers can be used by, for example, laying them parallel to each other on a window or wrapping them on a frame and coating with Nujol (an oil of low absorptivity in the regions of interest) to reduce the scattering. In addition to normal reflectance techniques, including attenuated total reflectance (ATR), diffuse reflectance (DRIFT) techniques have also proven useful. The photoacoustic spectroscopy (PAS) technique can also be used and is rapidly growing in use for samples, such as fibers, that have high scattering. It can depth-profile changes in compositions, such as induced by oxidation. A particularly useful recent advance is the development of FTIR

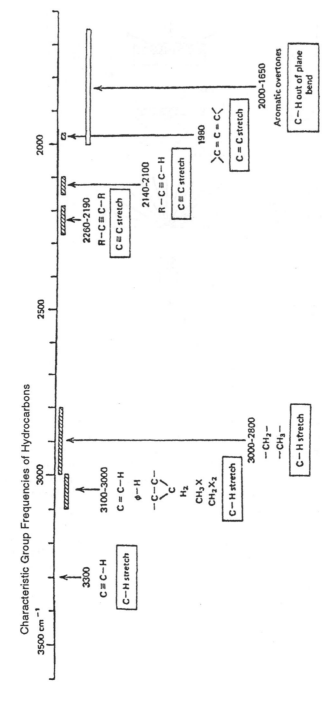

Figure 2 Characteristic group frequencies of hydrocarbons. The various types of groups shown absorb IR in the spectral ranges shown, and the position of the peaks in the ranges is a function of attached groups and the local environment. (From Ref. 10a. More extensive charts are given in Ref. 10b.)

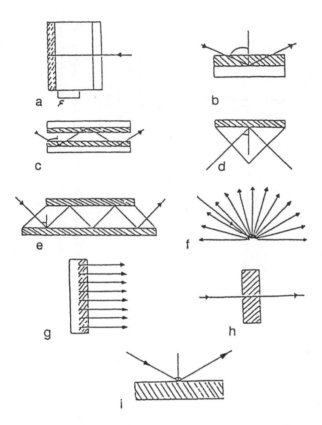

Figure 3 Diagrams of FTIR sampling techniques (a) Photoacoustic spectroscopy (PAS). Pressure fluctuations induced by the incident IR are detected by a sensitive microphone. (b,c) Specular or external reflection spectroscopy (RA). An IR beam passing through the sample at a steep angle (70–89.5°) is reflected from a mirror-like optical substrate one (b) or more (c) times. Useful for coatings. (d,e) Internal reflectance or attenuated total reflectance spectroscopy (IRS or ATR). The beam is transmitted through a high-refractive-index, low-absorptivity crystal (e.g., Ge) and is reflected once (d) or several (e) times at the crystal-sample surface. The method samples a depth of 0.5 μm of the sample's surface and can be used for opaque samples. (f) Diffuse reflectance spectroscopy (DRIFT). The scattered light is reflected from mirrors to the detector. Useful for fibers. (g) Emission spectroscopy. The emission spectrum of a sample is the mirror image of its absorbance spectra. Useful for metals, opaque samples and fibers for $v > 2000$ cm^{-1}; sample needs to be at thermal equilibrium. (h) Transmission spectroscopy. For films, or powdered samples dispersed in KBr pellets or suspended in an inert (to IR) oil on a window. (i) Spectral reflectance spectroscopy. Reflectance from the surface of the sample with angle of reflection equaling angle of incidence. Not useful for fibers. (From Ref. 8.)

microscopy, that is, the attachment of an IR microscope to an FTIR unit. This permits the observation (by visible light) and characterization of an individual fiber.

The other primary method of characterizing the substituent group composition is nuclear magnetic resonance (NMR) spectroscopy (and FT-NMR) [see Refs. 1 (Chap. 17), 8, and 13)]. In NMR the sample is placed in a slowly varying strong magnetic field H_0 while an oscillatory smaller field of frequency v_1 and constant maximum strength H_1 is applied at right angles (or H_0 and H_1 can be held constant while v_1 is varied). Energy is absorbed from the H_1 field when

$$H_0 = \frac{Ihv_1}{\mu} \tag{1}$$

where

I = the spin quantum number
h = Planck's constant
μ = the magnetic moment of the nucleus involved

For polymers the nuclei of interest, ^1H, ^{13}C, ^{19}F, and ^{17}O, all have spin (I) = 1/2 and thus two energy states, with the magnetic moments parallel (low energy) or antiparallel (high energy) to the applied (H_0) field.

Consider a single proton (^1H) in the sample. If isolated it will absorb energy when H_0, as it increases, becomes equal to Ihv_1/μ (in actuality a small number of the excess nuclei in the lower energy state "flip" their spins to the higher energy state). However, if neighboring bonded or nonbonded nuclei have a magnetic moment ($I > 0$) their fields (H_{loc}) will add (vectorially) to the applied H_0 field. If the internuclei vectors are stationary, as at low temperature, the result is absorption over a range of values of H_0, each nucleus absorbing when the field at its position is

$$H^*_r = H_0 + H_{loc} = \frac{Ihv_1}{\mu} \tag{2}$$

That is, a broad energy absorption line is observed. On the other hand if the internuclei vectors rapidly randomize in space, as in solutions or melts, the local field due to the neighboring nuclei averages to zero and a narrow line would be observed at the characteristic value of H_0 except for the influence of the magnetic moments of the electrons around the nucleus.

The electrons surrounding the nucleus produce a local field H_{loc} proportional to the applied field, and in the opposite direction

$$H_{loc} = -\delta H_0 \quad \text{and} \quad H_r = H_0 (1 - \delta) \tag{3}$$

where δ is the shielding constant and is a measure of the chemical (bonded) environment of the nucleus. For example, the ^1H devoid of electrons, as in $-\overset{\overset{\displaystyle C}{\|}}{C}-OH$ groups, have small δ and thus absorb at applied fields close to (but above) that given by Eq. (1), that is, when $H_r^* = H_r = H_0(1 - \delta)$. On the other hand the ^{1}H in $>CH_2$ groups have a large δ, the electrons shielding the nucleus, and therefore a larger H_0 is required before H_r^* is reached (i.e., resonance is obtained). The resonant field is "shifted."

Figure 4 shows the characteristic chemical shifts (in parts per million of the applied field since the shift is proportional to the field) with tetramethylsilane, in which the ^{1}H are more shielded than in any other organic compound being used as a standard. Its shift is defined as 10 ppm (τ scale) or 0 (σ scale), with all other resonances occurring at lower values of H_0.

With the development of FT-NMR it became possible to use ^{13}C despite its much lower abundance (relative to ^{12}C and ^{1}H). The advantage lies in the much larger shift, ≈ 600 ppm, versus <20 ppm for ^{1}H, permitting higher resolution. The general ^{13}H chemical shift chart is shown in Figure 5, ^{13}C spectroscopy is generally being used these days. Again, computer libraries aid spectra characterization.

Measurements of the chemical shifts for a polymer permit determination of its group composition, with the energy absorbed by a group (peak area) being proportional to its concentration. (For examples see Figs. 12–16.)

The chemically shifted lines in modern, high-resolution (large H_0) instruments are further split by spin-spin coupling (see later discussion) permitting determination not only of the constituent groups but also of their neighbors along the chain. So-called narrow-line NMR spectra (i.e., permitting resolution of both the chemical shift and spin-spin coupling) are most often obtained from solutions of the polymer, where the solvent should be devoid of the groups being measured in the polymer. This often requires deuterated (^{2}H) solvents for ^{1}H NMR and purified ^{12}C solvents for ^{13}C NMR.

Despite the broadening of the resonance band in solids, narrow-line spectra can be obtained by a combination of so-called "magic angle spinning" (MAS), in which the solid (fiber) sample is spun rapidly about an axis oriented at 57.4° to H_0, dipolar decoupling (DD), and cross-polarization (CP) (see Fig. 16a) techniques. For further description of this so-called "grand" experiment and its limitations (e.g., the chemical shifts may be affected by the conformation of the group involved) and of other special editing techniques (applications of various pulse sequences) applied to solution samples to enhance the resolution (over and above that resulting from higher H_0) and permit assignment of the peaks to the appropriate chemical groups the reader is directed to Ref. 8 or 15.

Ultraviolet (UV) and visible spectroscopy have more limited usefulness; many polymers show no absorption in the visible or near-UV range. Table 1 lists the groups whose absorption can be measured; in many cases these groups are formed by degradation during processing.

Group composition on surfaces of polymer samples, including fibers, can be determined by a number of techniques in which the surface is irradiated with

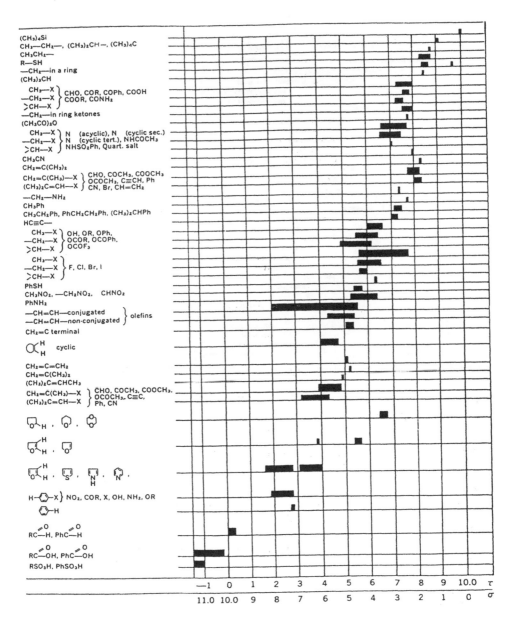

Figure 4 Characteristic NMR chemical shift spectral positions for ^1H in organic materials. Scale is in parts per million shift relative to $(CH_3)_4$ Si (top of list). (From Ref. 14b.)

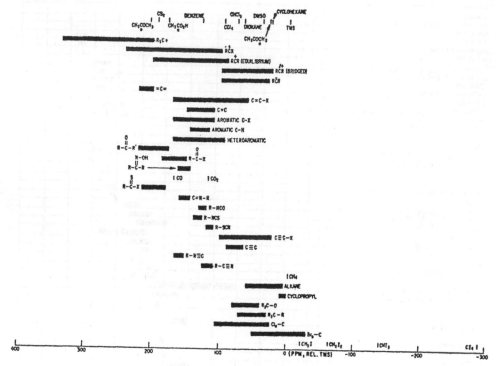

Figure 5 Characteristic ¹³C NMR chemical shift spectral positions; the zero is relative to (CH₃)₄ Si (TMS). (From Ref. 10a.)

Table 1 UV Absorption Ranges of Important Groups

Group (bond)	Absorption range (nm)	Transition energy (kJ/mol)
C-N (Amide)	250–310	478.8–386.0
C=O	187; 280–320	640.1; 427.5–373.9
C–C	195; 230–250	613.8; 520.4–478.8
O–H	230	520.4
C=C (isolated double bond)	180	665.0
-[CH=CH]ₙ- (conjugated double bond)		
n = 2	230	520.4
n = 3	275	435.3
n = 4	310	386.0

Source: Ref. 7.

photons (usually x-rays), electrons, or ions, with the energy of the resulting emissions being atom and group dependent. These techniques are of particular interest for the characterization of coatings, degradation effects, and surface treatments [see Refs. 1 (Chap. 24), 7 and 16]. For instance, in electron spectroscopy for chemical analysis (ESCA, also known as XPS, x-ray photoelectron spectroscopy) the sample is irradiated with monochromatic x-rays, knocking out electrons whose energy (which is measured) is characteristic of the atom type, electron shell from which removed, valency, and neighboring bonded atoms (i.e., group) (Figures 6 and 7). Other examples are Auger electron spectroscopy (AES), in which the excitation is by electrons, with resulting electron energies being measured, and secondary-ion mass spectroscopy (SIMS), in which ions are used to knock out ions whose mass is measured by a mass spectrometer. Depths characterized vary from a few atomic layers for SIMS to <30 layers for AES and <50 for ESCA. ESCA is particularly useful for polymers since, due to the use of x-rays, charging does not take place.

B. Configuration

Considered here is the arrangement along the chain of the constituent groups, that is, the microstructure of the chain. Included are sequence arrangement in copolymers, tacticity, cis–trans isomerism, and branches and cross-links. IR and NMR

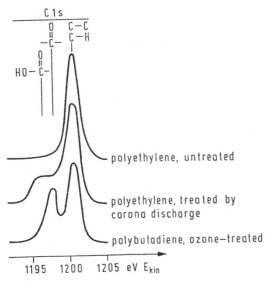

Figure 6 C_{1s} ESCA photoelectric spectra from the surface of the samples listed. (From Ref. 17.)

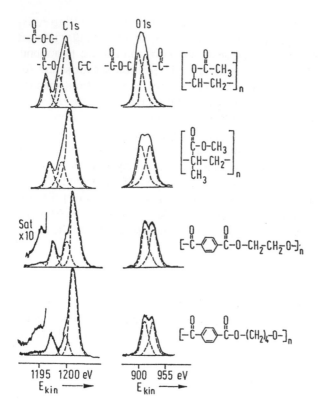

Figure 7 C_{1s} and O_{1s} ESCA spectra of polyvinyl acetate, PMMA, PET, and PBT. (From Ref. 16.)

are again the primary techniques, with the absorption frequencies of the constituent groups being shifted slightly, depending on their neighbors.

In IR we have already noted the shift in the CH_2 δ (bending) band as a result of its neighbors. The γ, τ, and ρ bands are particularly sensitive to their local environment. Figure 8 shows the shift in frequency of the CH_2 rocking mode, while Figure 9 shows how it can be used to determine the configuration of an ethylene-propylene copolymer. In practice only a sequence distribution can be obtained, averaged over all of the chains in the sample. Table 2 lists the important IR group frequencies used for structure determination, while Figure 10 shows a detailed correlation table for the $C{=}O$ ν vibration. As an example of IR's use for tacticity determination, Figure 11 and Table 3 show its use for predominantly isotactic and syndiotactic PMMA. All of the IR sample preparation techniques shown in Fig. 3 can be used. Although Figs. 8–11 are all examples of IR's use for characterizing configuration, differences in conformation (e.g., trans vs. gauche and different he-

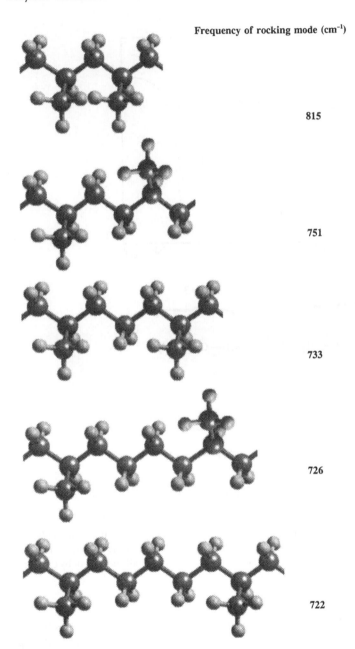

Frequency of rocking mode (cm⁻¹)

815

751

733

726

722

Figure 8 Frequency of the rocking mode as a function of number of CH₂ groups between "anchors" as determined from model compounds (data from Ref. 18). For more than 5 CH₂s absorption is also at 522 cm⁻¹.

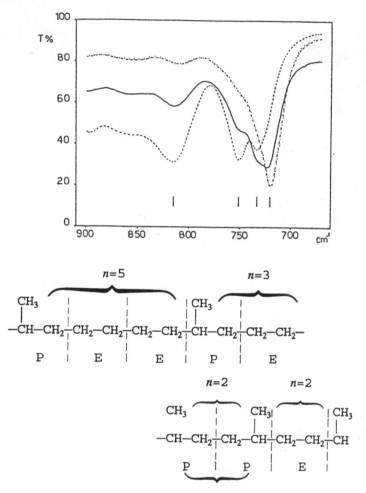

Figure 9 IR spectrum of ethylene propylene copolymers of 25% (--·-), 50% (—), and 75% (----) propylene composition in the CH_2 rocking region of the spectrum (compare with Figure 8). In pure PP the CH_2 rock band is 815 cm^{-1}. The presence of the 752 cm^{-1} band indicates reversal of PP chain direction, either by propylene tail-to-tail addition or by ethylene insertion between propylene heads. (Graph from Ref. 18.)

lical forms as well as crystalline vs. amorphous, all as described in Section II) can also cause changes in absorption band position. Separating these effects is of concern and requires complementary use of other techniques.

Branches (from either the end groups or branch points and crosslinks) can also be identified by IR. An example is shown in Figure 12 for polyethylene (PE)

Table 2   Important IR Group Frequencies

Average band position (cm^{-1})	Assignment	Remarks
3500	OH stretching	Broad band
2950, 2890	CH$_2$ stretching	Strong polymodal band
2260	NCO stretching (isocyanate group)	Strong band
1850–1650	C=O Stretching (carbonyl group)	Very strong band, position varies according to environment (e.g., ketone aldehyde, ester)
1620	C=C stretching	
1600	NH bending	Important indication of polyamides
1450	CH$_2$ bending ("scissoring")	
1100	C-O stretching (ether group)	
1000	C-C stretching (skeletal vibration)	
900–700	Benzene ring, CH out-of-plane vibrations	Two to three usually strong characteristic bands, positions depend on type of substitution
815	triazine ring out-of-plane vibration	Very sharp band
700	Carbon–halogen stretching	

Source: Ref. 7.

(BPE = branched or low-density polyethylene vs. LPE = linear or high density polyethylene). The CH$_3$ 1378 cm^{-1} band is due to CH$_3$-capped branches. It is, however, overlapped by several CH$_2$ bands (1304, 1352, and 1368 cm^{-1}), making the determination of the background difficult. Determination can be made either by subtraction of a (thickness-corrected) LPE spectrum or by deconvolution of the bands. However it must be noted that the absorbances of the CH$_3$ groups depend on the length of the branch to which they are attached (1.55:1.25:1 for CH$_3$ on methyl, ethyl, and longer branches). Thus the IR measurement is only a weighted average; high-resolution NMR can be used (see below) but is too expensive for general use, and thus the IR scale is commonly used.

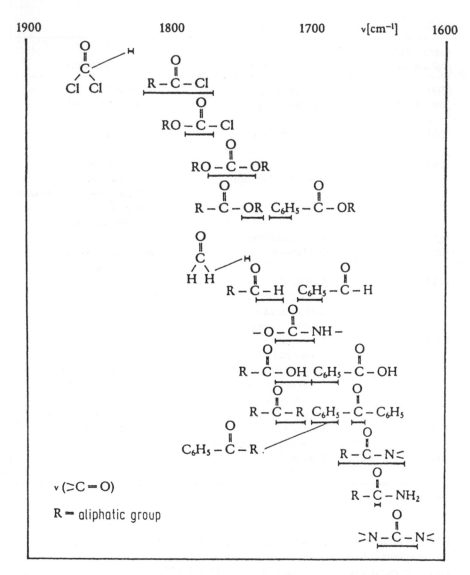

Figure 10 Correlation (v range in which observed) table of C=O stretching vibrations as affected by neighboring attached groups. (From Ref. 19.)

NMR is, in general, an even more sensitive sequence characterization technique. In addition to the chemical shift effect described earlier, the electrons around a nucleus produce additional, smaller (spin–spin coupling) shifts due to the effect of the magnetic moments of neighboring, nuclei whether or not the internuclear bonds are stationary or rapidly randomizing in space. The electron in

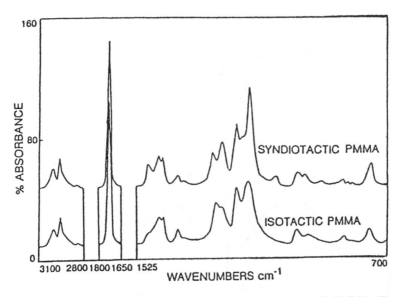

Figure 11 IR spectra of predominantly isotactic and syndiotactic PMMA. (From Ref. 20.)

Table 3 Wavenumbers and Assignments of IR Bands for Isotactic and Syndiotactic PMMA

Isotactic	Syndiotactic	Assignment
2995		$\nu_a(C-H)$
2948		$\nu_s(C-H)$
1750		$\nu(C=O)$
	1485	$\delta_a(\alpha\text{-}CH)$
1465	1450	$\delta(CH_2), \delta_a(CH_3-O)$
	1438	$\delta_s(CH_3-O)$
	1388	$\delta_s(\alpha\text{-}CH_3)$
	1270	$\delta_s(\alpha\text{-}CH_3)$
1260		$\nu_a(C-C-O)$
1252		coupled with
	1240	$\nu(C-O)$
1190	1190	skeletal
1150	1150	
996	998	$\gamma_r(CH_3-O)$
950	967	$\gamma_r(\alpha\text{-}CH_3)$
759	749	$\gamma(CH_3) + $ skeletal

Source: Ref. 20.

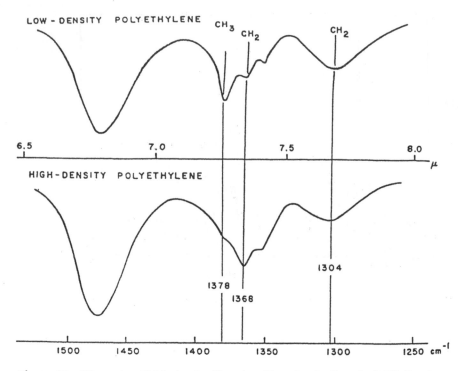

Figure 12 IR spectra of high-density (linear) and low-density (branched) PE. For determination of the degree of branching the 1378 cm^{-1} CH$_3$ band should be deconvoluted from its neighbors. (From Ref. 9.)

an atom having a neighbor with a spin ½ can see it with ±½; thus its chemically shifted line is split in two. As an example, consider CH$_2$Cl–CH$_3$ (Figure 13). Each ^{1}H on the CH$_2$ can be parallel or antiparallel to the H_0, yielding four possible arrangements with the total spins 1, 0, 0, and −1 as seen by the electrons on the CH$_3$ ^{1}H atoms. As a result the chemically shifted CH$_3$ line is split into three, with a ratio of 1:2:1. Likewise, the three ^{1}H on the CH$_3$ can have total spins of ±½, or ±1 ½, yielding four CH$_2$ lines with an area ratio of 1:3:3:1. This spin-spin interaction is over and above the chemical shift effect, but does not occur between ^{1}H atoms on equivalent groups (as in CH$_3$–CH$_3$) or in the same group. The amount of the shift is the same on both sets of ^{1}H atoms that are coupled (i.e., interaction of the CH$_2$ and CH$_3$ in Fig. 13), is independent of the field, and is affected by differences in chemical shift of the two groups. Thus one can determine the group involved by δ and the neighbors by the spin-spin coupling. With increasing field strength H_0, particularly with the advent of superconducting magnets, increasing resolution has permitted separation of spin–spin effects due not

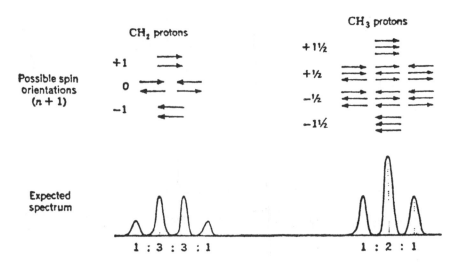

Figure 13   Spin–spin (^{1}H) coupling for ethyl chloride. (From Ref. 21.)

only to neighboring groups but also to groups several backbone bonds along the chain. Examples, for predominantly i- and s- PMMA, are given in Figures 14–16 for two different values of H_0.

Consider first the CH_2 and CH_3 (backbone) groups at low H_0 (low resonance frequency) (Figure 14). In s-PMMA the CH_2 lines should be a singlet; the CH_3 groups are on opposite sides of the chain so that the ^{1}H atoms on the CH_2 are equivalent (the unit is racemic).

CH₃ H R

$$-\overset{|}{\underset{|}{C}}-\overset{|}{\underset{|}{C}}-\overset{|}{\underset{|}{C}}-$$

R H CH₃

r

syndio-PMMA

$$R = \overset{\overset{O}{\|}}{C}-O-CH_3$$

CH₃ H CH₃

$$-\overset{|}{\underset{|}{C}}-\overset{|}{\underset{|}{C}}-\overset{|}{\underset{|}{C}}-$$

R H R

m

iso-PMMA

In i-PMMA (*meso*) the CH_3 are on the same side of the chain, and thus the two ^{1}H on the CH_2 are not equivalent and four peaks (as in $CH_2Cl - CH_3$) are seen. Note that there is some fine structure to these bands and that neither form is pure. For

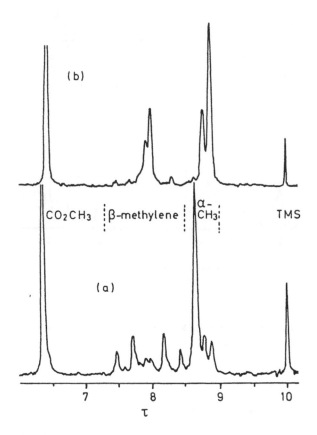

Figure 14 60-MHz NMR spectra of 15% solutions of predominantly (a) syndio and (b) iso-PMMA. The ester (β) methyl resonance is near 6.5τ, the β methylene resonance is from 7.5 to 8.5τ, and the α-methyl protons between 8.5 and 9.0τ. (From Ref. 22.)

the α-CH$_3$ group (on the backbone) the spin–spin coupling shift varies with the configuration of the neighboring CH$_3$, here considered as dyads.

```
  CH3                CH3            CH3
   |    H   R   H     |              |    H   R   H   R
   C — C — C — C — C                 C — C — C — C — C
   R   H   |   H   R                 R   H   |   H   |
           CH3                               CH3     CH3
       r       r                        r       m
     Syndio-PMMA                      hetero-PMMA
```

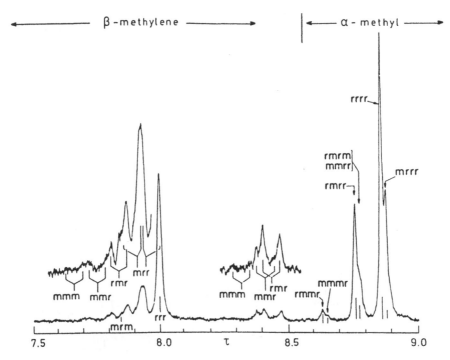

Figure 15 The β-methylene ¹H-220 MHz spectrum for predominantly syndiotactic PMMA in chlorobenzene at 120°C. The vertical markers for the spin–spin coupling positions for the triads (e.g., mmm) are joined by slanting lines that intersect at the positions (center) for the chemical shifts; tetrads can be resolved in the α-methyl region. (From Ref. 23.)

and

$$
\begin{array}{ccccc}
CH_3 & & CH_3 & & CH_3 \\
| & H & | & H & | \\
C-&C-&C-&C-&C \\
| & H & | & H & | \\
R & & R & & R \\
& m & & m & \\
\end{array}
$$
iso-PMMA

The three CH₃ peaks in Figure 14 are due to rr (*syndio*, at the highest field), rm (*hetero* or atactic), and mm (*iso*, at the lowest field) dyads. From the areas under the peaks the relative amount of the various types of dyads can be determined.

Figures 15 and 16 show the corresponding "high-resolution" (220 MHz, with current NMR machines operating at as high as 360 MHz) spectra for the same

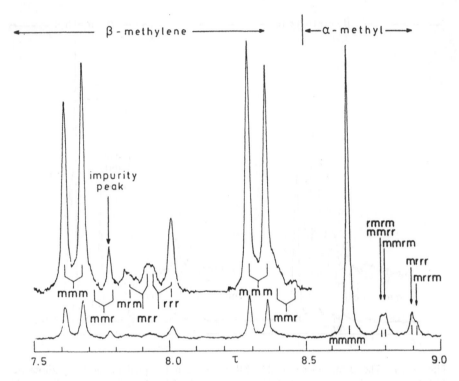

Figure 16   The ^{1}H 220 MHz spectrum for predominantly isotactic PMMA, as in Figure 15. (From Ref. 23).

types of samples. In both figures, the relative numbers of various types of triads can be determined from the CH_2 chemical shift region, while in the α-CH_3 region tetrads are identified. Similar techniques can be applied to isomerism, copolymer sequence, and so forth. The last example (Figure 17) shows the high-resolution scans used for determination of branch length in low-density PE.

Figure 18 shows the results of applying the various techniques utilized for producing narrow-line NMR spectra from solid samples; in this case ^{13}C spectra [25]. In (a), only scaler decoupling is applied, with no useful spectrum obtained. With dipolar decoupling (DD) and cross-polarization (CP), as in (b), broad resonances are obtained. With magic-angle spinning (MAS) and dipolar decoupling (c) the lines are narrowed but with considerable noise in the spectrum. In (d), the "grand" experiment (MAS + DD + CP), the various individual ^{13}C atoms can be resolved (not ^{1}H as in Figs. 14–16), but clearly tacticity could not be characterized.

Various other techniques, already discussed, can also be used for configuration characterization. In ESCA, for instance, the binding energies of the elec-

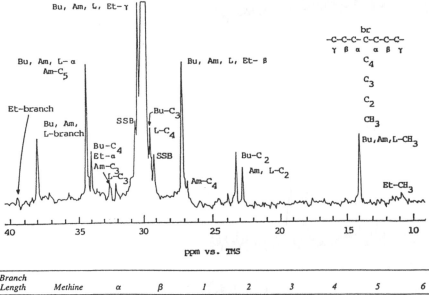

Branch Length	Methine	α	β	1	2	3	4	5	6
1	33.3	37.6	27.5	20.0					
2	39.7	34.1	27.3	11.2	26.7				
3	37.8	34.4	27.3	14.6	20.3	36.8			
4	38.2	34.6	27.3	14.1	23.4	–	34.2		
5	38.2	34.6	27.3	14.1	22.8	32.8	26.9	34.6	
6	38.2	34.6	27.3	14.1	22.8	32.2	30.4	27.3	34.6

Figure 17 ^{13}C FT-NMR spectrum at 25.2 MHz for a low-density PE obtained in trichlorobenzene at 110°C with 9500 scans. The structure at upper right illustrates the C positions; the branch carbons are numbered from the methyl (1) end toward the methane C branch point. The table below indicates the corresponding chemical shifts. Branches longer than three carbons yield identical spectra. Terminology in the figure indicates the length of the branch (Am, amyl, C_5; Bu, butyl, C_4; Et, ethyl, C_2; L, long C_6 and longer), position of branches on the branch (e.g., Bu-C_2, butyl branch on C_2 of primary branch), and SSB, spinning side band (see texts, e.g., Ref. 8, for discussion). All values given in parts per million (±0.1) downfield from TMS. The solvent was 1,2,4-trichlorobenzene, and the temperature was 125°C. (From Ref. 24.)

trons emitted depend not only on the atom involved and its closely associated (bonded) atoms, but also on their neighbors. An example is shown in Figure 19 for polyethylene–polytetrafluoroethylene (PE–PTFE) copolymers for which it is claimed [26] that pentad sequences can be distinguished. Likewise, UV-visible spectroscopy can, on occasion, be used, with the absorption frequency again being dependent on neighboring groups.

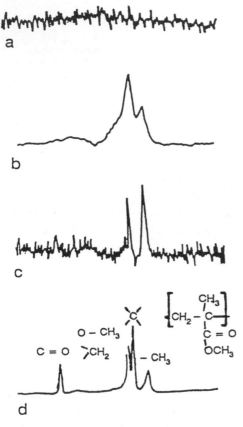

Figure 18 C^{13} solid state NMR spectra of PMMA: (a) scaler decoupling only; (b) DD+CP; (c) MAS+DD; (d) MAS+DD+CP. (From Ref. 25.)

C. Molecular Weight

Measurements based on colligative properties, dependent on the number of molecules in a given weight of sample (i.e., $\overline{M}_n$ the number average molecular weight), include boiling-point elevation, freezing-point depression, osmotic pressure (the most sensitive), and vapor-phase osmometry [27]. For $\overline{M}_w$ light scattering has long been used, with low-angle laser light scattering being a modification reducing the effects of dust particles and the need for angle extrapolation [27]. In general these methods have been replaced, since the 1960s, by GPC. This method (see, e.g., Ref. 28) permits determination, with suitable calibration, of $\overline{M}_n$, $\overline{M}_w$, $\overline{M}_z$, in fact the full molecular weight distribution. Since the method effectively measures the spatial size or hydrodynamic volume distribution of the molecules in a solvent, it is af-

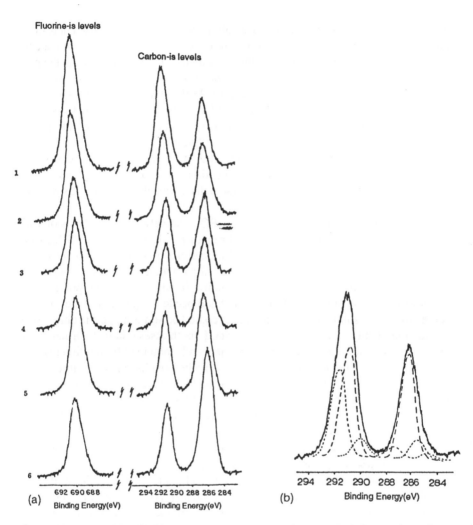

Figure 19 F_{1s} and C_{1s} ESCA levels for a series of copolymers of ethylene and tetrafluorocthylene (a) with the deconvoluted C_{1s} peaks for samples 1 ($\sim$ 62 mol % C_2F_4) in (b). (From Ref. 26a.)

fected by branching, copolymer sequence, and composition as they affect chain stiffness and other properties. It also suffers from the cost of the equipment required and the lack of suitable (safe) solvents for some polymers.

Chromatography processes, in general, involve transfer of solute between two phases, one stationary and the other moving. In GPC both phases are liquid (and the same, the solvent), the stationary phase being the liquid filling the pores in microporous gel particles packed in a column through which the solvent is flow-ing. As a "plug" of polymer solution is introduced into the solvent stream, a con-centration gradient is set up between the solution outside the particles and the pure solvent in the particles, resulting in polymer diffusion into the pores, with the smaller molecules being more able to diffuse. As the "plug front" moves down the column, the larger molecules will be retained to a smaller degree, with the smaller molecules remaining in the pores for a longer time before diffusing out into the now less concentrated solution following the "plug end." With the concentration of the elution stream being measured (by index of refraction) a plot of concentra-tion versus time (or elution volume) of retention is obtained that, by calibration, can be converted to a molecular weight (actually molecular coil size) distribution (Figure 20).

Using large columns, in preparative GPC, relatively sharp fractions of vari-ous molecular weight can be obtained. Prior to the development of GPC, frac-tionation was based on variations in solubility, either by adding a nonsolvent to a polymer solution, with the least soluble (e.g., highest molecular weight) molecules precipitating out first, or by differential elution in which a swollen polymer gel (swollen due to the presence of the solvent and needed to permit transfer of mol-ecules from the "solid" to the liquid phase) coated on some support is eluted with liquid of gradually increasing solvent power. In both cases relatively broad distri-bution fractions are obtained. Improvements can be obtained by fractionally pre-cipitating the polymer onto the column packing before elution and/or imposing a temperature gradient on the column in addition to the variations in solvent power. In the latter case, if the sample is present initially only at the top of the column it will undergo a series of precipitations and dissolution steps as it moves down the column. This is the basis of the relatively new TREF (temperature rising elution fractionization) technique used for fractionating linear low-density polyethylenes by both molecular weight and branch content [29].

Somewhat simpler, but less informative, techniques than GPC for molecular weight characterization are based on solution and melt viscosity [27]. The latter can be directly related to the melt spinnability of the polymers, and both are often used in technical data sheets for specific grades of a polymer. Solution measurements yield a "molecular weight" whose value (M_v) lies between $\overline{M}_n$ and $\overline{M}_w$ [30], while $\overline{M}_w$ can be obtained from the melt viscosity [usually it is the intrinsic viscosity, sometimes the inherent viscosity (discussed later,) or from the melt index (MI) or melt flow rate (MFR) that is given, with MI and MFR directly related to η (melt)].

In the simplest situations, solution viscosities are measured at appropriate temperature using either Ubbelohde or Ostwald–Fenske viscometers. Both permit measurement of the flow rate of the solution [time of flow for a given quantity of solution (t) or solvent (t_0)] through a calibrated capillary, with the measurements being made as a function of concentration. Extrapolation to zero concentration re-moves effects due to interaction of the molecules, as the viscosity due to the size of individual, noninteracting molecules is desired. Definitions of the relevant

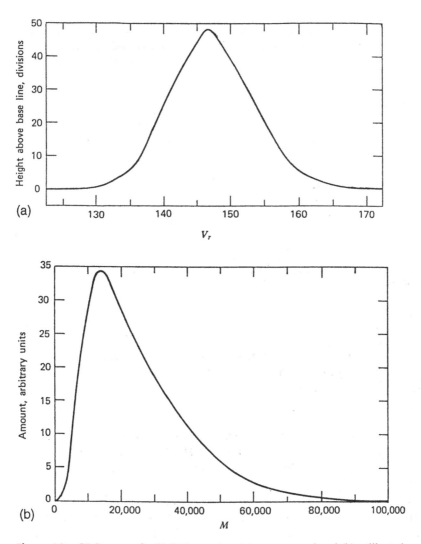

Figure 20 GPC curves for PMMA samples: (a) as measured and (b) calibrated molecular weight distribution. (From Ref. 28a.)

terms are given in Table 4. The last relation is related to the empirically determined expressions

$$\frac{\eta_{sp}}{c} = [\eta] + k' \, [\eta]^2 c + \ldots \tag{4}$$

$$\frac{\ln \eta_r}{c} = [\eta] + k''[\eta^2]c + \ldots \tag{5}$$

Table 4 Definition of Solution Viscosity Terminology

Common name	IUPAC recommended name[a]	Symbol and defining equation
Relative viscosity	Viscosity ratio	$\eta_r = \eta/\eta_0 \simeq t/t_0$
Specific viscosity	—	$\eta_{sp} = \eta_r - 1 = (\eta - \eta_0)/\eta_0 \simeq (t - t_0)/t_0$
Reduced viscosity	Viscosity number	$\eta_{red} = \eta_{sp}/c$
Inherent viscosity	Logarithmic viscosity number	$\eta_{inh} = (\ln \eta_r)/c$
Intrinsic viscosity	Limiting viscosity number	$[\eta] = (\eta_{sp}/c_{c=0} = [(\ln \eta_r)/c]_{c=0}$

[a]The IUPAC names are almost never used. They were proposed to eliminate the inconsistency that the common names do not have units of viscosity.
Source: Ref. 27.

where Eqs. (4) and (5) are the Huggins and Kraemer relations [31,32] and $k' + (-k'') = \frac{1}{2}$ if additional terms are not needed, $[\eta]$ is related to M_v through the Mark–Houwink relation [33],

$$[\eta] = KM_v^a \tag{6}$$

where K and a depend on polymer, solvent, and temperature, which are tabulated for many polymer–solvent pairs in the Polymer Handbook [4]. The solution viscosity can also be measured with various types of rheometers (capillary flow, cone and plate, couette, etc.) permitting determination of both $[\eta]$, if desired, and the more concentrated η of direct concern to fiber spinning from solution (see, e.g., Ref. 34). Note that the relation between $[\eta]$ and M_v applies to linear polymers since $[\eta]$ is actually a measure of molecular size (hydrodynamic volume); application to branched polymers of varying degrees of branching is inappropriate.

Melt viscosity, to obtain the MI or MFR (the term used depends on the polymer), is obtained using a simple capillary viscometer, a specified (e.g., ASTM D-1238) load being applied to a polymer melt at a specified temperature. The weight of polymer extruded through the specified die in, usually, 10 min is the MI or MFR; the higher η, the lower the amount. With Newtonian flow occurring in the system, $\eta = k\overline{M}_w^{3.4}$, with k again being tabulated for various polymers [4]. Again various rheometers can be used to characterize η, yielding such further information as effects of variations in shear rate, normal stress effects, etc. [34,35]. These are of particular concern in melt spinning, with high shear rates being desired for both economics and molecular orientation (see, e.g., Ref. 36).

II. PHYSICAL STRUCTURE

The physical structure of polymers in general, and fibers in particular, occurs over a range of size scales, thus requiring a range of techniques for characterization. Of concern are structures on (1) the size scale of the individual molecular segment (e.g, their conformation), (2) the relative number of regular and random conformations and their arrangement in space (i.e., degree of crystallinity and orientation), (3) the size and shape of the crystalline and amorphous regions, including defects and disorder in the crystals, and (4) the organization and interaction of these crystals in larger structures (e.g., spherulites and fibrils for crystallizable polymers, and domains of various shapes in block copolymers and blends.) The size scales of concern overlap, extending from 1–2 nm for (1) to as large as 100 μm for (4). Techniques of primary concern include x-ray, neutron, and electron diffraction, infrared spectroscopy, and transmission and various types of scanning electron and optical microscopy, but with a wide range of other techniques being of collaborative usefulness.

A. Molecular Conformation

Polymer conformation means the relative physical arrangement in space of the atoms of a molecular or molecular segment as affected by both intra- and intermolecular atomic interactions (e.g., covalent, hydrogen and van der Waals bonds). Although the overall dimension of the molecule, as characterized by the mean square end-to-end distance $<r^2>$, or radius of gyration, R_g, in either amorphous or crystalline regions could be considered here, we postpone its discussion to the next section. Here only the local segmental conformation is considered. In crystalline regions this is most often determined by x-ray (or electron) diffraction, that is, the determination of the shape of the molecule in the unit cell. For molecular conformation in the amorphous regions, and supplementing the diffraction studies for crystalline regions, infrared spectroscopy is the primary tool. Both of these, other than for recent developments in diffraction, rely on traditional techniques but with continued improvement in instrumentation and application of computer-based calculations and simulations.

Characterization of the unit cell of a polymer crystal requires determination of the spatial position of all of the atoms, from one or more molecular segments, each one physical repeat unit long, within the unit cell—that is, the conformation of the physical repeat unit and their relative positions. While this can be done by standard x-ray diffraction techniques, their application to polymers is made difficult by (1) the complexity (number of atoms and low symmetry) of the unit cells,

(2) general lack of availability of macroscopic single crystals permitting application of x-ray single-crystal techniques, (3) disorder of various forms in the crystalline regions that can be produced, and (4) the C, H, N, O atomic composition (low scattering power) of most polymers. See Ref. 37 for discussions of applications of standard x-ray techniques of crystal structure (lattice and unit cell) determination to organic nonpolymeric molecules; although there are numerous texts on crystal structure determinations for inorganic crystals, the problems involved in the determination of the crystal structure of organic molecular crystals are much closer to those involved in the determination of the crystal structure of polymers.

X-ray unit cell characterization of polymers generally takes advantage of two features relatively unique to macromolecular systems, the helical conformation of most polymers (synthetic and biological) in crystalline regions (Figure 21) and the ability to produce fibers in which there is an alignment of the molecular (backbone) axes along the fiber axis [39]. The latter isolates the Bragg reflections from one of the axes of the unit cell (usually **c**) along the meridian of the diffraction pattern,* and the former somewhat simplifies the determination of the molecular conformation from that pattern. For instance, the layer line spacing corresponds to the length of the physical repeat unit; the number of layer lines and distance to the layer line with the first meridional reflection correspond to the number of chemical repeat units per physical repeat and the axial distance per chemical repeat unit. The type of helix, that is, number of turns of continuous helix per physical repeat unit, can also be determined from the form of the pattern. Although these do not determine the relative spatial positions of each atom in the repeat unit, chemical or physical, knowledge of this type is of major assistance. Furthermore, this and a knowledge of the dimensions of the lattice are sufficient to determine the density of the perfect crystal needed, discussed later. Determining the positions of the atoms is generally aided by knowledge of the chemical structure and typical bond lengths and angles, as determined from low-molecular-weight "model compound" crystal structures. It is "confirmed," in the end, by comparison of calculations of the structure factors or intensities of the various reflections based on the positions and scattering of the individual atoms in all of the physical repeat units in the unit cell (as related by the symmetry of the lattice) and the experimental intensities. Details of the determination of the helical conformation and the unit cell are discussed in several texts [39,40], with that by Vainshtein [40] particularly recommended. Examples will be drawn from the use of electron diffraction, as described later. As pointed out in Ref. 39e, drawing, as well as defects, crystal thickness, and, in particular, temperature can have a significant effect on the unit cell parameters. For example, Bhatt et al. [41] have suggested that the angle between **c** and **b** (α)

---

*In triclinic and some monoclinic unit cells **c*** is not parallel to the fiber axis; see electron diffraction patterns in Fig. 37. This makes the determination more difficult.

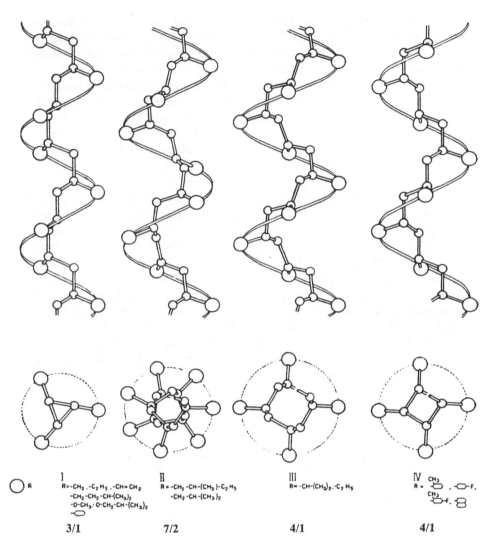

Figure 21 Representative helical conformation of isotactic polymers. The same pitch helix (different radius and starting point) can be drawn through any identical atom in the chemical repeat units (e.g., the backbone C containing the side group). The side groups, represented by R, corresponding to each helix are listed. A number of other helices are also found, such as 11/3 as well as 3/1 and 4/1 for polybutene (R = C_2H_5), 15/7 and 13/6 for PTFE, 9/5, or 28/16 for polyoxymethylene, etc. (From Ref. 38.)

in polyethylene terephthalate (PET) increases linearly with draw ratio, based on the separation of the $10\bar{5}$ peaks. The value of 98.5° usually reported corresponds to a draw ratio of 6–6.5. On the other hand, the separation may be due to the known tilt of the **c** axis relative to the fiber [45], varying with the amount of draw.

As indicated in Section I, infrared (and Raman) absorption spectra result from molecular vibrations that cause changes in the dipole moment (and polarizability) of the molecule; these spectra are unique to each molecule. They are used for configuration determinations but also depend on the conformation and molecular packing as they affect the intra- and intermolecular forces, respectively. For both IR and Raman, the normal mode vibrations of small atomic groups, such as CH_2 and $C=O$, are of concern, as affected by their local environment.

Conventional Raman spectroscopy, even though yielding equivalently useful information, has been used much less widely than IR for the study of polymers (see Refs. 8 and 42). Estimates are that it fails to produce usable spectra in more than 50% of normal samples [8], primarily due to fluorescence and self-absorption of the Raman photons. FT-Raman, with an attachment that can be added to some FTIR instruments in the near-infrared region, has improved the situation. Figure 22 is an example of the use of Raman for characterizing changes in the conformation of polyethylene terephthalate as it is annealed [43]; complementary studies by DSC and IR are described in Ref. 44 and Figs. 23 and 24.

It is well known that amorphous PET will crystallize when heated above T_g; at the heating rates used in a DSC it crystallizes at temperatures near 110°C. Figure 22(A) shows the Raman spectra in the region 750–1050 cm^{-1}; the peaks at 998 and 886 cm^{-1}, respectively, are assigned to trans (O-CH$_2$ stretch) and gauche conformations of the ethylene glycol segments. The 795 cm^{-1} band, shown to be independent of thermal treatment, was used as an internal thickness band. In IR the corresponding bands used were 973 (O-CH$_2$ stretch, trans), 898 (CH$_2$ rock, gauche) and 793 cm^{-1} (internal thickness). From the ratio of intensities of these bands, that is, $I(998)/I(795)$ and $I(886)/I(795)$, the fraction of trans and gauche conformations of the flexible -O-CH$_2$-CH$_2$-O segment (shown to be trans in the crystal by x-ray diffraction [45] can be obtained (Figure 22(B)). In agreement with DSC results (Figure 23(A)); crystallization (conversion of gauche to trans conformers) begins above T_g—for the annealing conditions used here, at ~90°C. Note that the initial bulk sample, an injection molded plate 3 mm thick, with the back-scattered Raman spectra being obtained, has a degree of crystallinity as determined by DSC of ~30% (Figure 23(B)). With a portion of the trans conformation known to occur in the amorphous regions, there is a clear discrepancy with the 25% trans conformer obtained by Raman. The "discrepancy," indicative of the power of the technique, is attributed to a skin-core morphology for the sample. The skin, being more rapidly quenched, is initially more amorphous than the core, with DSC giving a bulk crystallinity value and Raman sampling only, in this case, less than the 500 μm thickness observed by optical microscopy for the skin. Determination of the trans conformer content in a glassy amorphous sample (~15%) permitted separa-

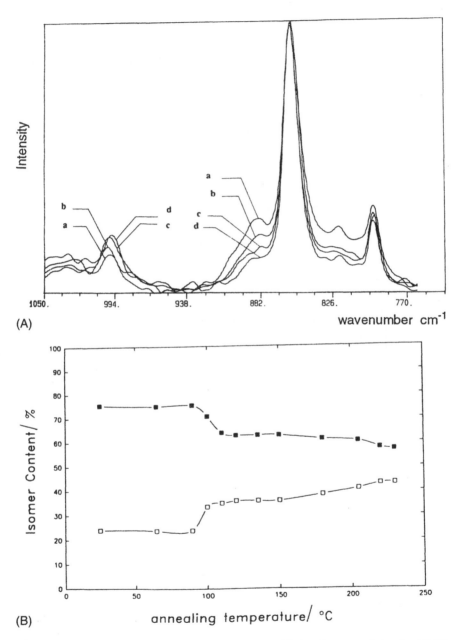

(A)

(B)

Figure 22 (A) FT-Raman spectra of PET: a) as molded and as annealed at b) 110°C, c) 150°C and d) 205°C for 1 hour. (B) Plot of distribution of trans (□) and gauche (■) conformers in PET based on Figure 22(A). (From Ref. 43.)

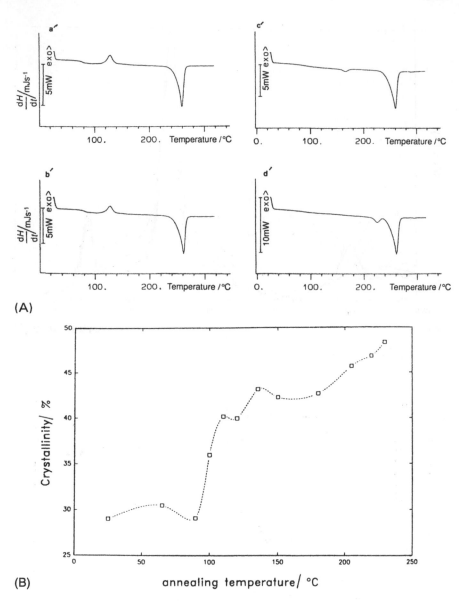

(A)

(B)

Figure 23 (A) DSC scans of PET as quenched (a′) and annealed at (b′) 90°C, (c′) 150° and (d′) 205°C for 1 hr. The exothermic peak in (a′) and (b′) results from crystallization during heating. (B) Plot of the DSC-determined crystallinity of an injection molded PET sample as a function of annealing temperature (1 hr). The crystallinity was calculated based on the ΔH_f measured relative to a 28.1 cal/g ΔH_f for 100% crystalline PET. (From Ref. 43.)

tion of the distribution of gauche (amorphous), trans (amorphous), and trans (crystalline) conformer as a function of annealing temperature (Figure 24(A)), with Figure 24(B) showing the (excellent) correlation between the DSC-determined skin-layer crystallinity and the crystalline trans content determined by Raman spectroscopy. Figure 25 show similar correlations obtained by IR, using the 973, 898 and 793 cm^{-1} bands (see also Figure 50).

As shown in Figure 22(B), in addition to the step change at about 100°C, there is a change in slope at ~140°C. Similar step changes and changes in slope are seen, for instance, for the width of the 858 cm^{-1} [complex mode consisting of ring C-C and C(O)-O stretching] and 1725 (C=O stretch) cm^{-1} bands (see original paper [43]). The step change is again related to conformation changes accompanying the crystallization above T_g, while the change in slope at ~140°C is attributed to the development of sufficient segmental mobility to permit further segmental rearrangement and an increase in crystallinity. We suggest the initial crystallization constrains the mobility of the remaining amorphous segments, with 140°C corresponding, in terms of Boyer's double-T_g model [46], to the higher T_g.

As an additional example of the application, in this case of IR, to details of the conformation of PET, we take an even earlier report by Koenig and Hannon [47]. Using solution-crystallized and annealed, molded as well as etched (fold surface removed) samples, in comparison with glassy and molten amorphous samples as extremes, they attributed the 988 cm^{-1} (and 1380 cm^{-1}) band to the (tight) gauche conformation of an adjacent reentry fold. With the 973 cm^{-1} band as a measure of the degree of crystallinity (trans content, discussed earlier) and the 793 cm^{-1} band as an internal thickness band, they could then follow the changes in numbers of tight folds and crystallinity with thermal treatment in both molded and drawn film (fiber-like) samples. As shown in Figures 25(A–D), the number of tight folds is zero in the melt and quenched amorphous samples, near zero in as-drawn films, but increases with annealing in both types of samples. As noted by the authors, the absence of the 988 cm^{-1} band does not necessarily indicate the absence of folds; its presence is interpreted as being due to the particular sequence of gauche conformations in a tight, adjacent reentry fold.

The preceding experimental methods have been supplemented (and in some views replaced) in recent years by semiempirical force-field calculations and molecular orbital calculations. Commercially developed programs (e.g., Ref. 48) permit determination of minimum energy conformations both for single molecules and as they are packed in a crystal (Figure 26). The Cerius^2 program, from Molecular Simulations, Inc., is of particular interest in that it also, based on the crystal packing module, permits display of the corresponding powder, fiber, and electron diffraction patterns for direct comparison with experimental results (Figure 27; see later discussions for ED pattern simulation and comparison). However, even within the same program, the use of different force fields for the description of inter- and intramolecular interactions yields different results.

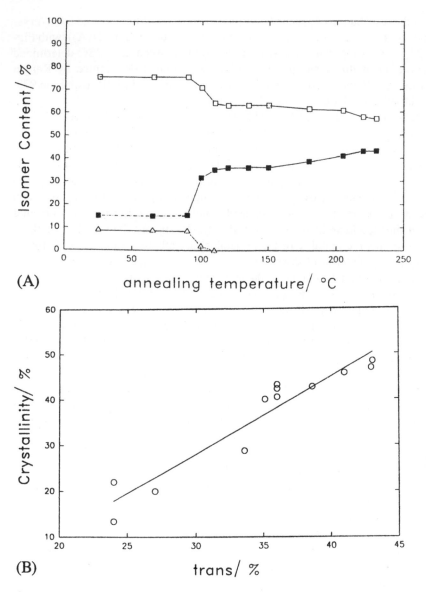

Figure 24   (A) Corrected distribution of trans and gauche conformers in injection molded PET: (□) gauche; (■) crystalline trans; (Δ) amorphous trans.   (B) Correlation of crystallinity in the skin layer as determined by DSC and the Raman-determined crystalline trans fraction. (From Ref. 43.)

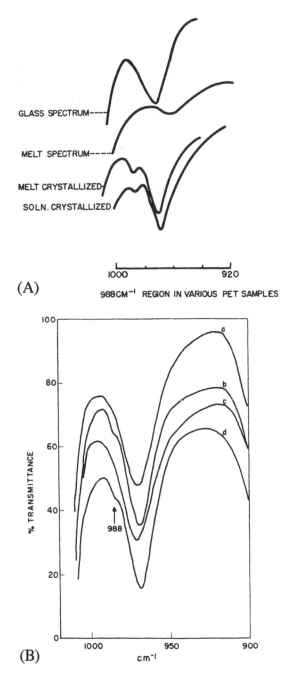

(A)

GLASS SPECTRUM----

MELT SPECTRUM----

MELT CRYSTALLIZED

SOLN. CRYSTALLIZED

1000   920

988CM⁻¹ REGION IN VARIOUS PET SAMPLES

(B)

% TRANSMITTANCE

100

80

60

40

20

0

a

b

c

d

988

1000   950   900

cm⁻¹

Figure 25 (A) 988 cm⁻¹ IR band in various PET samples. (B) IR spectra (900–1000 cm⁻¹) of an amorphous PET sample (a′) stretched 430%, (b′) annealed taut at

Sample	$\dfrac{A_{973} \text{ (cryst.)}}{A_{795}}$	$\dfrac{A_{988} \text{ (fold)}}{A_{795}}$	$\dfrac{A_{988}}{A_{973}}$
Melt-crystallized	2.94	0.44	0.15
PET high-IV (0.89) melt-crystallized	2.85	0.52	0.18
Crystallized from			
2-(2-butoxy ethoxy) ethanol (slow cool)	2.75	0.17	0.06
PET high-IV (1.60) melt-crystallized	2.65	0.41	0.16
Crystallized from			
2-(2-butoxy ethoxy) ethanol (190°C)	2.42	0.25	0.10
Crystallized from			
dimethyl phthalate	2.04	0.24	0.12
PET high-IV (0.89) as polymerized	1.66	0.10	0.06
(C) PET high-IV (1.60) as polymerized	1.64	0.16	0.09

Sample	$\dfrac{A_{973} \text{ (cryst.)}}{A_{795}}$	$\dfrac{A_{988} \text{ (fold)}}{A_{795}}$	$\dfrac{A_{988}}{A_{973}}$
(a) Amorphous film	0.33	0	0
(b) Sample (a) stretched 430%	1.28	0	0
(c) Sample (b) annealed at 140°C, 1/2 hr	1.49	0.08	0.05
(d) Sample (c) stretched 50%	1.76	0	0
(D) (e) Sample (d) annealed at 140°C, 1/2 hr	1.89	0.10	0.05

Figure 25 (continued) 140°C, 30 min (c′) part (b′) restretched 50% and (d′) reannealed at 140°C, 30 min.   (C) Table of ratios of absorbances of the 973 cm^{-1} (trans), and 988 cm^{-1} (fold) bands to the internal thickness 795 cm^{-1} band for the samples in Figure 25(A). Note the ratios needed to be corrected by the molar extinction coefficients in order to determine the "actual" degree of crystallinity and numbers of folds.   (D) Absorbance ratios for the drawn samples in Figure 25-B. (From Ref. 47).

Although, as indicated at the beginning of this section, x-ray diffraction characterization of polymer conformation (in crystalline regions) is aided by the use of fibers, it still requires development of a model from which calculated intensities of the reciprocal lattice "points" are compared with those measured. The x-ray fiber pattern corresponds to the rotationally averaged projection of the reciprocal lattice, with rotation about **c** (rather than **c***, which complicates determination of crystal structures of triclinic and monoclinic unit cells). We have contended for some time [50] that if appropriate samples are available, electron diffraction rather than x-ray diffraction is a better (simpler to interpret) method of determining crys-

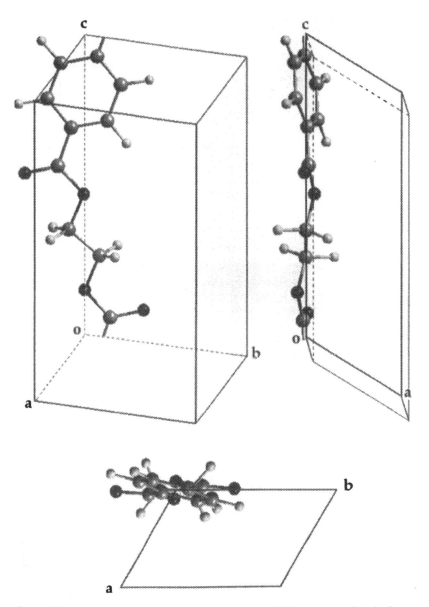

Figure 26 Simulation of the conformation of a single PET chain (one chemical repeat unit shown) and their packing in the crystal lattice. Additional segments are at each corner of the cell. The simulation was done using the Cerius^2 program [48a] (Universal force field) and the unit cell parameters described in Ref. 45. (From Ref. 49.)

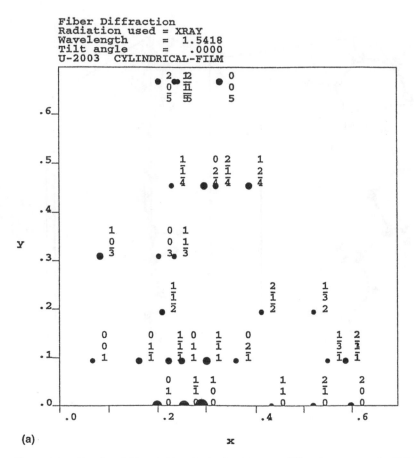

(a)

Figure 27 Simulated fiber (a,b) and powder (c) x-ray diffraction patterns for PET based on the crystal packing shown in Figure 26. $10\bar{5}$ at 42.5°C in the powder pattern is normally used for measuring the orientation of PET fibers. It is not seen in the "spot" fiber pattern (a) for which perfect orientation was assumed. In (b) a 5°C distribution (Gaussian, value at half maximum) was assumed for the simulation. (From Ref. 49.)

tal structure either by itself or at least in conjunction with x-ray diffraction. The potential of using ED has been greatly enhanced recently by Dorset's development [51] of *ab initio* direct phasing of the ED patterns permitting direct calculation of electron density projections. For general discussions of the application of electron diffraction to crystal structure determination see Ref. 52, with a number of recent papers on its application to synthetic and natural macromolecules in Ref. 53.

The advantage of electron diffraction is that, if appropriate samples are available, single-crystal or, if necessary, fiber diffraction patterns can be obtained. The single-crystal patterns are displays of the intensity distribution on a single plane

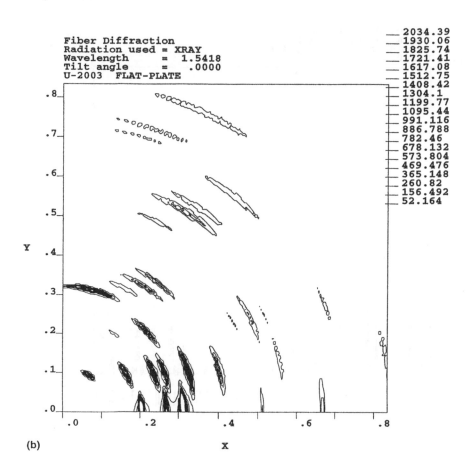

Fiber Diffraction
Radiation used = XRAY
Wavelength   = 1.5418
Tilt angle   =   .0000
U-2003 FLAT-PLATE

— 2034.39
— 1930.06
— 1825.74
— 1721.41
— 1617.08
— 1512.75
— 1408.42
— 1304.1
— 1199.77
— 1095.44
— 991.116
— 886.788
— 782.46
— 678.132
— 573.804
— 469.476
— 365.148
— 260.82
— 156.492
— 52.164

(b)

(or zone, due to the flatness of the sphere of reflection resulting from the small value of λ) of the reciprocal lattice from which, using Dorset's technique, the electron density of the unit cell projected on the corresponding real space plane can be calculated. On the other hand, with the simulation techniques the calculated intensities can be compared directly with the measured ones, with the results of small changes in conformation, packing, and various types of disorder being "immediately" visualized.

The problems in utilizing electron diffraction primarily are due to the need for "appropriate" samples:

1. The samples are preferably single crystals with lateral dimensions on the order of at least 1 μm (to yield sufficient intensity within the lifetime of the crystal as determined by beam damage and to permit isolation by selected area apertures).

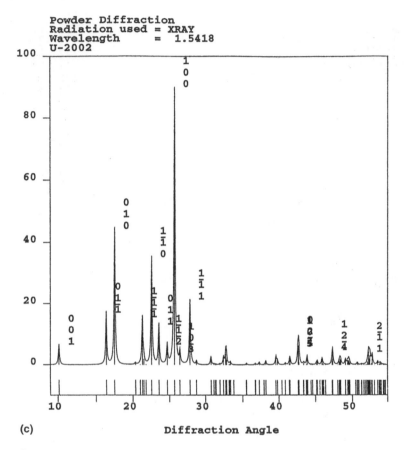

Figure 27   (continued)

2. They should be on the order, preferably, of 100 Å thickness to reduce or elim-
    inate dynamic (double) diffraction. The scattering power for electrons is suf-
    ficiently high that diffracted beams in thicker samples are rediffracted one or
    more times, resulting in "incorrect" intensity distributions.

3. If single crystals are not available, fibers (usually sheared films) can be used.
    These generally are annealed to develop as high crystallinity as possible, but
    still yield only rotationally averaged patterns. They still have advantages over
    x-ray patterns in that the flatness of the sphere of reflection permits obtaining
    00l reflections (meridian), as well as the averaged hk0 and hkl (equator and
    quadrant) reflections, without tilting the sample.

4. Assurance that the samples used are representative (same crystal structure
    as the bulk samples [fibers, films, or molded objects] of interest). This can
    be done by comparison with x-ray results. As shown later, the single-crystal

samples are often more perfect and may have different packing. For instance, parallel chain packing in crystals grown by simultaneous polymerization-crystallization, antiparallel chain packing in folded chain crystals, and statistically random chain direction in drawn fibers and films are to be expected, with different chain conformations in all three.

Sample preparation, as suggested earlier, is probably the critical factor in obtaining ED patterns, with it being in many cases more an art than a science. In the following we consider the methods, representative results, and potential difficulties.

1. *Crystallization from Dilute Solution.* As is well known (and discussed later), almost all polymers that are crystallizable can be crystallized from dilute (<0.1%) solution as lamellar, ~100Å thick, single crystals composed of folded (and therefore antiparallel or at least statistically random chain packing) molecules. This is independent of ongoing debates over the adjacency and regularity of the folding. These crystals yield hk0 ED patterns (Fig. 28), which are, in a sense, images of the hk0 plane of the reciprocal lattice permitting calculation of the electron density of the crystal projected on that plane in real space (Fig. 29). Interpretation of the electron density map, particularly if only the single hk0 map is available, in terms of the chain conformation is not straightforward; as in the case of interpretation of x-ray diffraction, it is usually compared with predictions from modeling.

2. *Epitaxial Crystallization.* For a reasonable number of polymers, epitaxial crystallization from dilute solution (see, in particular, papers by Wittman, Lotz and co-workers [56]) on various substrates (e.g., alkali halide or low-molecular-weight organic single crystals, or oriented polymer films) results in lamellar crystals in which the molecular axes are parallel to particular directions on the substrate surface, with, in many cases, one of the hk0 planes parallel to the surface. The lamellae grow out from the surface (Figure 30). Resulting ED patterns, with a greater potential for double diffraction due to the sample thickness parallel to the beam, permit determination of **c** axis spacings and may yield single, crystal patterns (or single-crystal patterns "crossed" at angles determined by the substrate lattice) of an hk0 zone. If so, electron density maps in a projection direction normal to the chain axis rather than parallel (as in crystallization from dilute solution) can be obtained.

3. *"Practical" Samples.* Much more difficult is the obtaining of appropriate (thin) samples from objects of practical interest, that is, bulk melt crystallized samples, fibers, films, etc. In a few cases crystallization of thin films (i.e., <1000 Å) from the melt on, for example, glass slides has yielded single crystals, or spherulitic structures in which the lamellae in a large enough area yield a common lattice orientation, suitable for ED [58,59]. Dynamic diffraction is of more concern than for crystallization from dilute solution. In other cases slow crystallization (i.e., at low supercooling) of normal-thickness samples

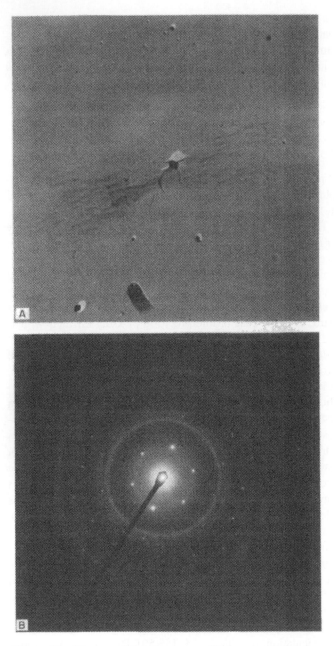

Figure 28   Electron diffraction pattern (B) from a solution grown single crystal of poly-ethylene sulfide (A). (From Ref. 54.)

(a)

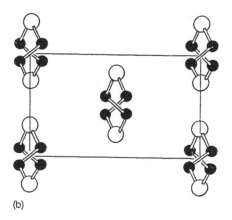

(b)

Figure 29 (a) Electron density of the polyethylene sulfide crystal projected on the hk0 plane and (b) corresponding projection of the unit cell. (From Refs. 54 and 55.)

with free surfaces will yield samples with few tie molecules. When single stage replicas are made, one or a few lamellae may adhere to the replica, permitting electron diffraction. It must be recognized, however, that in "thick" samples, particularly with free surfaces, low-molecular-weight and less crystallizable polymer molecules may exude to the surface during crystallization, causing concern over the representativeness of the patterns obtained.

Fibers often have a skin–core as well as a fibrillar morphology, with weaker cohesion between the skin and core and between the fibrils than between the molecules within the structural units. R. Scott showed many years ago [60] that it was

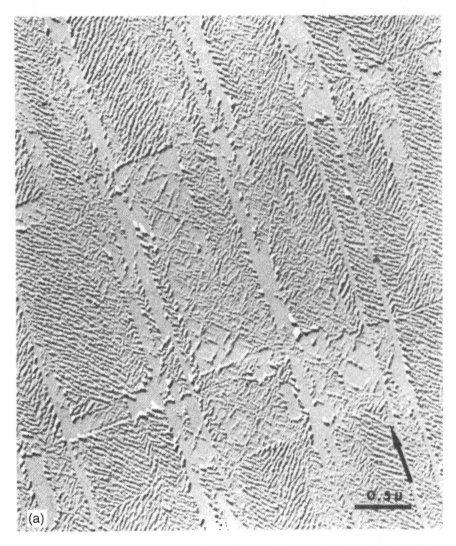

Figure 30   (a) Lamellar crystals of LPE crystallized from xylene solution at 108°C on a cleaved NaCl crystal. Arrows indicates an NaCl [100] direction. (b) Corresponding ED pattern from a thicker sample crystallized at 88°C. (From Ref. 57.)

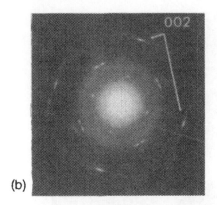

(b)

possible, for some fibers, to nick the surface with a razor blade and strip off a thin "ribbon," thin enough in some areas to obtain ED patterns and dark-field micrographs even in normal (100 kV) TEMs. These yield, however, fiber rather than single-crystal patterns.

Another "neat" technique for producing drawn films suitable for ED was developed by Gohil and Petermann (for detailed description see Ref. 61). The films are "extracted" from a molten pool, being drawn to varying degrees in the molten state and cooled rapidly, resulting in crystallization from an oriented state. The resulting shish-kebab morphology before and after annealing yields fiber-type ED patterns.

Fiber patterns can also be obtained, and most often have been obtained, by shearing (or "smearing") a polymer melt (e.g., with the edge of a razor blade or glass slide) on a glass slide. Following cooling, usually rapidly to prevent relaxation, the films are often annealed below T_m to develop improved crystallinity. Examples are shown in Fig. 31 for poly(2,6-oxynaphthoate) (PONA); after annealing below T_{k-m}, the film was sheared above T_{k-m}, that is, in the liquid crystal state. This pattern, which is similar in the diffuseness of the equatorial reflections to that in x-ray patterns from fibers of the same polymer (with 006 only being seen by tilting), should be compared with the single-crystal whisker patterns in Fig. 37.

On occasion, preferred orientation of an hk0 plane relative to the substrate is obtained. For a family of polyether liquid crystal polymers, single-crystal patterns (h0l) were obtained, but this is not normally the case. The polyether patterns (Figures 32(a) and 32(b) demonstrate the problems arising from triclinic and monoclinic cells; although $\mathbf{c}$ (the molecular axis) is parallel to the shear direction, $\mathbf{c}^*$ is not. Rotationally averaging this type of pattern, as observed from fibers, makes determination of indices much more difficult.

Recently [64] we have described the confined thin-film melt polymerization (CTFMP) technique for preparing lamellar (single crystal and single disclination

Figure 31   CTFMP-polymerized PONA sheared at 280°C. (From Ref. 62.)

domain) samples suitable for ED. Initially applied to thermotropic liquid crystal polyesters, it appears more broadly applicable (for review see Ref. 65) to condensation polymers in general, including copolymers. The CTFMP technique involves the simultaneous polymerization-crystallization (in some cases in the liquid crystal form initially) of the monomers in a thin film between glass cover slips; the result is extended chain lamellae, which, for reasons unknown as yet, are often ∼100 Å thick, regardless of polymerization times and temperatures. An example, for poly(2,6-oxynaphthoate) (PONA) [34a] is shown in Figure 33, with ED patterns (phase I and phase II) in Figure 34. The availability of patterns such as this, even if only they are available, greatly simplifies indexing of x-ray or ED fiber

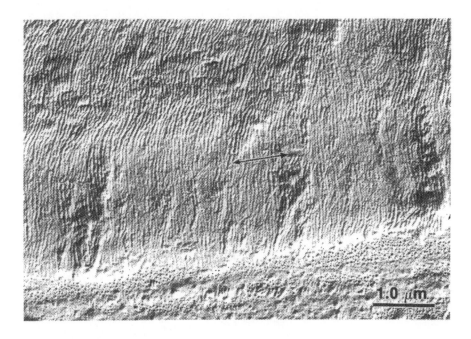

(a)

Figure 32 (a) TEM micrograph of a sample of an aromatic aliphatic azomethine polyether (C_{10} aliphatic segment) sheared at 195°C and annealed at 150°C, 2 hr ($T_{k-m} = 175°C$). The thickness of the lamellae, which are normal to the substrate and shear direction, is ~400Å; they consist of extended chains.

patterns. This is particularly true if two (or more) phases are present simultaneously in the "fibers."

Figure 35 shows micrographs and ED patterns of PET crystals prepared by the CTFMP technique. PET has a triclinic unit cell; thus if the **a** − **b** (hk0) real space plane is parallel to the substrate, **c** will lie at an angle (here substantial) to the beam direction (the normal to the substrate) as well as **a*** and **b***. This appears to be the case for the PET crystals polymerized-crystallized at 200°C, and only a few or no reflections can be obtained without tilting the sample in the microscope. On the other hand, for the crystals in Fig. 35c, **c** is parallel to the beam, **a*** and **b*** lie in the plane of the substrate, and hk0 ED patterns with several orders of reflections are obtained (Fig. 35d). Unfortunately, the thickness of these crystals, which consist of numerous (twinned) lamellae lying at an angle to the substrate, is such that dynamic diffraction is occurring; the relative intensities of the various reflections are "incorrect," and normal kinematic diffraction equations for the relation between intensity (amplitude) and atomic positions cannot

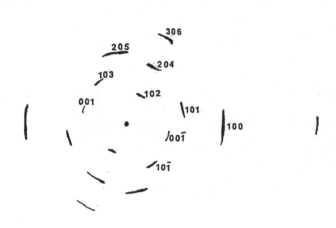

(b)

Figure 32 (continued) (b) Schematic diagram of an ED pattern from the sample in (a). This is a single crystal h0l pattern with the polymer having a triclinic unit cell (i.e. **c*** is at an angle to **c**). The fiber axis is vertical. (From Ref. 63).

be used. The "correct" distribution of intensities, based on the crystal structure described by Daubeny et al. [45], is shown in Figure 36. The Wilson test can be used to determine if the observed variation in intensities is affected by dynamic diffraction [67].

In a few cases hk0 zone (i.e., containing 00l reflections) single-crystal patterns have been obtained for polymers by both x-ray and ED. For the former solid-state polymerization (e.g., of POM from needle crystals of trioxane [68]) has been used. For the latter, whisker-type single crystals have been grown, by polymerization-crystallization, from solution for a number of thermotropic liquid crystal polymers (see Ref. 69 and references therein). An example, for PONA, is

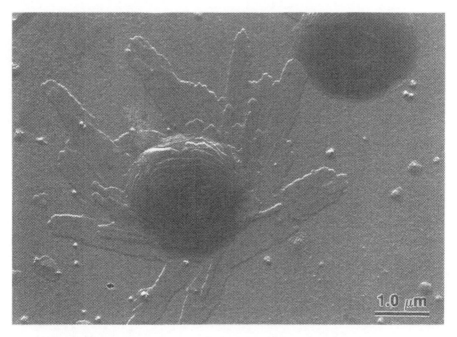

Figure 33 Lamellar single disclination of PONA, with protruding single-crystal lamellae, grown by the CTFMP technique at 180°C. (From Ref. 66.)

shown in Figure 37. Although x-ray powder patterns have been interpreted in terms of an orthorhombic unit cell [70], clearly these patterns (with the equatorial reflections corresponding to phase I, Figure 34) indicate a monoclinic unit cell. Crystal packing simulation, using Cerius^2, suggests the conformation and packing (parallel chains) shown in Figure 38, with the corresponding predicted patterns. The agreement, while close, is not totally satisfactory. Antiparallel (or statistical) chain packing is not expected, based on the polymerization process. The lateral streaks on some of the meridianal reflections is attributed to an axial disorder, for instance an occasional reversed chain direction.

Dark-field microscopy (Figure 37a, insert) shows a clear center line in the whiskers (also seen in PpABA whiskers); it is assumed this is the location of the H-bonded region of the original dimers, with growth in both directions on the acetoxy ends of the monomers, that is, parallel chain packing in each half of the whisker but in opposite directions. In the CTFMP lamellae the carboxy ends are presumed to attach to the glass surface (resulting in excellent adhesion), the chain direction being the same (acetoxy upward) in all of the lamellae in a given crystal or disclination domain. This permits rapid end-linking (chain extension), as is observed when the samples are heated above their T_{k-m}.

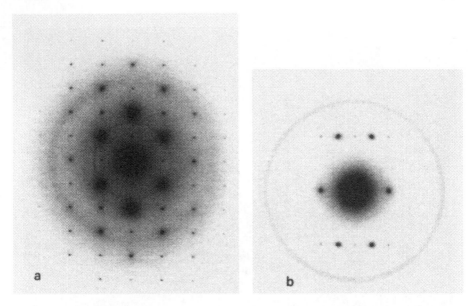

Figure 34 ED patterns of PONA phases I (a) and II (b) from CTFMP single crystals. (From Ref. 66.)

Both x-ray and ED can be used to follow changes in the molecular packing and conformation with temperature. Again ED has an advantage, if the samples are suitable, of being able to follow the changes in a single crystal (patterns). Examples, for PpOBA phase I and II (similar to the PONA lattice packings in projection along **c**), are shown in Figure 39. Although in most polymers beam damage is sufficiently high that different crystals are used for patterns at successively higher temperatures (or longer times at a given temperature), in these aromatic ring-containing polymers several patterns could be obtained from the same crystal as it was heated. Clearly there is a gradual change from the low-temperature crystal form to the high-temperature form (here over a 50°C temperature range) within an individual crystal, with diffuse scattering streaks (representing disorder) connecting pairs of reflections, a superlattice type structure at intermediate temperatures (shown by the satellite reflections), and considerable order in the high-temperature forms. The proposed transformations are shown in Figure 40-a. Furthermore, on cooling, there was shown to be a memory effect, of both phase and lattice orientation, with none of these effects being observable by x-ray diffraction, which averages over the entire illuminated volume. On the other hand, it is possible some of the observed ED effects are caused or affected by the adhesion to the substrate, with this adhesion preventing free rotation of the chains. The latter has been suggested to occur in the high-temperature form, at least segment-wise (phenyl rings and carbonyl groups), based on dielectric and broad-line NMR

Figure 35 CTFMP single crystals of PET polymerized at (a) 200°C, 10 hr and (b) its ED pattern obtained by tilting the crystal 27°C, (c,d) Micrograph and ED pattern (no tilt) of a PET crystal grown (polymerized) at 225°C, 6 hr. (From Ref. 49.)

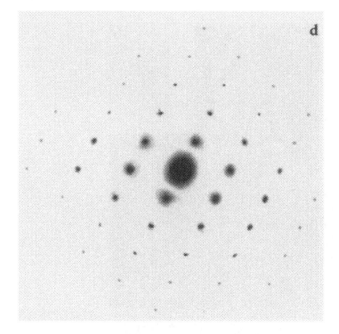

Figure 35 (continued)

measurements [72,73]. In our samples, in which the high-temperature forms (Figure 40-b) are metrically hexagonal, but are both orthorhombic, the phenyl ring orientation is maintained, leading to a difference in intensity of the 200 and 110 reflections, despite their identical spacings. If this occurs in the bulk samples, it again could not be observed by x-ray diffraction.

An important factor to note in all of the foregoing discussions of interpretation of diffraction results is that while disagreement between the intensities predicted on the basis of particular model of molecular conformation and packing can prove the model incorrect, agreement does not prove it to be correct. Another model may also be found leading to similar or better agreement. Furthermore, even if disagreement occurs, caution is needed in rejecting a model; we have found that only slight shifts in axial position or molecular conformation can lead to substantially improved agreement.

There have been limited applications of high-resolution TEM to lattice imaging of polymers, with recent image processing techniques permitting resolution of individual molecules (for review see Ref. 74). Although they were initially thought limited to highly radiation-resistant polymers, such as poly(p-xylene) (PPX) and poly(p-phenylene terephthalamide) (PPTA), even PE single crystals have yielded lattice fringes (Figure 41-A).

Figure 41-B is an example of the application of image processing to a PPX single crystal. Figure 41-B (b') is the ED pattern from a β form crystal as shown

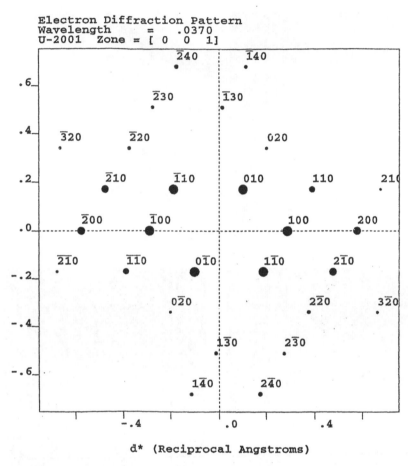

Electron Diffraction Pattern
Wavelength   =   .0370
U-2001 Zone = [0   0   1]

d* (Reciprocal Angstroms)

Figure 36 Simulated ED pattern based on the crystal diagrammed in Figure 26. Note the 020 and higher order 0k0 reflections are much weaker than in the pattern; this is tentatively attributed to dynamic diffraction. (From Ref. 49.)

in Figure 41-B (a'). Its high-resolution dark-field image, taken using the encircled 400 reflection and its satellites, is shown in Figure 41-B (c'); and 18-Å 400 planes can be seen in two of their three directions. In Figure 41-B (d') an optical diffraction pattern taken from the original negative (currently the negative would be scanned and the Fourier transform calculated using a computer) is shown; the similarity to that in (b') is clear. An optically filtered (mask with holes placed over the diffraction spots in the optical bench used to reform the image) image is shown in Figure 41-B (e'); in the image one is looking down the molecular axis. This corresponds to taking the inverse Fourier transform with only the "spots" of

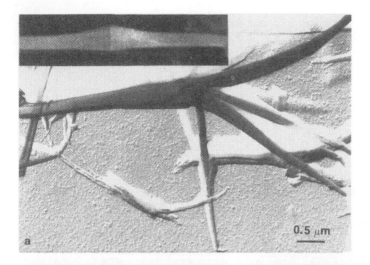

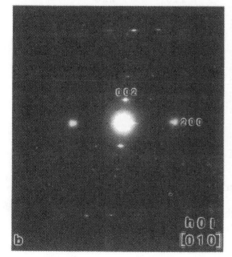
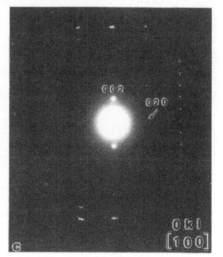

Figure 37 (a) PONA whisker with (b) [010] (h0*l*), and (c) [100] (0k*l*) zone ED patterns. A dark-field micrograph using hk0 reflections is inset in (a). (From Ref. 69a.)

the initial transform being used. Figure 41-B (f′) is a representation of the image, with (g′) being the unit cell as derived from a crystal structure analysis using WAXD [76b].

Figure 41-C is an example of the application of high-resolution TEM to fragments of a PPTA fiber. Using 006 and 110, 200 reflections of the ED pattern (a′)

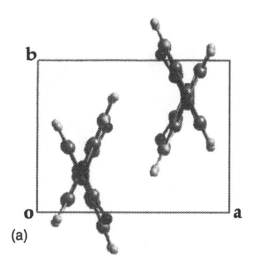

(a)

Figure 38   (a) Crystal structure (a,b,c) suggested by the Cerius [2] simulation program for PONA phase I with corresponding simulated ED patterns, (d) [001], (e) [100], and (f) [010]). (From Ref. 66.)

the dark-field images in (b′) and (c′) were obtained. The 006 image suggests alternating domains in and out of diffracting position; this can be related to the pleated sheet conformation of the fibrils and molecules in the fiber [76d]. Using the 110, 200 reflections, from planes parallel to the molecular axes, small crystallites are seen randomly everywhere along the axis. In Figure 41-D is shown a high-resolution image of the sample along with its optical diffractogram. The 4.3-Å lattice fringes correspond to the 110 planes, with the bright areas with fringes being of the same size as the bright spots in the dark-field image. Some fluctuation in the fringe direction can be seen in the image with a slight curvature of the image being seen near the arrow.

In all of the foregoing the concentration has been on the conformation (and packing) of the molecule in the crystalline regions of a polymer. Conformations in the amorphous regions, being, almost by definition, more random, are more difficult to determine. As already shown, IR and Raman are useful for local conformations, such as relative numbers of trans and gauche conformers. Mobility of these constituent groups can be characterized by "broad"-line NMR, including spin–lattice and spin–spin relaxation-time measurements permitting determination of the temperatures (relaxation or T_g) at which the various types of groups can undergo rotation (see, e.g., Ref. 77). For characterization of the molecular conformation in amorphous regions, particularly of totally amorphous polymers, on a

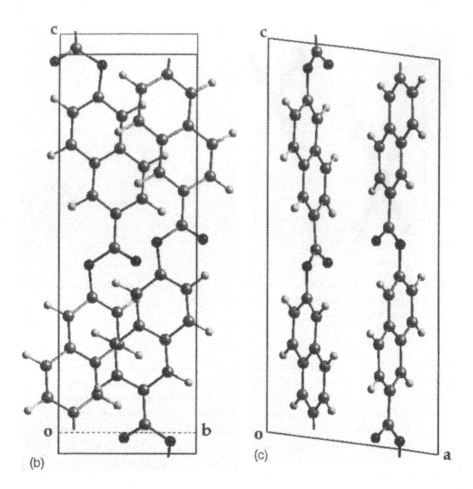

Figure 38 (continued)

larger scale, normally stated in terms of the mean square end-to-end distance, $<r^2>$, or radius of gyration, R_g, neutron scattering has been the primary tool. It depends on the difference in scattering power of ^1H and ^2H. In practice a dilute solution of deuterated polymer in hydrogenated (or vice versa) is prepared. The resulting (small-angle) neutron scattering pattern, interpreted in the same way as the light-scattering patterns from dilute solutions of a polymer in a solvent, yields R_g and, by assumption of a random coil, $<r^2>$. The interpretation also assumes the deuterated and hydrogenated molecules are thermodynamically perfectly miscible, a feature that does not always occur. However, for glassy amorphous samples, which can be quenched from the melt, this is of less concern than when similar studies are used to describe the conformation of chains in the crystalline regions of partially crystalline samples.

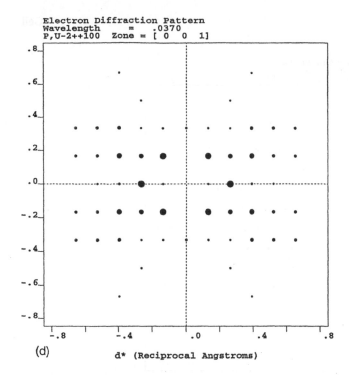

Electron Diffraction Pattern
Wavelength   =   .0370
P,U-2++100   Zone = [0   0   1]

(d)   d* (Reciprocal Angstroms)

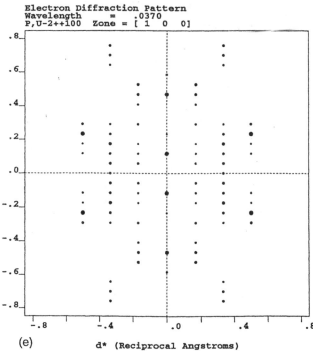

Electron Diffraction Pattern
Wavelength   =   .0370
P,U-2++100   Zone = [1   0   0]

(e)   d* (Reciprocal Angstroms)

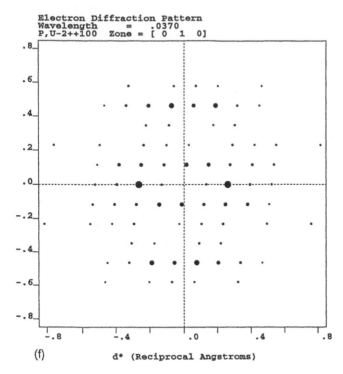

(f)   d* (Reciprocal Angstroms)

Figure 38   (continued)

In a recent development Wunderlich and co-workers [78] have used a full-pattern (two-dimensional Reitveld) wide-angle x-ray diffraction refinement [79], in conjunction with DSC, SAXS, and thermal mechanical analysis, to characterize the amorphous "structure" in a PET fiber. Based on a model of the crystallite structure, a three-dimensional diffraction pattern was calculated and then "rotated" to produce a two-dimensional fiber pattern for comparison with that observed at each point for which data was collected using a four-circle diffractometer. The latter collects data as for a pole figure. The model used incorporates parameters describing the average size of the crystallites in three dimensions, the distribution of orientation of the crystallites relative to each other and the average orientation relative to the fiber axis, the unit cell parameters, atomic positions (determined using a rigid body model), a temperature factor, a paracrystallinity matrix for defects of the second kind, and instrumental parameters.

Figure 42-A shows a contour plot of one quadrant of the observed fiber diffraction pattern, with Figure 42-B being the crystalline-phase diffraction pattern based on the model, with the background removed. Figure 42-C is the residual fiber scattering after subtraction of the crystalline scattering [i.e., (b) − (a)]. Assuming

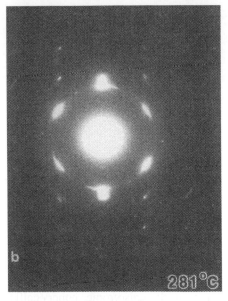

Figure 39 ED patterns of phase I PpOBA CTFMP crystals at (a) room temperature, (b) 281°C, and (c) 346°C with schematic (d) of 346° pattern. ED patterns of phase II ppOBA CTFMP crystals at (e) room temperature; (f) 282°C with (g) schematic, and (h) 346°C with (i) schematic. (From Ref. 71.)

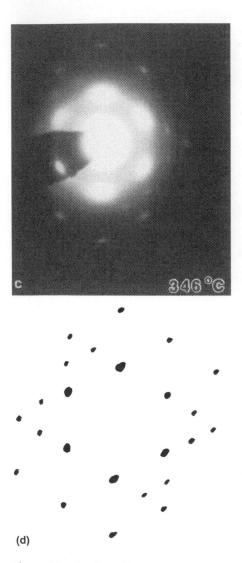

(d)

Figure 39 (continued)

the presence of an isotropic amorphous phase with a truly liquid-like structure results, when it is subtracted, in the pattern from the anisotropic, noncrystalline
phase shown in Figure 42-D.

This anisotropic phase, defined as the intermediate phase, is clearly oriented,
with the broad peaks at 0.2 and 0.28 Å^{-1} on the equator being near the 010 and
100 reflections. Wunderlich and co-workers suggest the initial modulus of the PET
fibers is primarily determined by the amount and orientation of the intermediate
phase. The tenacity also is strongly dependent on this phase. They suggest it is

mainly present between the fibrils, with the "truly" amorphous regions being between the crystallites within the fibrils.

DSC measurements of the heat of fusion (yielding w_c), and the change in specific heat at T_g (yielding w_a) permit determination of the rigid (rigid above T_g) amorphous fraction, that is, $w_r = 1 - w_c - w_a$. This rigid amorphous fraction is presumed related to the intermediate phase. Figure 42-E shows a comparison of the three phases as determined by x-ray and DSC.

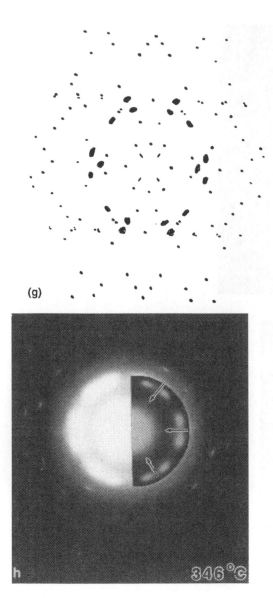

(g)

Figure 39 (continued)

B. Degree of Crystallinity and Orientation

1. Crystallinity

Determination of the degree of crystallinity, from which a single number representing the fraction of molecular segments in crystalline regions is desired, are, perforce, based on a two-phase model; that is, the segments (of lengths down to a

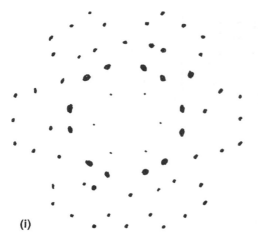

(i)

single C atom) are to be assigned to either a crystalline or amorphous region. Clearly, in reality this cannot be done, since, as will be seen, the assignment depends to some extent on the characterization technique used and the two-phase model ignores the possibility of intermediate degrees of order (and disorder).

Standard (traditional) techniques for characterizing the degree of crystallinity include x-ray diffraction, density, heat of fusion, and IR, with x-ray being, in one sense, the fundamental technique. All of these, however, characterize different aspects of the degree of order in the sample: for example, x-ray characterizes long-range order; density characterizes overall packing; heat of fusion characterizes changes in entropy as well as enthalpy; and IR characterizes short-range order. It is thus to be expected that the various methods will give different answers for a given sample even if the methods have been standardized (i.e., ASTM or ISO) for particular samples to give the same answers. For the density method, x-ray diffraction (or ED) is used to determine the volume and number of physical repeat units in the unit cell, permitting calculation of the density of a 100% crystalline sample. Likewise, heat of fusion measurements are calibrated against either density or x-ray measurements if 100% crystalline samples, as is usually the case, are not available. IR is, in some cases, calibrated similarly, with x-ray characterization of unit cell and conformation also aiding in the determination of appropriate bands to use. For details of the techniques, readers are directed to various general [27] and specific texts (for x-ray, Ref. 39; specific heat, Refs. 7 and 80; and for IR, Refs. 6–8).

In the simplest methods of x-ray crystallinity characterization, diffractometer scans of compression molded plaques (total thickness usually 1/16–1/8 in) are used. Various methods (see Ref. 39) can be used to draw in a line separating the background (air and incoherent scatter) from the polymer scattering over some pre-selected range of scattering angle 2θ plus another line separating the crystalline

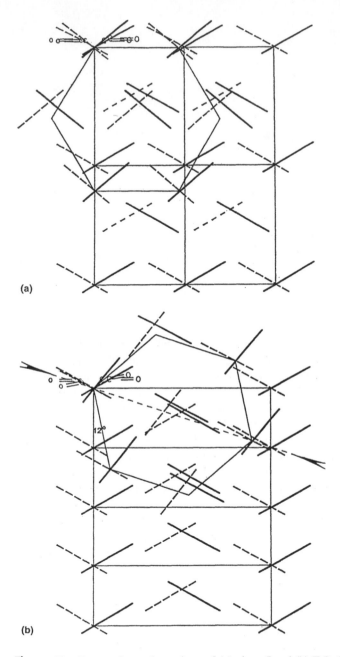

Figure 40 Proposed transformations of (a) phase I and (b) II PpOBA unit cells. Recent simulation (66) suggests the carbonyl groups (C=O) are closer to the planes of the phenyl rings in the room-temperature forms. Diagrams of the proposed orthorhombic high temperature (c) I (HT) and (d) II (HT) crystal structures. They are metrically hexagonal. The solid and dashed lines represent successive phenyl rings along the chain. (From Ref. 71.)

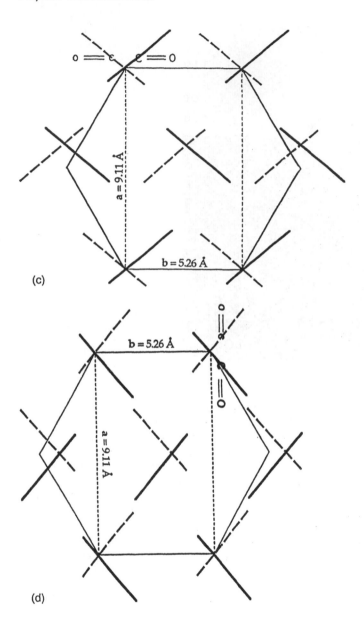

(c)

(d)

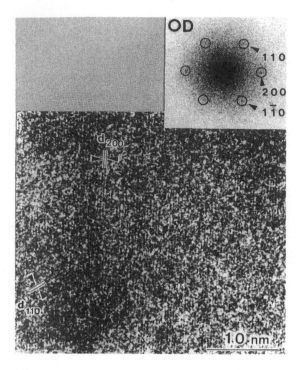

(A)

Figure 41 (A) Lattice image of a PE single crystal taken at a magnification of 90,000 at 4.2°K and 160 kV (to reduce beam damage). Inset is the optical diffraction pattern from the negative indicating the 110 and 200 planes are resolved in the micrograph (From Ref. 75.) (B) (a′) PPX single crystal.   (C) (b′) ED pattern of a β form. (c′) high-resolution dark field image taken using the encircled 400 reflection poly(p-xylylene) (PPX)   (D) (d′) Optical diffraction pattern from a negative taken at 500 kV, (e′) lattice image obtained by optically filtering the image, (f′) model structure based on ED, (g′) structural model showing orientation of the phenyl rings; the molecules at the corners statistically take one of the three equivalent orientations shown. The bold line on each phenyl ring represents the upper part of the ring. (From Refs. 76a,b.)   (C) (a′) ED pattern and (b′,c′) dark-field micrographs of the PPTA sample used for (D). In (b′), using the 110 and 200 reflections, small crystallites are seen, of approximately the same size as the domains showing the lattice fringes in (D). (From Ref. 76c.)   (D) High resolution of a PPTA torn fragment from a Kevlar fiber originally annealed at 400°C. The molecular axis is vertical, with the insert pattern being the optical diffraction pattern from the negative. (From Ref. 74.)

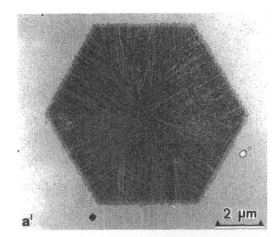

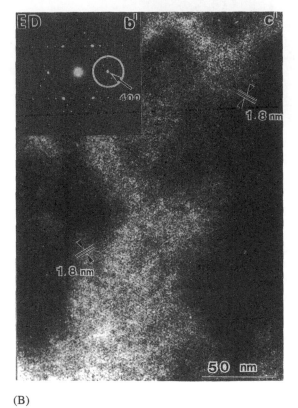

(B)

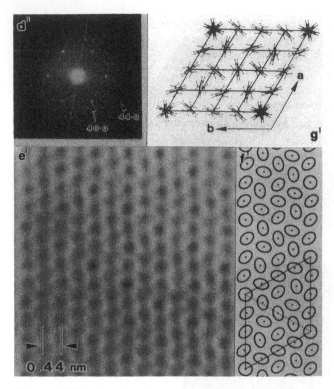

Figure 41(B) (continued)

from the amorphous scattering (e.g., Fig. 43). The fractional crystallinity is then given by the ratio of the area of the crystalline scattering to the total polymer scattering, with the crystals having to have lateral (hk0) diameters of some 8–10 molecules to be considered crystalline. While the different methods give different absolute values, all (should) extrapolate for polyethylene to a 100% crystallinity at a density of 1.00 g/cm³, corresponding to the density of the unit cell but to different densities for the (unattainable) 100% amorphous material. In our view, since it is known the two-phase model is incorrect, the relative values obtained by any of the methods can be used to predict variations in properties, with appropriate recognition of potential problems. The method can be used even for samples having more than one crystalline phase, either for individual crystallinities or overall values.

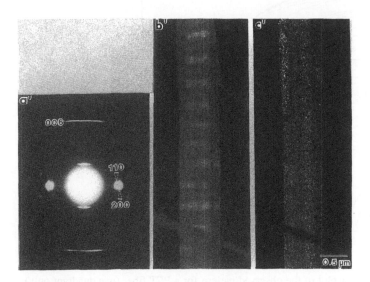

Figure 41(C)

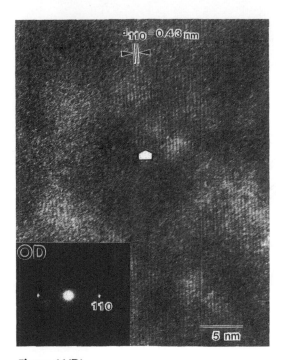

Figure 41(D)

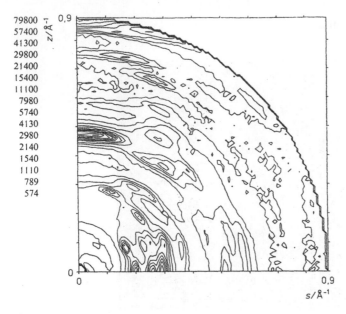

79800
57400
41300
29800
21400
15400
11100
7980
5740
4130
2980
2140
1540
1110
789
574

z/Å⁻¹ 0,9

0

0   0,9

s/Å⁻¹

Figure 42-A Contour plot of the observed pattern for a PET fiber (one quadrant only). The intensity values of the contours, starting from 0, are shown at the upper left. The pattern has been corrected for absorption, polarization, and Compton scattering.

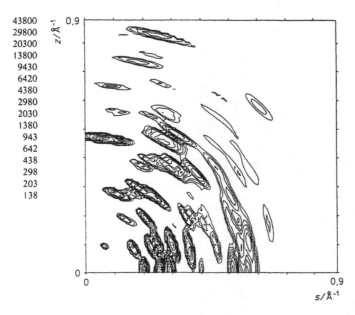

43800
29800
20300
13800
9430
6420
4380
2980
2030
1380
943
642
438
298
203
138

z/Å⁻¹ 0,9

0

0   0,9

s/Å⁻¹

Figure 42-B Contour plot of the crystalline phase diffraction. This can be compared with the plot in Figure 27-b.

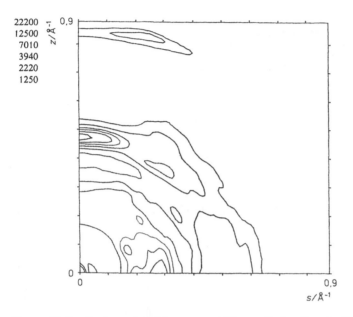

Figure 42-C Contour plot of the noncrystalline scattering (i.e., A−B).

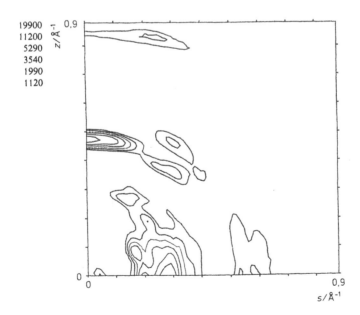

Figure 42-D Contour plot of the anisotropic, noncrystalline (i.e., intermediate phase) scattering obtained from Figure 42-C by subtraction of an isotropic scattering.

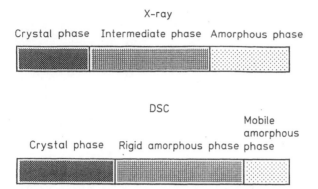

Figure 42-E Comparison of the relative amounts of the three phases as defined by x-ray scattering and DSC. (From Ref. 78b.)

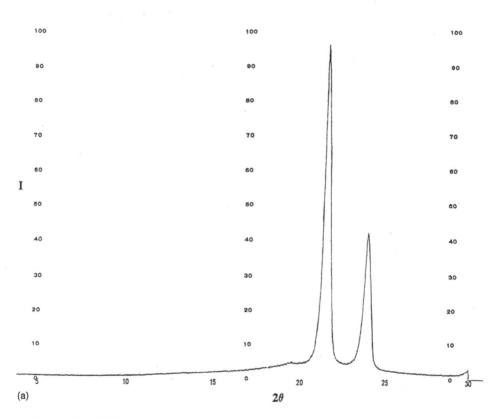

Figure 43 Wide angle x-ray diffraction patterns of PE samples of (a) 80%, (b) 55% and (c) 37% crystallinity.

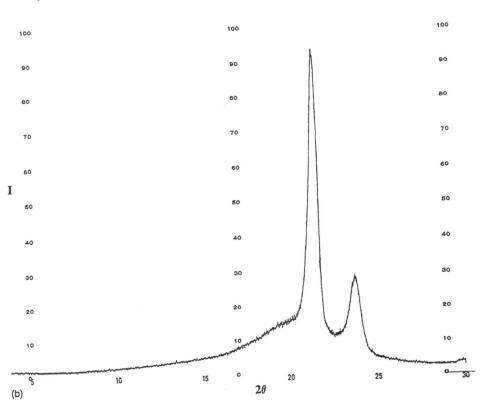

(b)

2θ

For fibers a major problem in x-ray determinations of crystallinity is the need for random crystal orientation in the x-ray beam. By some means, such as chopping and mixing, the fibers need to be randomized in space. This is often difficult and thus other methods are more appropriate, with x-ray patterns from molded samples being used for calibration. In addition, if filled samples (e.g., pigment, plasticizers, fibers, etc.) are to be measured, diffraction from the additives must be subtracted.

Density measurements can be made using density-gradient columns or, more precisely, if thought useful, by flotation. Fibers, films, or parts of molded objects can be used, with the major problem being the presence of voids or fillers in the samples. In addition, if two crystalline phases are present, with different crystal densities, there is no way by density alone to separate their contributions or determine an overall value.

Thermal analysis, by differential thermal analysis (DTA) or DSC, can be used to measure the heat of fusion of small ($\sim$10 mg) samples. DSC measurements are more often used, with the peak areas being directly related to the amount of heat

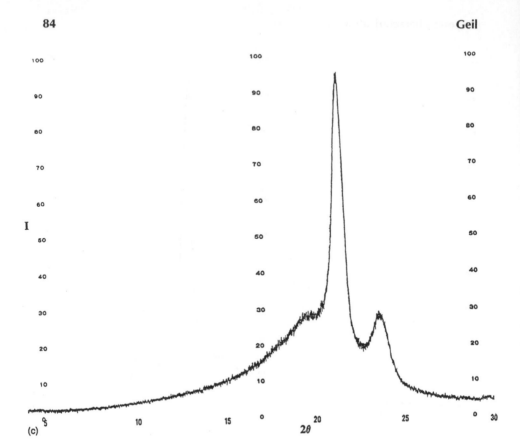

Figure 43   (continued)

absorbed by the sample as it melts, whereas DTA requires calibration. For traditional methods see Refs. 7 and 80. Required is either knowledge of the heat of fusion of a crystalline sample, or extrapolation as a function of crystallinity measured
by x-ray or density. As in the case of x-ray measurement, a baseline needs to be
drawn in, with ΔH_f being related to the area under the peak (see Fig. 44).

Problems arise if crystallization or annealing effects accompany the melting.
Similar problems, for fibers, could result from relaxation of the orientation. A
recent development [81], modulated DSC, permits separation of the reversible
(heat capacity related) and nonreversible (kinetic) effects occurring during heating of the sample, with the difference giving the initial crystallinity of the sample
(Fig. 44). In making these measurements, a small sinusoidal temperature variation
(here $\pm 0.53°$C/40 sec) is superimposed on the 5°C/min heating rate. The total
curve in this figure represents the normal DSC scan of a quenched PET sample.
The T_g is seen as the change in baseline at 75°C. Crystallization from the glass
(exothermic) is seen as the peak at 134°C (extrapolated onset 126°C), with a total
ΔH of 36.40 J/g, while melting occurs with a peak at 253°C (extrapolated onset

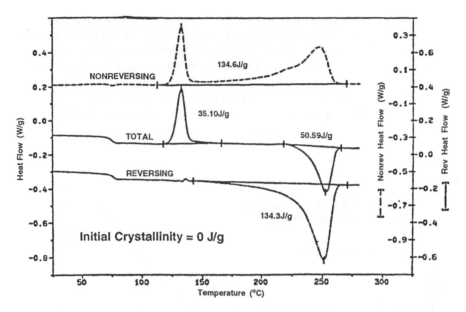

Figure 44 Modulated DSC scan of a quench-cooled PET film measured at a 5°C heating rate with a ±0.53°C amplitude modulation, 40 sec periodicity. The total scan is that normally measured with crystallization occurring at ~135°C, melting at 250°C. The "nonreversing" curve shows both crystallization at 135°C and further crystallization from ~175 to 240°C. The "reversing" curve represents the total heat of fusion of all the crystals formed during heating; because it is equal to the nonreversing ΔH_c it indicates there was no initial crystallinity. (From Ref. 81c.)

239°C) having an area of 50.59 J/g. This suggests some crystallinity was present in the initial sample, an amount that could be quantized if the ΔH_f for the crystal were known. However, the authors [49] claim x-ray indicates no crystallinity in similarly quenched samples. The reversing component of the total curve, with an area of 134.3 J/g and a single, broad peak at 250°C, represents the ΔH_f of both the initial crystallinity and the melting of all the crystals that formed during heating. The nonreversing component represents the exothermic crystallization processes occurring during heating. The cold crystallization peak at 134°C is the same as in the total scan, but additional ordering is seen to occur in two steps, at ~220 and 240°C. The total area is 134.6 J/g, the same as the heat absorbed during melting; that is, the sample had no initial crystallinity.

There are, in general, a number of structure-dependent IR (and Raman) bands; these include both those that are conformation dependent and those split due to the symmetry of the crystal. A list of these structure-dependent bands is given, for example, in Ref. 82. Since the conformation and the splitting are local-order, short-range effects, crystallinity measurements by IR can differ significantly from measurements by other techniques. For instance, for PET, the ratio of trans CH_2 conformers to total ethylene glycol units (Figure 22), even though all segments in the crystalline regions have a trans conformation, does not represent the degree of crystallinity; as shown, trans conformers also occur in the amorphous region.

Another example is polypropylene (PP). In both the hexagonal and monoclinic crystal forms the molecules have a 3/1 helical conformation. The 1003 cm^{-1} band has been attributed to segments with this conformation, while the 1028 cm^{-1}

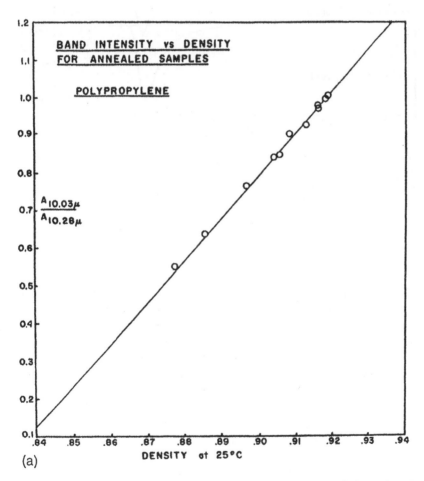

(a)

Figure 45 Plot of $A(1003)/A(1028)$ versus density for (a) annealed samples of PP and (b) quenched samples as they are annealed. (From Ref. 83.)

band can be used as an internal thickness band. As shown in Fig. 45a, a plot of $A(1003)/A(1028)$ versus crystallinity as measured by density, for slowly cooled samples, gives a linear plot, suggesting its use; presumably one could characterize the sum of the α and β form crystalline phases. However, as shown in Fig. 46, a quenched and a slowly cooled PP sample have nearly the same IR scans, whereas the x-ray patterns differ drastically. Figure 45b shows the changes in the $A(1003)/A(1028)$ ratio versus density as the quenched samples are annealed; they

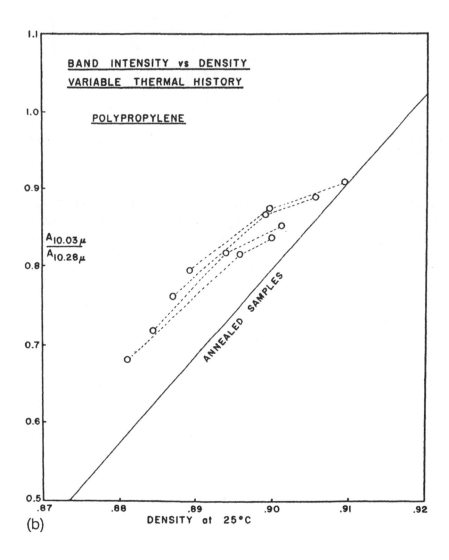

(b)

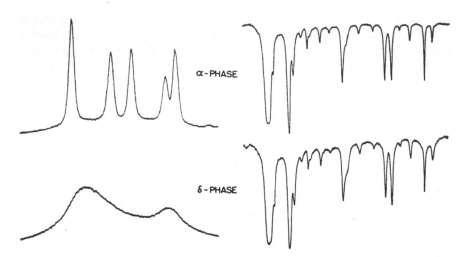

α-PHASE

δ-PHASE

Figure 46 X-ray and IR scans of annealed (upper) and quenched (lower) samples of PP. (From Ref. 83.)

approach the line obtained for the slowly cooled samples. The "problem" is attributed to the attempt to relate the 3/1 conformation to the degree of crystallinity; when PP is moderately quenched it "crystallizes" in the so-called smectic form [84]. Although details of the molecular packing are still being debated, it consists of domains of aligned 3/1 helical molecular segments. This short-range order is seen as "crystalline" by IR but nearly amorphous by x-ray (and somewhere in between by density and ΔH_f).

As Koenig points out [8], although crystal field splitting can be observed for PE (for the 720 methylene rock and 1460 cm^{-1} methylene bands), permitting their use for crystallinity measurements, IR is generally not suitable for other polymers. In PP it is the intramolecular helical splitting, not the intramolecular crystalline splitting, that is observed: "Unfortunately, the spectral results for PP are typical for most semicrystalline polymers. Generally the chains are simply too far apart and the intermolecular forces are so small that the crystal field splitting is not observed" [8]. As he suggests, and we, density is a simple, rapid, and cheap method of measuring crystallinity, while x-ray serves as the "final test."

All of the foregoing methods are based on the two-phase concept. In actuality, polymer crystals are expected to have a variety of defects and/or disorders. For instance, polyethylene single crystals grown from solution have a crystallinity, as measured by x-ray or density, of only 80%, with only part of the amorphous component being attributable to the folds [85]. Furthermore, in bulk crystallized polyethylene and other polymers a crystalline–amorphous interphase of varying degrees of mobility is proposed, based, for example, on heat capacity, NMR, and x-ray measurements [78].

Hosemann has described one method of characterizing the possible disorder in crystalline regions (see Refs. 6, 39, and 86) (Fig. 47). Disorders of the first kind retain the long-range order of the lattice. However, either (1) the lattice points are statistically displaced small amounts from their proper positions (as in thermal vibrations), (2) there is a statistical distribution of up and down helices or helix hand, or (3) there is statistical substitutional disorder, as in random copolymers of like size units. In distortions of the second kind each lattice point varies statistically in position relative to its nearest neighbors; that is, there is short-range but not long-

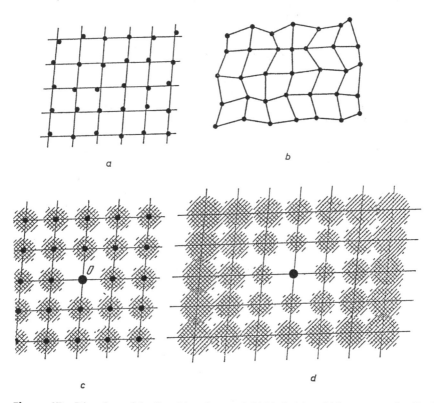

Figure 47 Disorders of the first (a) and second (b) kind; (c) and (d) represent the distribution functions (average positions of the atoms relative to any center) for (a) and (b). The disorder in (a) could also involve a statistical distribution of up and down or right- and left-handed helices, or of a substitutionally disordered lattice of equal size units; (b) would result if the substitutional units have a different size [as in random P(ONA)p(OBA) copolymer]. (From Ref. 40.)

range order. This can occur in random copolymers of unlike size units. The effects of these disorders on the diffraction patterns (in x-ray diffraction and ED) is described in Table 5. From the decrease in intensity of increasing order of a given reflection and the broadening (of all reflections if due to size, or of increasing orders if due to distortions of the second kind) the degree of disorder can be characterized. The statistical and substitutional disorders result in diffuse scattering (streaks if x-ray fiber or ED single crystal patterns are observed, as in Figure 37). The presence of the disorders clearly complicates the measurement of the crystallinity. The thermal vibrations, for example, subtract intensity from the crystalline peaks and add it to the diffuse background, suggesting an increase in amorphous content. If the crystals are small enough (<10 repeat distances, i.e., ~50 Å), the breadth of the reflections may be so large they are difficult to distinguish from amorphous scattering; this is one interpretation of the scattering from the "smectic" PP.

Table 5   Types of X-ray Scattering by Various Types of Crystals.

System	Diffraction	Diffuse Scattering
Ideal crystals without disorder	Sharp	None
Distortions of the first kind		
Statistical disorder and		
substitution type	Sharp	$N(<f^2> - <f>^2)$. continuous, gradually varying
Frozen-in thermal vibration		
type	Sharp, decrease in I with θ	$N<f>^2 (1-D^2)$, yields temperature diffuse scatter, increasing background with θ
Effect of crystallite size	Broadened equally for all orders of given reflection, inverse to N in the lattice direction	same as above if present
Distortions of the second kind	Broadened increasingly with θ, proportional to degree of disorder in lattice direction	same as above if present

Note: f = structure factor for unit cell; $<f^2>$ = mean square value, perfect or imperfect; D = Debye factor = $\exp(-8\pi^2 <\mu^2> \sin^2\theta/\lambda^2)$ with $<\mu>^2$ = mean square amplitude of vibration (of atoms); and N = number of unit cells (in given lattice direction).
Source: Modified from Ref. 6.

Ruland [87], with modifications by Vonk to permit easier calculation using a computer [88], developed a method to characterize both the crystallinity and the degrees of both types of disorder. As described in Ref. 39e, in particular, it involves separation of the incoherent background scattering and the amorphous scattering from the crystalline scattering, for various increasing ranges of 2θ. Comparison of the integrated crystalline scattering, corrected by adjustable terms characterizing the two types of disorder, to the total integrated intensity should be a constant, the degree of crystallinity, regardless of the range of angles chosen. Thus both crystallinity and disorder can be obtained. An example, for PP, is shown in Figure 48 and Table 6.

2. Orientation

Characterization of the degree and type of orientation, of particular concern in fibers and films but also of interest in, for example, injection-molded (in which variations through the thickness and with distance from the gate affect the properties) and extruded objects, is relatively straightforward for the molecular seg-

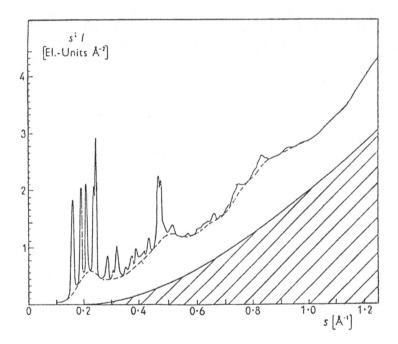

Figure 48 Plot of Is^2 versus s ($s = 2 \sin \theta/\lambda$) for a quenched sample of PP annealed 30 min at 160°C. The rise at large angles is due to the multiplication by s^2. The shaded area is the incoherent scattering. (From Ref. 87a.)

Table 6 Crystallinity of PP Samples as a Function of k (Disorder Factor) and Range of Angles of Integration

Interval	Polypropylene							
	Sample 1		Sample 2		Sample 3		Sample 4	
$s_o - s_p$	$k = 0$	$k = 4$	$k = 0$	$k = 4$	$k = 0$	$k = 4$	$k = 0$	$k = 4$
0.1–0.3	0.270	0.329	0.353	0.431	0.546	0.666	0.120	0.146
0.1–0.6	0.159	0.294	0.222	0.411	0.333	0.616	0.078	0.144
0.1–0.9	0.105	0.305	0.145	0.421	0.220	0.638	0.044	0.128
0.1–1.25	0.067	0.315	0.095	0.447	0.145	0.682	0.029	0.136
xcr		0.31		0.43		0.65		0.14

Note: For all samples a value of $k = 4$ gives the "same" crystallinity regardless of the range of integration. Since all k are identical regardless of thermal history, it is assumed k is due to thermal vibrations; disorders of the second kind would be expected to vary with thermal history.
Source: Ref. 87a.

ments in the crystalline regions but is more complex if crystalline and amorphous orientation is desired. Yet the latter may significantly affect the properties. For measurements of the orientation of crystalline segments, x-ray diffraction is the primary technique, with IR dichroism also appropriate. The latter is particularly applicable for determining, separately and jointly, the crystalline and amorphous orientation if either absorption bands due to both conformations can be separately identified (as for the trans and gauche PET bands described earlier, with the caveat that some of the trans conformers are in the amorphous regions) or by subtraction if only a crystalline band can be isolated (see later discussion). Birefringence is also used for characterizing overall orientation, with amorphous orientation being obtainable by subtraction of crystalline orientation obtained by another technique.

Fibers are the simplest systems for which to characterize orientation. In general, they have cylindrical symmetry about the fiber axis, for both the crystalline and amorphous regions. On the other hand there may be a skin–core effect, depending on the method of preparation and diameter, in which there is either a skin of different degree of orientation than the core, or a gradation of orientation from surface to center. Ignoring the latter effects, a relatively simple method to obtain an indication of the degree of crystalline orientation is to take a flat-plate or cylindrical film x-ray pattern, the fiber (or bundle of co-aligned fibers) being oriented along the **z** axis (cylinder axis) of the film. Typical flat-plate WAXD photographs are shown in Figure 49 (in this case, however, of films). Desired would be an indication of the distribution of the molecular axes in the crystal, (**c**), at angles to the fiber axis, with this distribution presumably having cylindrical symmetry about the fiber axis. Although the 00l reflections, whose

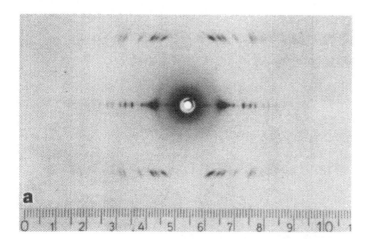

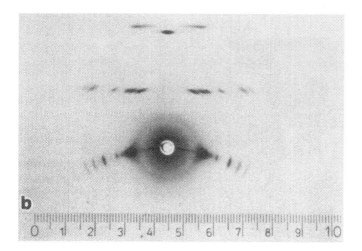

Figure 49 Cylindrical WAXD photographs of drawn samples of (a) PE, (b) PE tilted, (c) POM, (d) PTFE (15°C), (e) PP short exposure, and (f) PP long exposure. (From Refs. 89 and 92b.)

azimuthal distribution would yield this information, after correction for the increase in circumference of rings about the fiber axis with increasing deviations from the axis,* are not on the pattern (the fiber needs to be tilted for 00l to inter-

---

*The sphere of reflection intersects a decreasing fraction of the "ring" corresponding to crystals aligned off axis as the angle of misalignment increases (see Ref. 39).

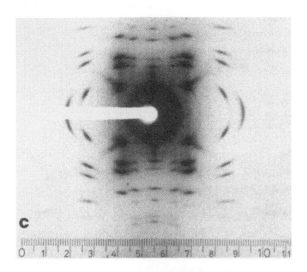

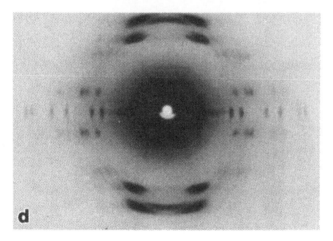

Figure 49   (continued)

sect Ewald's sphere), the same information can be obtained from the equatorial (hk0) reflections.

The degree of orientation can be expressed either in terms of the half (or full) width at half height of a microdensitometer azimuthal scan along one of the equatorial hk0 reflections (even for monoclinic and triclinic cells since **a*** and **b*** are perpendicular to **c**) or in terms of Hermans' orientation function [90]

$$f = \frac{3(\cos^2 \phi) - 1}{2} \tag{7}$$

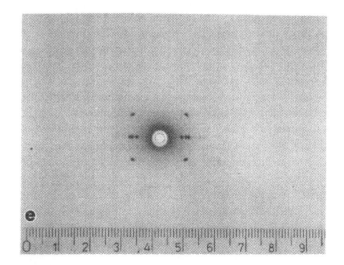

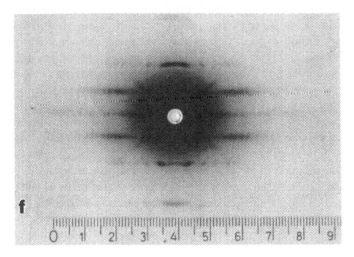

where ϕ is the angle of deviation from perfect alignment (here **c** is parallel to the fiber axis and/or **a*** and **b*** perpendicular. For reasonable degrees of alignment the correction mentioned earlier is not needed and, with ϕ being the angle between **a*** or **b*** and the equator:

$$(\cos^2 \phi) = \frac{\int_0^{2\pi} I(\text{hk0}) \cos^2 \phi \, d\phi}{\int_0^{2\pi} I(\text{hk0}) \, d\phi} \tag{8}$$

For a "perfectly" oriented fiber, f would be 1 as defined here,* whereas it is 0 for random orientation and $-\frac{1}{2}$ for an orientation perpendicular to the fiber axis.

For more complex types of orientation, as found in drawn films of some samples (e.g., PET), biaxially oriented films, or the skins of injection molded objects, more complex x-ray methods should be used. The simplest again consists of WAXD photographs, this time three patterns with the beams in three mutually perpendicular (machine, transverse and thickness, as earlier) directions. An example for a drawn film of PET is shown in Figure 50a with the method of sample preparation shown in Figure 51b. [91]. If the film had cylindrical symmetry the transverse and thickness patterns should be identical while the machine direction pattern should consist of rings; the observed pattern is attributed to preferential alignment of the phenyl rings parallel to the surface, with the molecular axes, as expected, being aligned in the machine direction. A similar set of patterns is shown in Figure 50(b) for a "pushtruded" shot gun shell casing LPE sample (see Figure 51). The casing is produced by forcing an LPE tube over a mandrel at $\sim 120°C$, that is, below T_m.

As noted above, 00l (planes normal to **c**) reflections are not observed on flat plate photographs with the beam perpendicular to **c** (i.e., the draw direction in fibers), since the Ewald's sphere does not intersect their reciprocal lattice spots. Although they can be observed if the fiber is tilted, such patterns are not that useful for orientation measurements. Representations of the entire distribution of crystal unit cell axes can be obtained by using pole figures (or inverse pole figures) with details of the method given in standard texts [6,39e]. A pole figure is, in essence, a two-dimensional stereographic projection plot of the three-dimensional distribution of a given reciprocal lattice vector (e.g., for PE, [200]* (**a**), [020]* (**b**), or [002]* (**c**)). From two such plots (Figures 51c, d) one can obtain a visualization of the distribution of the unit cell axes. On the pole shown in Figures 51a, b, the numbers indicate the number of counts detected per unit time when the sample is oriented that the normal (i.e., the pole, hence the name of the plot, and here the axis involved, **b** and **c**) to the plane being examined lies in the given direction. The Figures 51a, b show the raw date while for Figures 51c, d they have been smoothed and plotted as if a hemisphere was drawn about the sample, with the machine direction aligned E–W, the transverse direction aligned N–S, and

---

*We have defined ϕ as the angle of deviation of the hk0 reflection from the equator; normally it is defined as the deviation of **c** from the meridian, in which case the equation becomes

$$\langle \cos^2 \phi_m \rangle = \frac{\int_0^{2\pi} I(hkl) \cos^2\phi \, \sin \phi d\phi}{\int_0^{2\pi} I(hkl) \sin \phi d\phi} \tag{9}$$

with $\sin \phi$ accounting for the corrections mentioned earlier. In this case, perfect alignment for [001] is 1, for an [hk0] it is $\frac{1}{2}$ (see Refs. 39 and 90).

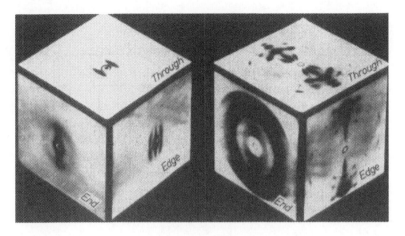

(a)

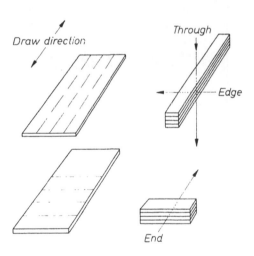

(b)

Figure 50 (a) Flat-plate SAXD and WAXD patterns of drawn films of PET with the beam in the thickness, transverse and draw direction. (b) Diagram of method of obtaining the patterns. (From Ref. 91a.) (c) Similar patterns of a pushtruded LPE sample. (From Ref. 92.)

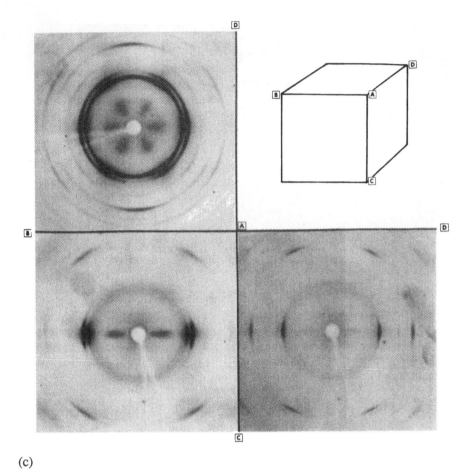

(c)

Figure 50   (continued)

the thickness direction in the center. Values of equal intensity are connected with "contour lines," as on a topographic map, to display (here) the unit cell axes' distribution (Fig. 51c, d) Note that the axes (MD, ND, and TD) have been rotated between Figures 51 a, b and c, d. These pole figures (with [110] and [200] also being obtained) were interpreted as indicating an alignment of **c** in the machine direction, indicated by the high concentration of (002) poles at the E and W poles. In combination with SAXS patterns taken with the beam at various angles to the sample (Figure 51e), they were interpreted in terms of lamellae oriented as in Figure 51f. The difference in the two-point SAXS spacings with beam orientation agrees with the lamellar tilt suggested. With chain folding in LPE occurring on 110 and 200 planes, the fold planes align in the transverse direction (Figure 51g) and prevent

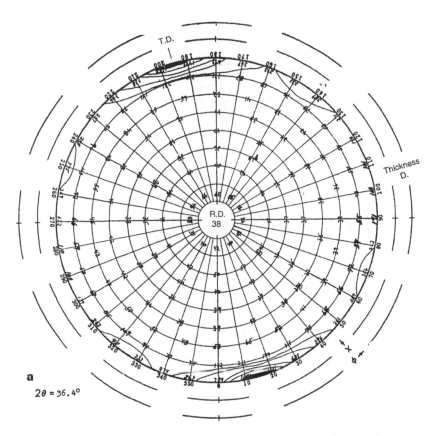

a

$2\theta = 36.4°$

Figure 51 (a,c) 020 and (b,d) 002 pole figures for drawn, rolled LPE. The raw data are shown in (a) and (b), with contour lines shown in (c) and (d). Sample directions: R.D., roll direction; T.D., transverse direction; and Th.D., thickness direction. Usually a Wolff plot projection would be used rather than that shown here. SAXS patterns obtained with the beam oriented at various directions to the sample are diagrammed in (e), with a two-point pattern being seen with the beam normal to the sample and a four-point pattern with the beam in the transverse direction. The proposed model (f) has a roof-like arrangement of the lamellae, with the chain axes in the machine direction and the ridge of the roof in the transverse direction. The fold planes (110 and 200) are aligned in the transverse direction, parallel to the surface (g). (From Refs. 92a,b.)

the axial splitting often seen in highly drawn films. This orientation also leaves the close-packed 110 and 200 planes parallel to the surface. Pole figures for fibers having cylindrical symmetry (Figure 51a,b) would consist of uniform bands of intensity on the equator (outer ring) for [hk0]* directions, for example, [200]* (a* or **a**), [110]* or [020]* (b* or **b**) for PE, the azimuthal breadth along the lines of longitude (radial lines in Figures 51a,b) being directly related to the half width, half height

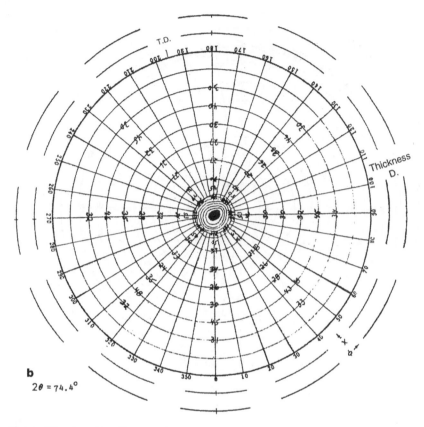

b

$2\theta = 74.4°$

Figure 51   (continued)

value of Hermans' f cited earlier, while [00l]* (**c**)* would either be a "spot" at the center (N pole) (as in Figure 51b) for orthorhombic, tetragonal, and hexagonal cells, or a ring of some breadth about this direction for monoclinic and triclinic cells.

3.   Amorphous (and Overall) Orientation

One of the simplest ways to measure overall orientation (if the crystal orientation is known, "subtraction" permits determination also of the "amorphous orientation") is to determine the change in dimensions as a sample is permitted to freely relax while being heated [93, 94]. We have found that free relaxation can be permitted by using a liquid stable to suitably high temperature and noninteracting with the polymer, with the method being useful for both fibers and more complex types of orientation. Figure 52 is an example for a cross section of two types of PE pipes heated on a dish of glycerin; the preferential orientation of the external skin in one type of pipe (externally cooled) versus skins on both surfaces (cooling inside also) results in the differences in the shrinkage.

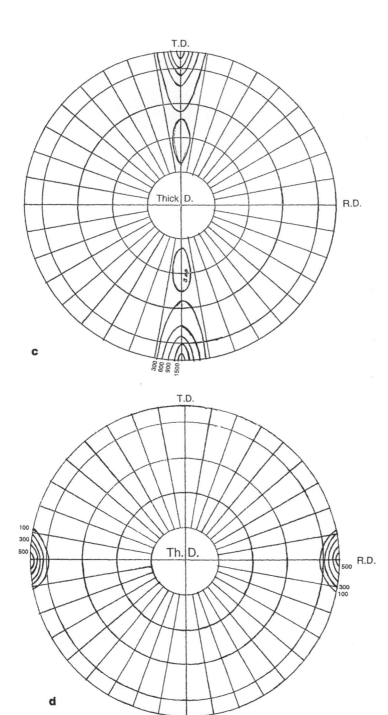

c

d

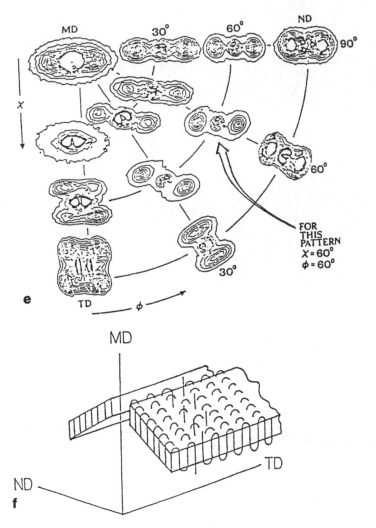

Figure 51 (continued)

Birefringence (Δn), as measured in a polarizing microscope using either a compensator or the Becke line method using immersion oils of varying index of refraction [95], is another relatively simple method of determining the overall orientation of a sample. Letting $\Delta n_y = n_z - n_x$, where n_x and n_z are the indices of refraction in the $\mathbf{x}$ and $\mathbf{z}$ (fiber axis) direction, for light propagating in the y direction, Hermans f_{cz} (for the angle between $\mathbf{c}$ and $\mathbf{z}$) is given by

$$f_{cz} = \frac{\Delta n_y}{\Delta n_y^\circ} \tag{10}$$

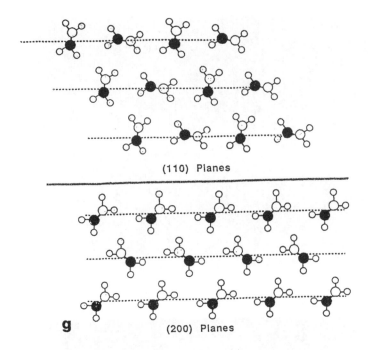

(110) Planes

(200) Planes

g

where Δn_y° is the maximum possible birefringence, that is, all chains perfectly aligned. For semicrystalline polymers of crystallinity fraction ψ,

$$\Delta n_y = \psi \, \Delta n_y \, (c) + (1 - \psi) \, \Delta n_y \, (a) + \Delta n_y(f)$$
$$= \psi f_{cz} \, (c) \, \Delta n_y^{\circ} \, (c) + (1 - \psi) f_{cz}(a) \, \Delta n_y^{\circ}(a) + \Delta n_y(f) \qquad (11)$$

where $\Delta n_y(c)$, $\Delta n_y(a)$ are the Δn for the crystalline and amorphous regions in the fiber, $\Delta n_y^{\circ}(c)$, $\Delta n_y^{\circ}(a)$ are the values for completely oriented crystalline and amorphous regions, and $f_{cz}(c)$ and $f_{cz}(a)$ are the respective Hermans f and $\Delta n_y(f)$ a small contribution from form birefringence. Furthermore, $\Delta n_y^{\circ}(c)$ and $\Delta n_y^{\circ}(a)$ differ only due to differences in density, that is $\Delta n_y^{\circ}(c) = \rho_c/\rho_a \cdot \Delta n_y^{\circ}(a)$. Thus either the overall orientation or, if $f_{cz}(c)$ is known from x-ray diffraction (it would equal the f described above), $f_{cz}(a)$ can be determined. The latter, on which the modulus of fibers is strongly dependent, is usually less than $f_{cz}(c)$, but approaches it in the extended chain fibers of PE [96]. Values of $f_{cz}(c)$ up to 0.999 and $f_{cz}(a)$ have been obtained [96a]. For x-ray methods of determining $f_{cz}(a)$ see Ref. 96b.

Infrared dichroism is another, related method of determining f_{cz} and, if $f_{cz}(c)$ is known, $f_{cz}(a)$. From knowledge of the orientation of the transition moment (direction of dipole oscillation) of the IR vibration (or polarization direction for polarized Raman scattering) relative to the chain axis, measurements of the absorption intensity with parallel (to the fiber axis) and perpendicularly polarized

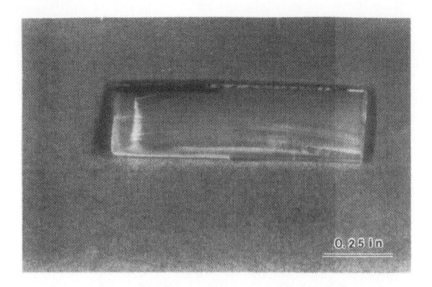

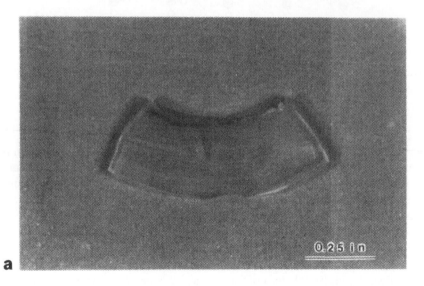

Figure 52 Longitudinal sections of extruded LLDPE pipes originally cooled (a) on the external surface only and (b) on both surfaces, before (top) and after (bottom) heating to T_m on a glycerine surface. The external surface is upper in each photograph. (From Ref. 93.)

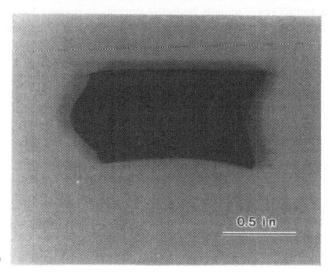

b

incident IR give the dichroism $R = A_{\parallel}/A_{\perp}$ which can be used by itself, or converted to Hermans' f:

$$f = \frac{(R - 1)(R_0 + 2)}{(R + 2)(R_0 - 1)} \qquad (12)$$

where R is the experimentally measured dichroism ratio and R_0 is the dichronic ratio for a perfectly uniaxially oriented sample. If the transition moment makes an angle ϕ with respect to the molecular axis, $R_0 = 2 \cot^2 \phi$, with R_0 varying from ∞ to 0 as ϕ varies from 0 to 90°. No dichroism ($R_0 = 1$) is observed if $\phi = 54° \, 44'$, the so-called magic angle (IR); for NMR the magic angle is 57.4° (3 $\cos^2 \phi' - 1 = 0$), with ϕ' being the angle between axis of rotation of the sample and the applied field H_0. Details of methods of determining ϕ and thus R_0 are given in texts (e.g., Refs. 6 and 8). Examples of transition moment directions relative to the chain axis for various IR bands and the use of the 1220 cm^{-1} band to determine f for PP films drawn to various draw ratios are shown in Table 7 [97].

A major advantage of the IR technique is its ability to detect and separate the orientation of the conformers in the amorphous regions from those in the crystalline regions, or of two components in blends or other multiphase systems, again, however, with the caveat that some fraction, for example, of the "crystalline" conformers (i.e., trans in PET) may actually be in the amorphous regions. Recent x-ray methods (described in Section II.A) have permitted separation of crystalline, intermediate and amorphous (assumed random) orientations [78].

Although not expected in fibers, nonaxially symmetric orientation is frequently found in uniaxially drawn films (e.g., PET, as shown in Fig. 50a). For such samples, as well as biaxially oriented films, injection molded objects, etc., the absorbance of the various IR bands in the three directions needs to be characterized. For this purpose A_z, the absorbance in the third (thickness) direction, as well as A_y ($A_{\parallel}$) and A_x ($A_{\perp}$), need determining. From A_z, obtained by sample tilting (see Ref. 8),

$$A_0 = \frac{A_x + A_y + A_z}{3} \qquad (13)$$

and the orientation parameters A_x/A_0, A_y/A_0 and A_z/A_0 can be determined. Figure 53 shows results for a PET film drawn to varying degrees. The conformer composition was determined by comparison with predicted spectra for pure trans and gauche conformers, with the separation into trans (crystalline) and trans (amorphous) similar to that described for Figure 22.

As indicated x-ray, birefringence, and IR are the most common techniques for characterizing orientation, with the three methods characterizing different aspects of the orientation, and with relaxation being an additional, even simpler technique. Electron and neutron (discussed later) diffraction yield results similar to x-ray, with ED restricted to specially prepared (thin) samples. Other, less used techniques include NMR [99], permitting determination of the orientation of mo-

Table 7 (a) Dichroic ratio (R), Orientation Function (F), and Average c Axis Orientation Angle (θ) for Drawn Films of PP of Various Draw Ratios Based on the Dichroic Ratio for the 1220 cm^{-1} Band at Band Angle 90° and (b) PP Bands and Their Vibration Direction (Transition Moment) Relative to the c Axis

(a)

Draw Ratio[a]	R	F	θ°
1	1.015	−0.010	55.1
1.5	0.745	0.186	
4	0.536	0.366	40.6
7	0.056	0.918	13.5
8	0.51	0.926	12.9
9	0.036	0.947	10.8

(b)

Frequency (ν)	α_v°	Frequency (ν)	α_v°
928 cm^{-1}	90°	1220 cm^{-1}	90°
973 cm^{-1}	18°	1256 cm^{-1}	0°
998 cm^{-1}	18°	1307 cm^{-1}	0°
1045 cm^{-1}	0°	1363 cm^{-1}	90°
1103 cm^{-1}	90°	1378 cm^{-1}	70°
1168 cm^{-1}	0°		

[a]The number shown is the ratio of the length of the drawn film to the initial length of the film.
Source: Ref. 97a and b.

bile segments (see Section III) and the use of fluorescent probes (e.g., Ref. 100). In the later case anisometric fluorescent molecules (e.g., dyes) are absorbed in the amorphous domains of the sample, before or after deformation, and their orientation is determined, for instance by polarized UV spectroscopy.

With increases in intensity of x-ray sources (rotating anodes and synchrotrons) and the development of FTIR (and FT-Raman) techniques, it has become possible to follow deformation and crystallization processes in real time. Modern FTIR instruments (as of 1993) permit obtaining and storing scans (interferograms) at rates up to nearly 100/sec, giving a time resolution approaching 10 msec. Although at low resolution, if the time frame of the phenomenon being examined is sufficiently long, (e.g., polymerization reactions, mechanical deformation, including dynamic mechanical spectroscopy, creep and stress relaxation, fatigue and crystallization; see, e.g., Ref. 8), single interferograms can be transformed into spectra essentially instantaneously to follow incremental changes. Interpretation is as described elsewhere in this chapter.

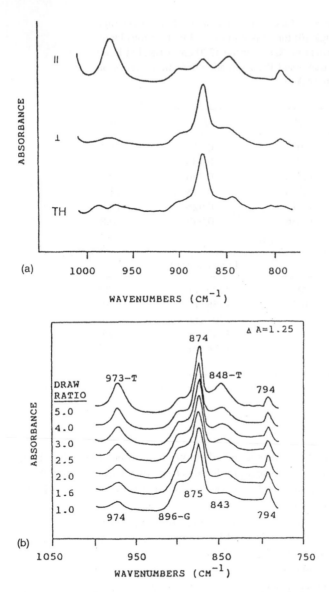

Figure 53 Effect of draw ratio on orientation of PET films. (a) Spectra of uniaxially drawn PET film with the beam polarized parallel and perpendicular to the draw direction and parallel to the thickness direction of the film. (b) Calculated isotropic spectra of uniaxially oriented PET as a function of draw ratio. The 973 and 848 cm⁻¹ bands (trans) increase while the 896 (gauche) band decreases. (c) Orientation (trichroic) functions for the 973 cm⁻¹ band as a function of percent elongation. Perfect parallel orientation has a value of 2. (d) Conformer composition (trans; total, crystalline, and amorphous) of drawn PET films. (From Ref. 98.)

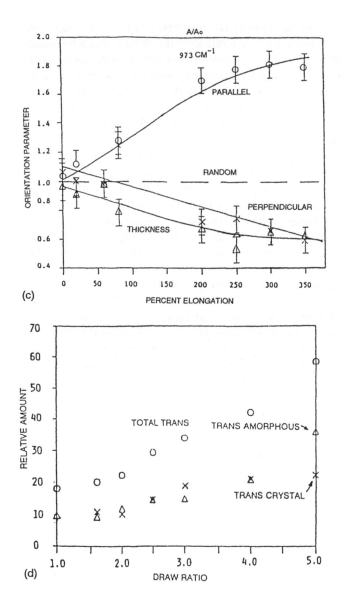

(c)

(d)

Rotating anode x-ray tubes, commercially available from several sources, are designed to permit dissipation of the heat induced by the electron-beam bombardment of the target generating the x-rays. Conventional sealed copper tubes, as often used, have a 1×10 mm focal spot (region from which x-rays are generated) and can be operated, typically, at 40 kV, 20 mA, with higher power being used in broad-focus (e.g., 2×10 mm) and lower in fine-focus (e.g., 0.4×8 mm) tubes. The fine focus, looked at on end from a slight angle, gives a high-brilliance "point" source, while the broad focus looked at normal to the long direction yields an intense line focus. With a normal focus sealed tube, a wide-angle flat-plate point-source pattern of a 3-mm-thick (near optimum for maximum intensity) PE sample takes 5–10 min at a sample-to-film distance of 5 cm; a small-angle pattern (see below) of the same sample at a 17-cm distance takes nearly a day. Typical, reasonable-cost rotating anodes operate at 60 kV, 100, 200, or 300 mA, that is, with up to nearly 30 times the power, with correspondingly increased intensity. These units have tubes with diameter on the order of 6 in, rotating at 6000 rpm. Even higher power tubes are available: Rigaku Denki Corp. produces a unit operating at 1.5 A, 60 kV, with the tube, 12 in. in diameter, rotating at 9000 rpm. For in-lab applications for such tasks as characterization of orientation and crystallization at various positions along a spin line, such systems are ideal. However, even sealed tubes can be used for such purposes, using longer exposures, since it is assumed the physical structure changes are stable as a function of distance from the spinneret, although obviously changing within the fiber as it moves away from the spinneret [101].

Synchrotron radiation is increasingly being used as a high-intensity source [102]. The advantages are [39e]:

1. A continuous spectrum from infrared ($\lambda \approx 10^{-2}$ mm) to hard x-rays ($\lambda \approx 0.1$ Å).
2. High inherent collimation of the beam (0.1 mradian in the x-ray λ region).
3. A source intensity and brightness, due in part to the high collimation, several orders of magnitude higher than the largest rotating anode sources.
4. If desired, short (~ 100 psec) pulses with a frequency of a few nanoseconds.
5. Highly polarized radiation.

Applications have been made to crystallization and annealing (e.g., unoriented PET [103]) and to phase separation in blends and block copolymers, using small-angle x-ray diffraction (discussed later), and to crystallization, melting, and phase transitions, using wide-angle diffraction for a number of polymers (e.g., PP [104] and polyesters [105]). The latter includes simultaneous wide- and small-angle x-ray scattering and light scattering. General reviews are given in Ref. 102, as well as Ref. 39e.

Further enhancement of the rates of detection of x-ray scattering has come through the development of linear and two-dimensional position-sensitive detectors (PSDs) [39e]. These detectors permit scattering angle detection of x-rays scattered over a broad angular range simultaneously (as on film) but with the advantages of counter detection and pulse height (λ) discrimination. Solid-state linear PSDs as well as Vidicon tube two-dimensional detectors are more recent advances, enhancing the range of intensity (solid state, but at the expense of λ discrimination, which, however, is not as important for synchrotron radiation) and the resolution ($\sim$300 μm as compared to 1 mm for the two-dimensional PSD).

C. Domain Size, Shape, Organization, and Interaction

The subject of this section is generally called polymer morphology, that is, the external structure of the physical entities making up the sample. In Section II.A we discussed characterization of the internal physical structure of crystalline domains, that is, the unit cell and defect-disorder structure, while Section II.B described characterization of their relative amount and orientation. Pertinent to this section is the discussion there of crystal size effects, that is, x-ray (or electron) diffraction line (or spot) broadening. As shown there, the size of the reciprocal lattice spots is inversely related to the size of the corresponding crystal, with the width of observed reflections being the result of these spots, for one or more crystals, being sampled by Ewald's sphere of reflection. Unfortunately, for crystal size characterization, disorders of the second kind (but not the first) also broaden the reflections, with separation of the two effects normally requiring measurement of the breadth of at least two orders of the reflections from planes normal to the crystal dimension being probed. For instance, 002 and 004 are needed for determination of crystal size in the c axis direction of PE, with even 002 being of low intensity on diffractometer scans and absent on flat-plate photographs from fibers. For fibers, however, x-ray line broadening remains the most appropriate technique, with tilting of the fibers being used when necessary. Note, however, that the x-ray line broadening technique yields the size of the crystal, not the crystal–crystal spacing.

The basic physical structural units for polymers generally have at least one dimension on the order of 100 Å; this includes the thickness of lamellar crystals and all dimensions of fringed micelle-like crystals, fibril diameters, radii of gyration of random coils in amorphous polymers (glass and melt), and domain sizes in block copolymers. For structure on this size scale, small-angle (x-ray and neutron) scattering and transmission electron microscopy are the traditional techniques.

If the domain structure results in a difference in electron density, small-angle x-ray scattering (SAXS) can be used. Chapter 7 of the recent x-ray text by Balta-Calleja and Vonk [39e] gives an excellent description of its application to polymer morphology characterization. The requirement of measuring close to zero scattering angle (for CuK_α) radiation, as normally used, a 100-Å periodicity gives a Bragg peak at 0.88° 2θ) requires that a highly collimated beam be used, necessitating long exposures and/or the use of high-intensity sources and, for fibers, the newer two-dimensional detectors.

Descriptions of the applications (and methods of data interpretation) of SAXS to determination of the specific surface, width of the phase boundary, and radii of gyration of domains of one phase dispersed randomly or regularly in another phase are given in Ref. 39e. If the domains are periodically arranged, as the lamellae in crystalline polymers or the domains in some block copolymers, Bragg's law ($n\lambda = 2d \sin \theta$) can be applied to the peaks observed in the diffraction patterns (after appropriate corrections) to yield domain spacing. For dilute systems of particles, Guinier's law [106] can be applied:

$$i(\theta) = \frac{Nm^2}{V}e^{-16\pi^2\theta^2 R_g^2/3\lambda^2}$$

(14)

where m is the number of excess electrons per particle over that in the matrix ($+$ or $-$), N is the number of particles in the illuminated volume V, and R_g is the radius of gyration, defined by

$$R_g^2 = \frac{1}{m}\int v'r^2\,dm$$

(15)

v' being the volume of the particle. From plots of $i(\theta)$ versus θ^2 both R_g^2 (from the slope) and Nm^2/V (intercept) can be obtained. From knowledge of the composition and chemical structure Nm/V can be determined, permitting determination of m, a quantity directly related to the mass of the domain. R_g^2, on the other hand, can be related to the overall size and shape of the domain. Both of these measurements are often used for biological systems, for instance globular proteins and viruses in suspension, with relative values of R_g^2 for various shape particles being given in standard SAXS texts [106].

Representative schematic SAXS patterns from fibers are shown in Figure 54. Two-point and four-point patterns, elongated into streaks parallel to the meridian, equator, or at angles, are seen, often with diffuse scatter (as in Fig. 54k) superimposed on the center of the pattern. Similar patterns are obtained from uniaxially drawn films, with three-dimensional flat-plate patterns again being needed if the films do not have cylindrical symmetry (Figure 50).

The central streak is often interpreted in terms of voids; if the voids are elongated along the fiber axis (e.g., between fibrils), the streak is elongated (by the reciprocal relationship) along the equator. Such a streak may also be related to the crystalline–amorphous morphology (but a fibrillar morphology, densely packed, would not yield the streaks), with the presence of voids being confirmed by either the presence of light scattering (opaqueness not due to fillers) or loss of the SAXS scattering when the fiber is soaked in a liquid of electron density similar to that of the fiber. Compression of the fibers normal to their axes can also reduce void scatter.

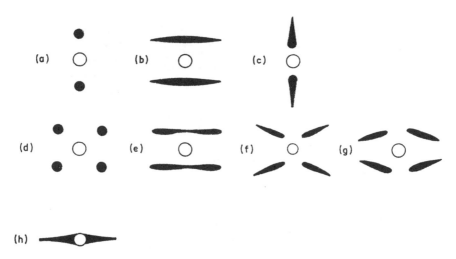

Figure 54 Schematic diagrams of various types of SAXD fiber patterns; the fiber direction is vertical, with (a–c) two-point patterns, (d–g) 4 point patterns, and (h) an equatorial streak. (From Ref. 39e.)

The two-point and four-point patterns (as well as the central streak) often undergo significant changes during annealing. Annealing taut, as in heat setting, usually leads to a perfecting (increase in crystallinity as shown by WAXD, sharpening of the SAXS pattern and decrease in the scattering angles of the maxima) of the morphology. (Reference 107 discusses applications of SAXS to fibers.) Annealing relaxed, on the other hand, can lead to major changes in the form of the pattern, with the fiber often losing nearly all of its original draw (the basis of the previously described shrinkage measurements of the degree of orientation).

The various patterns in Figure 54 are usually described in terms of morphological models, with, unfortunately, different models sometimes leading to the same type of pattern. This is representative of a feature of all diffraction patterns; while disagreement between the pattern calculated for a model and that observed indicates the model is incorrect, agreement does not indicate the model is correct; a different model may also lead to a similar pattern.

Interpretation of the two- and four-point patterns has been made based on both continuous lamellar structures (more or less normal to the fiber axis; e.g., Figure 55) and a microfibrillar structure based on either a fringed micellar model (Figure 56a) or a paracrystalline fibrillar model (Figure 56b). Bonart and Hosemann's model (Figure 55) can explain the patterns in Figure 54a, b with the width of the streak being related to the curvature of the lamellae and the meridional spacing being related to the center-to-center distance (periodicity) of the lamellae. As shown later, there is reasonable electron microscope evidence for this type of structure.

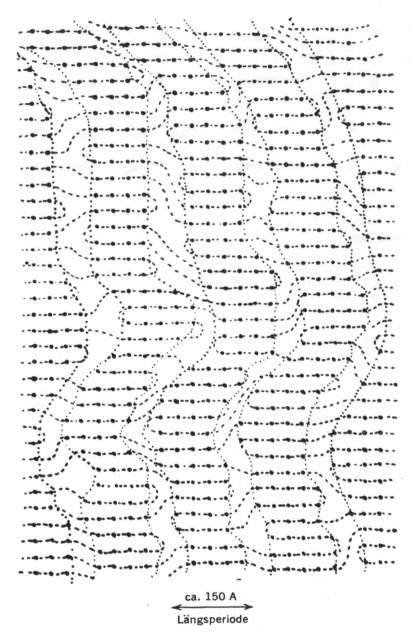

ca. 150 A
Längsperiode

Figure 55 Bonart and Hosemann's model for the arrangement of the molecular segments and the lamellae in an as-drawn fiber or film. (From Ref. 108.)

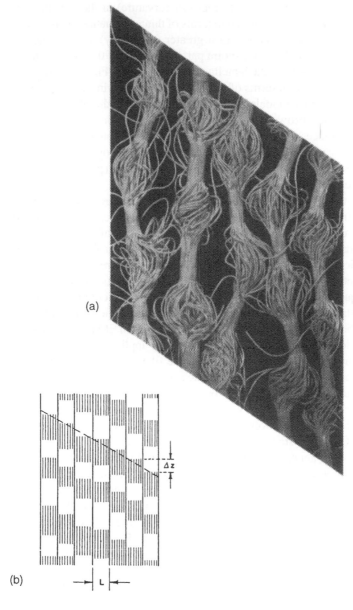

(a)

(b)

Figure 56 Paracrystalline (for lamellae) fibril models as proposed by (a) Statton [109] and (b) Hosemann [110] that would yield four-point patterns. (From Refs. 109 and 110.)

Annealing taut leads to a perfecting (decrease in curvature) of the lamellae, increase in spacing (sometimes interpreted in terms of thinner lamellae melting, segments adding to the thicker, resulting in a greater periodicity) and decrease in streak width. Interpretation of the four-point pattern in terms of this type of model requires a preferred tilt angle for the lamellae. In the models in Figure 56 the four-point patterns are explained in terms of axial shift of neighboring crystalline domains in the fibrils. In those models the broadening normal to the fiber axis, as in Figure 54e, is due to the narrow diameter of the fibrils. Further details can be found in Ref. 39e and references therein. The central streak, often found in association with one of the other patterns, is usually attributed to voids elongated parallel to the fiber axis. Other methods of characterization of the void (pore) structure in fibers are discussed in Chapters 7 and 8.

Considering the dimensions of the morphological features of a fiber, transmission electron microscopy (TEM) would be expected to be an ideal technique. Unfortunately, sample preparation problems (with thicknesses of less than 1000 Å required for normal TEM) have greatly restricted its use. Typical fibers are even too thick for application of high-voltage microscopy; furthermore, the micrographs are always projections of the density of the sample through its entire thickness. Standard techniques for preparing thin samples from typical commercial fibers include surface stripping [111], sections cut by ultramicrotome from embedded samples, and replicas of etched samples (for descriptions of TEM polymer sample preparation techniques see Ref. 112); examples of their application to drawn films are given in Figures 57–62. Direct imaging of the fiber surface, by replication, would not be expected to be informative, as the surface is affected by the process of spinning. Etchants (using, for instance, ions or $KMnO_4$–H_3PO_4) chosen to attack amorphous regions in preference to crystalline regions might be expected to reveal the morphology (Figure 58). However, because of their diameter, fibrils are difficult to replicate. Only two-stage (reduced resolution) replicas are possible. Annealing alone can also "generate" a visible morphology (Figure 58a, 59).

The use of cut or stripped sections, or drawn thin films, permits the use of dark-field (imaging of the electrons diffracted into one or more ED reflections) and phase-contrast microscopy (see Ref. 61). As shown in Figure 60, the crystals (here lamellae) are imaged (see also Figure 41). These micrographs can readily be interpreted in terms of Bonart and Hosemann model (Fig. 56). If drawn normal to the original draw direction the films readily break down into fibrils, with, I suggest, the separate fibrils resulting from deformation of initial individual lamellae. The crystals in neighboring fibrils coalesce into the lamellae, with the lateral order perfecting during annealing. The retention (or regeneration by localized melting–recrystallization at the microneck edges) of folds is shown by the result of further draw (Figure 61), with even the annealed samples breaking down into fibrils under the action of lateral stresses. Tie molecules between the original

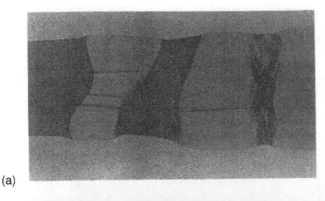

(a)

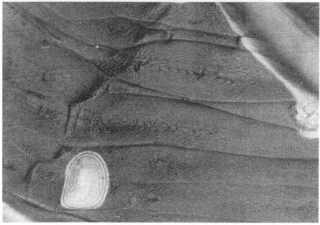

(b)

Figure 57   (a) Photograph of a partially drawn LPE film. The drawn regions are opaque due to the development of voids. Shear bands are seen at the right as a neck develops. (b) Optical micrograph (reflected light) of the necking zone of an LPE fiber originally crystallized with a free surface. (From Ref. 89.) (c) TEM micrograph of a microneck at a POM spherulite boundary. (From Ref. 113.)

lamellae, in bulk samples, are proposed to form intrafibrillar links in the drawn samples. This breakdown into fibrils is particularly obvious in simultaneously or uniformly biaxially drawn samples (Figure 62).

As indicated by the preceding micrographs, the use of thin films to study the morphology of uniaxially drawn polymers is much simpler than the corresponding study of fibers. While the observed morphology is believed similar to that in the fibrils, caution in extrapolation is indicated.

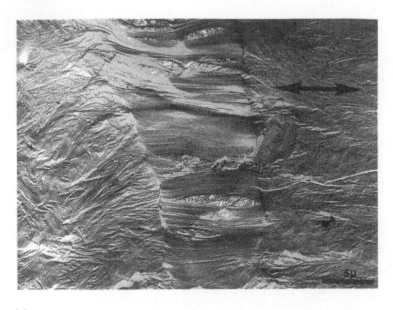

(c)

Figure 57 (continued)

TEM is also the method of choice for observing the organization of the crystalline lamellae in unoriented samples and of the domains in block copolymers. In both cases, however, the need for thin samples and sufficient image contrast requires special preparation techniques. Etching, sectioning, staining, and replication are often used (see Ref. 113).

Characterization of the morphology of amorphous polymers (the size and shape of random coils) utilizes small-angle neutron scattering (see Ref. 118). This is possible because of the difference in scattering power of ^1H and ^2H, so that a

Figure 58 (a) LPE drawn 8 times at 60° C, annealed 1 hour at 110°C. (b) LPE drawn 13.5 times at 60°C, annealed at 120°C for 7 hr and treated with fuming HNO_3 for 3 hr. The macro fibrils (linear domains of various uniform lamellar spacing) were suggested to correspond to different degrees of initial draw. A small angle electron diffraction pattern is inset (From Ref. 114.)

(a)

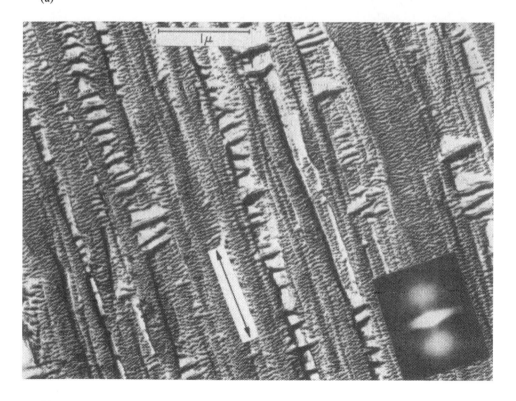

(b)

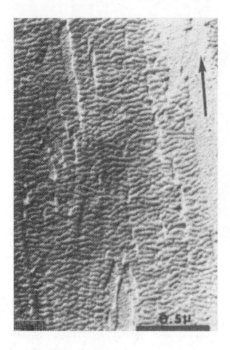

Figure 59   LPE thin film drawn and annealed on a substrate at 125°C for 1 hr. (From Ref. 115a.)

deuterated polymer molecule in a matrix of hydrogenated molecules, or vice versa, scatters similarly to an isolated polymer molecule in solution. Random coil dimensions (R_g); including coil anisotropy in drawn samples, and molecular (particle) weights can be obtained. Figure 63 shows an example for polystyrene. Miscibility in blends can also be studied, because the labeled molecules in a truly miscible blend have a random coil R_g, whereas clustering, resulting in larger R_g (and M_w), occurs for poorly miscible or nonmiscible blends. Small-angle neutron scattering can also be used to follow changes in the gross conformation of molecules during deformation of bulk semicrystalline polymers.

The method has also been used to characterize the overall conformation (type of folding) of the molecules in the crystalline state. For instance, the R_g of regular, adjacent-reentry folded molecules (i.e., forming a plane on the growth face) would be considerably different from the near-random coil R_g expected for a switchboard type of conformation. These studies also receive contributions from wide angle neutron diffraction, with the crystalline segments of a labeled polymer in the adjacent reentry model being adjacent to each other (at an ~4.5 Å spacing in many

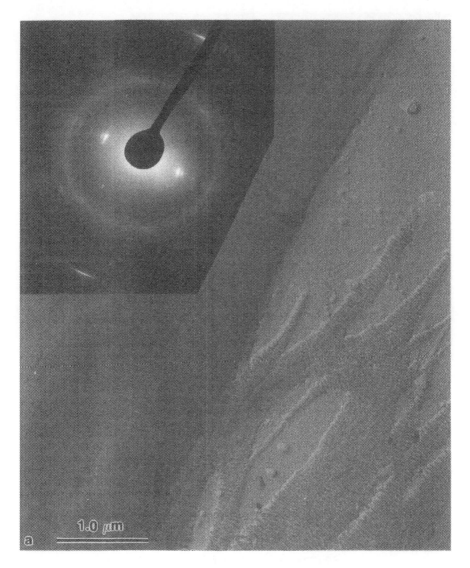

Figure 60 (a) Bright-field and (b) dark-field micrographs of a drawn annealed (130°C, 30 min) LPE thin film. (From Ref. 115b.) (c) Phase-contrast micrograph of a partially drawn, annealed film of LPE. (From Ref. 116).

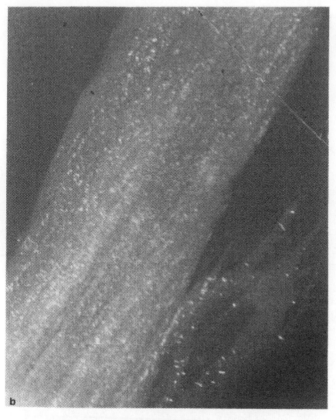

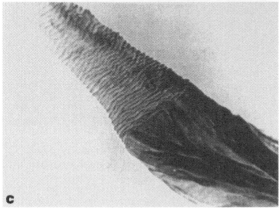

Figure 60   (continued)

Figure 61   TEM micrograph of a drawn, annealed LPE thin film redrawn across a crack in a carbon film. (From Ref. 115a.)

polymers) but being more widely and randomly spaced in the switchboard model. Caution in interpretation is needed, however, due to phase segregation of the components. For PE, for instance, slow cooling from the melt results in clustering of the minor component. Using quenched samples and the same data, various interpretations have resulted in values of essentially zero (less than three consecutive adjacent stems, with a 30% probability of escape to an adjacent lamella) [120] to 65% [121] adjacency (see also Ref. 118).

Although the basic structural units in polymers have a dimension on the order of 100 Å (lamellar thickness, fibril diameter, phase domain size, R_g), in crystalline polymers they are often organized into larger structures. These structures, termed spherulites, are often large enough to observe by optical microscopy (Figure 64). Using crossed polars it can be shown that the molecular axes in the lamellar crystals of which they are composed are tangentially oriented; TEM shows the

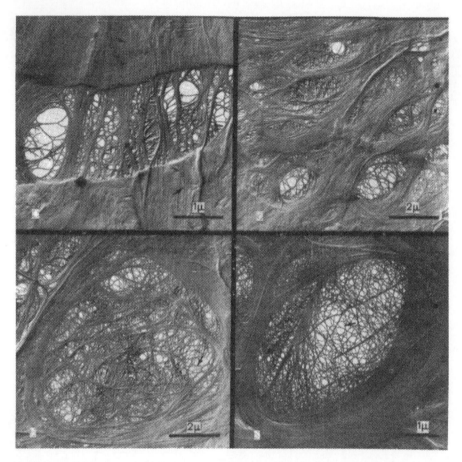

Figure 62 POM film biaxially drawn as a bubble. (From Ref. 117.) All samples drawn 1.25 × 1.25 except lower right (1.5 × 1.5). Arrows in upper left indicate short fibrils of edges of lamellae; on lower two figures branched fibrils.

lamellae grow outward from a central nucleus. For details of the growth and structure of spherulites see Refs. 89 and 122–125.

With spherulites being observed for samples of a wide range of crystallinity, the amorphous regions are incorporated within the spherulites, in between the lamellae, at the spherulite boundaries, and so forth. Tie molecules—molecules whose ends are incorporated in neighboring lamellae and traverse the intervening amorphous regions—were proposed to primarily determine the physical properties of semicrystalline polymers [126]; mats composed of solution-grown single crystals, in which few or no tie molecules are expected, flake apart readily like mica. The numbers of tie molecules are expected to increase with increasing

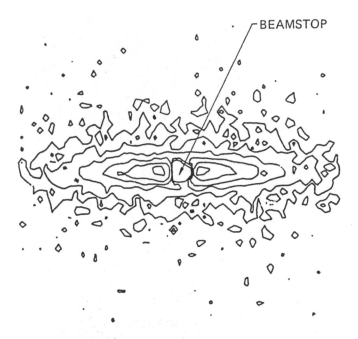

Figure 63 Two dimensional small angle neutron diffraction pattern from a drawn blend of deuterated polystyrene in normal polystyrene taken using an area detector. The asymmetry of the scattering indicates the (reciprocal asymmetry of the deuterated polystyrene coil dimensions. (From Ref. 119.)

polymer molecular weight and rapidity of crystallization, contributing to both the toughness of the sample and, as suggested earlier, to interfibrillar links if the polymer is drawn. Although we know of no way to "count" the number of tie molecules, evidence for their existence has been obtained, such as by swelling a sample, fracturing it in the swollen state, and then subliming the swelling agent from near the fracture surface [127]. Replication of this surface, as the separation of the lamellae in the swollen state is maintained by the residual swelling agent beneath the surface, reveals "interlamellar links" spanning pores between the lamellae (Figure 65). These links, on the order of 100 Å in diameter, were proposed to be made up of tie molecules and other interlamellar amorphous material (e.g., cilia and segregated molecules) drawn into the fibrils during the swelling. Similar interlamellar links are seen between lamellae in replicas of slightly drawn samples in which the lamellae are normal to the surface [113]. In both cases, if the lamellae are not separated too far, relaxation results in closure of the pores and a "disappearance" of the fibrils. A similar structure is observed between many of the lamellae in slightly drawn samples of, for example, PP and

Figure 64 Optical micrograph between crossed polars of monoclinic (dark) and hexagonal (bright) crystal structure PP spherulites. (From Ref. 89.)

POM, crystallized from oriented melts (with all lamellae perpendicular to the orientation direction and the surface). If annealed taut the interlamellar links crystallize, do not relax, and permit retention of the pore structure; the result is porous membranes whose cell size can be controlled by the original processing and amount of draw [128].

With the basic structural units having dimensions on the order of 100 Å, SEM has not been very useful in characterization of their organization. Although sample preparation is easier, polymers only needing to be coated with a conducting (metal) film for most observation, the resolution is sufficiently low that TEM observation of replicas is often more revealing; in both cases only surface structure is observed. The recent development of low voltage SEMs has permitted observations of noncoated, insulating samples; the resolution, however, is even lower.

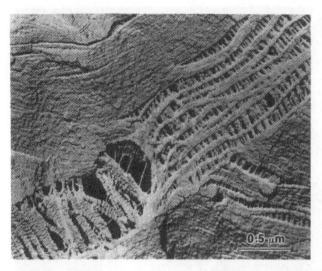

Figure 65 Fracture surface of swollen poly(4-methyl pentene-1). Interlamellar fibrils on the right can be seen that disappear (relax) when the swelling agent is removed; on the left of the lamellae themselves have been drawn. (From Ref. 127.)

On the other hand, due to its much greater resolving power and depth of field than an optical microscope, as well as the ability to select regions of the sample for characterization with x-ray dispersive analysis, SEM is ideal for the characterization of the surfaces of textile fibers and fabrics. Its use for these purposes is discussed in the next chapter.

In the past few years an alphabet soup of scanning probe microscopes, (SPMs) have been developed [129]. These include scanning tunneling (STM), atomic force (AFM), lateral force or friction force (LFM, FFM), and electrochemical force (ECFM) microscopy, with various microscopes having resolution sufficient to observe individual atoms (on appropriate samples), ease of sample preparation (often none) and ability to operate in air or liquids. Their use to date for polymer characterization, however, has been relatively limited. Lamellar thicknesses at the edges of solution-grown single crystals, for instance, have been measured, but the potential of observing the fold packing on their surfaces has not yet been fulfilled. Application to fibers is more difficult than to relatively flat specimens. Figure 66 shows a split PPTA fiber. The annealed fiber was embedded in epoxy and split longitudinally prior to observation. The epoxy surface is at upper right, and the skin of the fiber next, with the core having a microfibrillar morphology. Microfibrils cut during sectioning can also be seen.

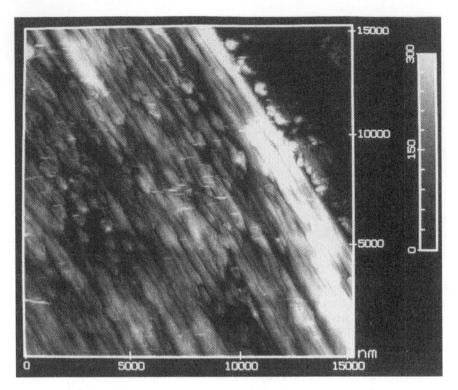

Figure 66   AFM image (scan size 15 μm of a PPTA fiber that was annealed at 450°C, embedded in epoxy and cleaved parallel to the fiber axis. (From Ref. 130.)

Figure 67 is an AFM image of a drawn film of PE. The lamellae of the shish-kebab structure, produced by the Gohil and Petermann method [61], are clearly seen. At higher magnification (Figure 68) the authors claim individual PE molecules can be seen. The spacing, 5.6Å, is larger than the expected 4.9-Å spacing of the 100 planes. Although most of the images published to date, in my opinion, have not been more informative than possible through TEM and SEM, the methods have considerable potential. For instance, for step heights less than 20–30 Å (as possible, e.g., on the as-polymerized polyester crystals [Figures 33 and 35]) they would be far superior to TEM; the latter is restricted to ~15Å visibility and ±10A accuracy due to the granulation of the shadowing material. Most representative of that potential is a recent publication by Stocker et al. [131b] on the direct observation of the packing of right and left handed helical s-PP molecules on the edges of lamellar crystals (Figure 69) grown by epitaxy. Figure 70 is a high resolution AFM micrograph, with the sample and microscope tip in water,

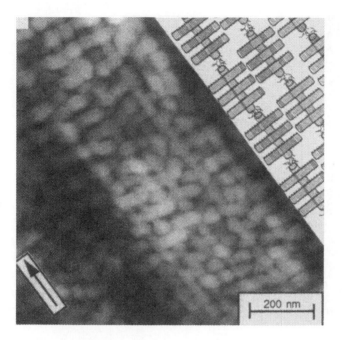

Figure 67 AFM image (in propanol) of film of polyethylene drawn from the melt by the method of Gohil and Petermann [61]. The lamellae of the shish-kebab structure protrude a few nanometers out of surface. A model, based on TEM studies, is at upper right. (From Ref. 131a.)

of a portion of the edge of one of the lamellae, the reduction of capillary forces making it possible to use imaging forces nearly an order of magnitude less than in air.

While the unfiltered images, as in Figure 70a, are noisy, the 2D FFT in Figure 70b clearly shows substantial periodicity is present and even in the unfiltered image one can clearly see orientation 10° to the vertical, as well as a periodic structure at an ∼45° angle to that orientation. In the filtered image, i.e., reconstructed using the maxima in the FFT, not only is the periodicity displayed more clearly, but the suggestion is that individual CH_2 and CH_3 groups are being resolved, as suggested by the model in Figure 70d, with the uppermost groups, visualized in the AFM, being darkened. As the authors note, the helices of one hand are better resolved than those of the opposite hand, here left handed better than right handed. This is due to the resolution in the scan direction being limited by a stick-slip process of the tip while the better resolution in the orthogonal direction is dependent on the density of the scan lines. For applications of the various methods the reader is directed to Ref. 129 and the manufacturer's literature.

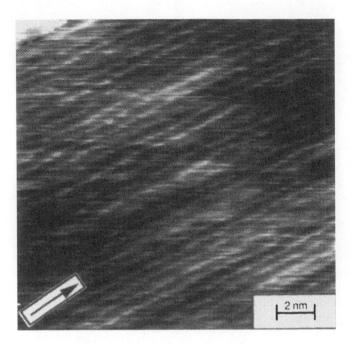

Figure 68 AFM image of a portion of one of the lamellar edges in Figure 67, at higher magnification. (From Ref. 131.)

III. PHYSICAL PROPERTIES

Because of their importance in practical textile use, characterization of single fiber tension, flexure, and friction properties, as well as the hand of textiles, is described in individual chapters in this volume. The chemical, biological, and thermal barrier properties of textiles are also described. Here we are concerned with physical properties of the polymers themselves that are of concern in their application as textile fibers.

With melt viscosity already described earlier, our primary concern is with their relaxation and transition behavior, that is, the temperatures of onset of particular types of segmental motion at the frequency of the measurement and the effects of stress on individual molecules. These are discussed next.

The primary techniques for relaxation measurements—creep, stress relaxation, and dynamic mechanical analysis (DMA)—are well known, with dielectric spectroscopy also being useful if the polymer contains polar groups, and DSC (as described earlier) being used for T_g (for discussions of method, results and interpretation see, e.g., Ref. 132). An additional, somewhat less used technique of in-

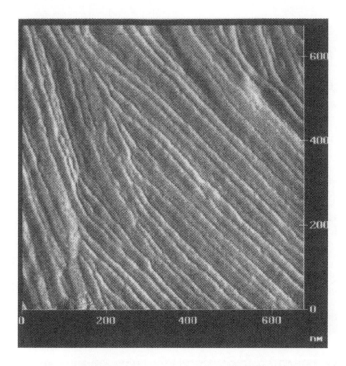

Figure 69   AFM image, in water, of a s-PP film crystallized epitaxially on an oligophenyl substrate. The lamellae are normal to the substrate permitting the high resolution images in Figure 70 of the molecules on the fold plane on their edge. (From Ref. 131b.)

terest is "broad-line" NMR. Changes in the second moment (width) of the NMR line can be related to the types of motion involved, with applications to oriented systems aiding in defining the types of motion. For these purposes measurements of spin–lattice and spin–spin relaxation times (T_1 and T_2) are also of value (see Refs. 7 and 8). An example of the use of the shape of the broad line is shown in Figure 71. The spectrum was divided into three peaks: L_s for the mobile protons in the amorphous regions, L_b for the immobile protons in the rigid amorphous (interphase) regions, and G for the immobile protons in the crystalline regions. The relative areas of the peaks are directly proportional to the fractional amounts in each type of region. The results are thus similar to those obtained by the "whole pattern" x-ray method described earlier. It is noted the measurement was made at 200°C, well above T_g and only slightly below T_m (215°C).

With the lower temperature relaxations generally having lower activation energies, their relaxation temperature increases faster with frequency of measure-

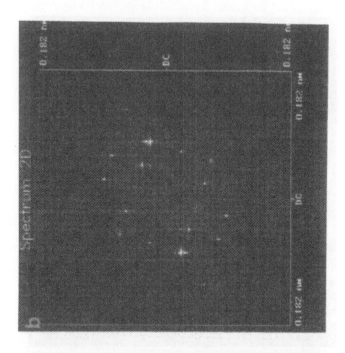

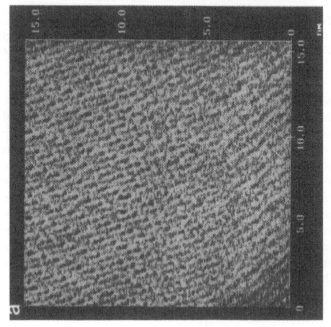

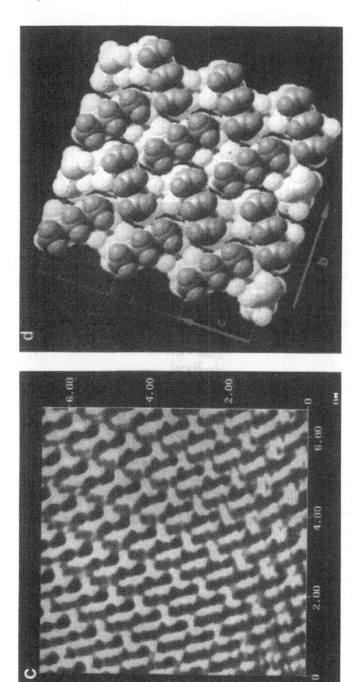

Figure 70 (a) High resolution AFM image of an area near the center of Figure 69. The chain axes are oriented at ∼10°C to the vertical. (b) Two dimensional fast Fourier transform of a portion of the image in (a). (c) Reconstruction of the image using the 20 pairs of spots in (b). (d) Schematic representation of the molecular arrangement in (c). The rows of three CH_2 correspond to "exposed" CH_3-CH_2-CH_3 groups; their alternate tilt in (c) and (d) confirms the proposed alternate left and right handed helical molecule packing in the crystal. (From Ref. 131b.)

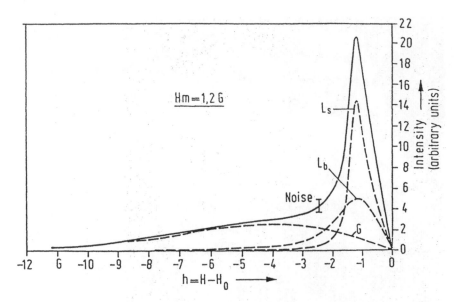

Figure 71   Broad line NMR spectrum of nylon 6 after annealing 20 hr at 200°C. (From Ref. 133.)

ment than the higher temperature relaxations; that is, the relaxations are closer together with higher frequency. Thus low frequencies (as in creep and stress relaxation) permit better resolution of the relaxations. Two relatively recent techniques
taking advantage of this effect are thermally stimulated creep and current. In thermally stimulated creep, which can be applied to oriented samples, a load is applied
to a sample at low temperature and the response is measured as the temperature is
slowly raised; changes in creep rate occur as various types of molecular motion
become activated. In thermally stimulated current (or thermally stimulated discharge current) an electric field is applied to a film with conductive coatings on
both surfaces at a temperature above that of any relaxation being probed, the sample is cooled to a suitably low temperature, the field is turned off, and the current
through an external circuit is measured as the film is heated. The elevated temperature field orients any mobile dipoles in the film, which remain frozen when
the field is removed at low temperature. When segmental motion occurs as the
sample is heated, the dipoles relax, and an image charge on the electrode is released; it is this that is measured. Effective frequencies are 10^{-3}–10^{-4} Hz, permitting excellent resolution of the relaxation processes. By polarizing the samples
over only a limited temperature range, distributions in relaxation times in individual peaks can be determined. References 134 and 135 are excellent reviews of the
techniques and their application.

Although the foregoing discusses primarily the measurement of relaxation temperatures, all can be applied to T_g and T_m measurements as well, with the qualification that for application of mechanical measurement the sample needs to maintain its integrity through the transition; for example, T_g would be measured by DMA for a semicrystalline polymer but not for a wholly amorphous one.

WAXD and IR have been used to characterize the effect of stress on individual molecules: WAXD for the modulus of crystalline regions, and IR for changes in vibration frequencies with stress in both crystalline and amorphous regions. The WAXD measurements (e.g., Ref. 6) rely on measurement of changes in **c** axis spacings with stress in well-oriented samples. The ratio of the stress to the **c** axis strain is the modulus of the crystalline regions. Table 8 is a comparison of observed and calculated values for a number of polymers. The single molecule values are obtained by multiplying the crystal moduli by the cross-sectional area of the molecule.

IR can also be used to study the effects of stress on polymers. For steady stress the goal is to obtain the molecular stress distribution, using initially highly oriented samples. Under stress small shifts in position and shape of some of the peaks are observed, with the shift depending on the initial draw ratio and thermal history of the sample. In other cases, where a phase change occurs under stress, there can be a change in absorbance of appropriate bands. Figures 72 and 73 show examples of these effects for PET and polybutylene terephthalate (PBT), respectively.

Zhurkov et al. [136] have shown that the frequency shift for many bands is approximately linear with applied stress, that is, $\Delta v = \alpha_x \sigma$, where α_x depends on the morphology of the sample and the temperature [138]. The results in Figure 72, for the 976 cm^{-1} band (trans, primarily crystalline), show that significant variations in α_x occur for various degrees of initial, controlled shrinkage during annealing (at 155°C for 17 hr in nitrogen) and that there is a threshold stress before shifting occurs. The latter is interpreted in terms of the disordered amorphous chains bearing the initial stress. The decrease in α_x with initial shrinkage is attributed to chain disorientation during annealing and to variations in the crystalline texture.

In PBT there is a reversible change in crystal structure under stress, with the aliphatic segments transforming from gauche-trans-gauche to all-trans. As shown in Figure 73, the decrease in CH_2 absorbance at 1460, beginning at 2–3% strain, accompanies the beginning of the phase change, while the 1485 cm^{-1} CH_2 band, attributed to the all-trans conformation, increases, both in a reversible manner. The 1510 cm^{-1} p-phenylene band was used as an internal thickness band to correct for changes in thickness.

Table 8 Comparison of Observed and Calculated (from Bond Energies) Single Chain Moduli

Polymer[a]	Tenacity (g/denier)	Tenacity (kg/mm²)	Elongation (%)	Macroscopic modulus (g/denier)	Macroscopic modulus (dyn/cm²)	Crystallite modulus[b] Obsd.[c] (dyn/cm²)	Crystallite modulus[b] Calcd. (dyn/cm²)	M_p (decomposition point) (°C)
Kevlar	25	330	5	850	111×10^{10}	153×10^{10}	182×10^{10}	(500)
PRD-49	15	200	3	1050	134	—	163	(500)
Nomex	5.5	68	35	82	10	88	90	(415)
PET	9.0	110	7	160	19.5	108	124[d]	260
it-PP	9.0	74	15	120	9.6	34	28[e]	170
PE	9.0	78	8	100	8.5	235	296[e]	135
Nylon 6 (α)	9.5	97	16	50	5.0	165	244[f]	225
PEOB (α)	5.3	64	25	75	8.9	5.9	2.4	225

[a]PEOB: poly(ethylene oxybenzoate).
Source: Ref. 6.

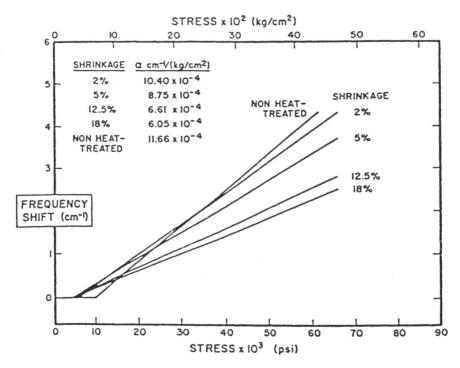

Figure 72 Frequency shift (of the peak) versus applied stress for PET films for the 976 cm^{-1} trans band. Controlled prior shrinkage was accomplished by annealing. (From Ref. 138 as reported in Ref. 137).

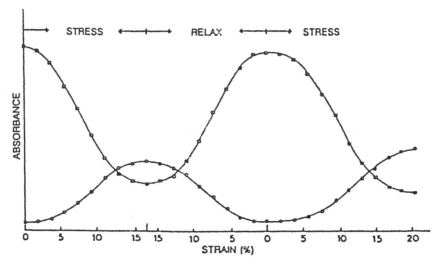

Figure 73 Absorbance of the CH$_2$ bands at 1460 cm^{-1} (□) relaxed form and 1485 cm^{-1} (o) stressed form as a function of reversible stress and strain. The absorbances were corrected for changes in sample thickness using a 1510 cm^{-1} aromatic ring band as an internal thickness band. (From Ref. 139.)

REFERENCES

1. C. Booth and C. Price, eds., *Comprehensive Polymer Science*, Vol. 1, *Polymer Characterization*, Pergamon Press, Oxford, 1992.
2. J. Haslem, H. A. Willis, and D. C. M. Squirrel, *Identification and Analysis of Plastics*. 2nd ed. Butterworth, London, 1972.
3. D. Braun, *Simple Methods for Identification of Plastics*, Hanser, Munich, 1982.
4. J. Brandrup and E. H. Immergut, eds., *Polymer Handbook*, 3rd ed., Wiley, New York, 1989.
5. (a) G. S. Beddard, Emission Spectroscopy, *Comprehensive Polymer Science*, Vol. 1, (C. Booth and C. Price, eds.), Pergamon Press, Oxford, 1992, Chap. 22.2 (b)G. Kampf, Characterization of Plastics by *Physical Methods*, Hanser (Macmillan), Munich (New York), 1986, Chap. 7.
6. H. Tadokoro, *Structure of Crystalline Polymers*, Wiley, New York, 1979.
7. G. Kampf, *Characterization of Plastics by Physical Methods*, Hanser (Macmillan), Munich (New York), 1986.
8. J. L. Koenig, *Spectroscopy of Polymers*, American Chemical Society, Washington, D.C., 1992.
9. J. L. Koenig, *Chemical Microstructure of Polymer Chains*, Wiley, New York, 1982.
10. (a) J. G. Grasselli and W. M. Ritchey, eds., *Atlas of Spectral Data and Physical Constants for Organic Compounds*, Vol. 1, 2nd ed., CRC Press, Cleveland, OH, 1975. (b) R. C. Weast, ed., Handbook of Chemistry and Physics, CRC Press, Boca Raton, FL (annual).
11. (a) L. J. Bellamy, *The Infrared Spectra of Complex Molecules*, Wiley, New York, 1975. (b) D. Dolphin and A. Wick, *Tabulation of Infrared Spectra Data*, Wiley, New York, 1977.
12. J. C. Hennicker, *Infrared Spectrometry of Industrial Polymers*, Academic Press, New York, 1967.
13. F. A. Bovey, *Nuclear Magnetic Resonance Spectroscopy*, 2nd ed., Academic Press, San Diego, 1988.
14. (a) F. A. Bovey, *High Resolution NMR of Macromolecules*, Academic Press, New York, 1972. (b) E. M. Mohacsi, *J. Chem Educ. 41*:38(1964).
15. R. A. Komarshi, ed., *High Resolution NMR of Synthetic Polymers in Bulk*, VCH, Deerfield Beach, Fla., 1986.
16. R. Holm and S. Storp, *Surf. Interface Anal. 2*:96 (1980).
17. R. Holm, Proc. 8th Int. Microchem. Symp., 1980, p. 257.
18. G. Bucci and T. Simonazzi, *J. Polym. Sci. C7*:203 (1964). See also: C. Tosi and F. Campbell, *Adv. Polym. Sci. 12*:87 (1973).
19. F. M. Schnepel, *Chem. Uns. Zeit 13*:33 (1979).
20. J. M. O'Reilley and R. A. Mosher, *Macromolecules 14*:602 (1981).
21. J. T. Arnold, S. Dharmatti, and M. E. Packard, *J. Chem. Phys. 19*:507 (1951).
22. F. A. Bovey, F. C. Schilling, F. L. McCrackin, and H. L. Wagner, *Macromolecules 9*:76 (1976).
23. R. C. Ferguson, *Macromolecules 2*:237 (1969).
24. J. C. Randell, *Polymer Characterization by ESR and NMR* (A. E. Woodward and F. A. Bovey, eds.), ACS Symp. Series 142, American Chemical Society, Washington, D.C., 1980.

25. J. Schaefer, M. D. Sefcik, E. D. Stejskal, and R. A. McKay, *Macromolecules 14*:188 (1981).
26. (a) D. T. Clark, W. J. Feast, I. Ritchie, W. K. R. Musgrave, M. Modena, and M. Ragazzini, *J. Polym. Sci., Polym. Chem. Ed., 12*:1049 (1974). (b) D. T. Clark, ESCA applied to polymers, *Advances in Polymer Science, Vol. 24*, (H. J. Cantow et al., eds.), Springer Verlag, Berlin, 1977.
27. F. W. Billmeyer, *Textbook of Polymer Science*, Interscience, Wiley, New York, 1962.
28. (a) E. A. Collins, J. Bares and F. W. Billmeyer, *Experiments in Polymer Science*, Wiley, New York, 1973. (b) J. V. Dawkins, Size exclusion chromatography, *Comprehensive Polymer Science*, Vol. 1 (C. Booth and C. Price, eds), Pergamon Press, Oxford, 1992, Chap. 12.
29. F. M. Mirabella, Jr., and E. A. Ford, *J. Polym. Sci., Polym, Phys. B25*:71 (1987).
30. J. R. Shaefgen and P. J. Flory, *J. Am. Chem. Soc. 70*:2709 (1948).
31. M. L. Huggins, *J. Am. Chem. Soc. 64*:2716 (1942).
32. E. O. Kraemer, *Ind. Eng. Chem. 30*:1200 (1938).
33. This equation is usually attributed to Mark, Houwink, and Sakurada; see Ref. 30.
34. T. G. Fox, S. Gratch, and S. Loshack, *Rheology*, Vol. 1 (F. R. Eirich, ed.), Academic Press, New York, 1956, Chap. 12.
35. S. Oka, *Rheology*, Vol. 3 (F. R. Eirich, ed.), Academic Press, New York, 1960, Chap. 2.
36. J. E. McIntyre, Polyester fibers, *Handbook of Fiber Science and Technology*, Vol. IV, *Fiber Chemistry* (M. Lewin and E. M. Pearce, eds.), Marcel Dekker, New York, 1985.
37. A. I. Kitaigorodski, *Organic Chemical Crystallography*, Consultants Bureau, New York, 1961.
38. G. Natta and P. Corradini, *Nuovo Cimento, Suppl. 15*:9 (1960).
39. (a) Ref. 6. (b) L. E. Alexander, *X-Ray Diffraction Methods in Polymer Science*, Wiley, New York, 1969. (c) M. Kakudo and N. Kasai, *X-Ray Diffraction by Polymers*, Elsevier, Amsterdam, 1972. (d) J. E. Spruiell and E. S. Clark, *Methods of Experimental Physics*, Vol. 16B (R. A. Fava, ed.), Academic Press, New York, 1980, Chap. 6. (e) F. J. Balta-Calleja and C. G. Vonk, *X-ray Scattering of Synthetic Polymers*, Polymer Science Library, Vol. 8 (A. D. Jenkins, ed.), Elsevier, Amsterdam, 1989.
40. B. K. Vainshtein, *Diffraction of X-rays by Chain Molecules*, Elsevier, Amsterdam, 1966.
41. G. M. Bhatt, J. P. Bell, and J. R. Knox, *J. Polym. Sci., B14*:373 (1976).
42. (a) R. J. Young, *Characterization of Solid Polymers* (S. J. Spells, ed.), Chapman & Hall, London, 1994, Chap. 6. (b) N. Schlotter, Raman spectroscopy, *Comprehensive Polymer Science*, Vol. 1 (C. Booth and C. Price, eds.), Pergamon Press, Oxford, 1992, Chap. 21. (c) P. Hendra, C. Jones, and G. Warnes, *Fourier Transform Raman Spectroscopy*, Ellis Horwood, Chichester, 1991.
43. J. C. Rodriguez-Cabello, L. Quintanilla, and J. M. Paster, *J. Raman Spectrosc. 25*:335 (1994).
44. L. Quintanilla, J. C. Rodriguez-Cabello, T. Jawhari, and J. M. Paster, *Polymer 34*:3787 (1993).
45. R. de P. Daubeny, C. W. Bunn, and C. J. Brown, *Proc. R. Soc. (Lond.) A226*:531 (1954).
46. R. Boyer, *Macromolecules 6*:288 (1973) and *J. Macromol. Sci., Phys. B8*:503 (1973).
47. J. L. Koenig and M. Hannon, *J. Macromol. Sci., Phys. B1*:119 (1967).

48. (a) Cerius², Molecular Simulations, Inc. (b) SYBYL, Tripos Associates, Inc. (c) Biosym Technologies, Inc.

49. T. C. Long, J. Liu, B.-L. Yuan and P. H. Geil, paper presented at American Physical Society Meeting, San Jose, Calif., March 1995, in press.

50. W. Claffey, K. Gardner, J. Blackwell, J. Lando, and P. H. Geil, *Phil. Mag. 30*:1223 (1974).

51. (a) D. L. Dorset, *Characterization of Solid Polymers* (S. J. Spells, ed.), Chapman & Hall, London, 1994, Chap. 1. (b) D. L. Dorset, *Macromolecules 25*:4425 (1992).

52. (a) Z. G. Pinsker, *Electron Diffraction*, Butterworths, London, 1953. (b) B. K. Vainshtein, *Structure Analysis by Electron Diffraction*. Macmillan, New York, 1964.

53. J. R. Fryer and D. L. Dorset, eds., *Electron Crystallography of Organic Molecules*, NATO ASI Series, Vol. C328, Kluwer, Dordrecht, 1991.

54. H. Hasegawa, W. Claffey, and P. H. Geil, *J. Macromol. Sci., Phys. B13*:89 (1977).

55. B. Moss and D. L. Dorset, *J. Macromol. Sci., Phys. B22*:69 (1983).

56. (a) J. C. Wittman, A. M. Hodge, and B. Lotz, *J. Polym. Sci., Polym. Phys. Ed. 21*:2495 (1983) (b) J. C. Wittmann and B. Lotz, *Prog. Polym. Sci. 15*:909 (1990).

57. S. H. Carr, A. Keller, and E. Baer, *J. Polym. Sci. A-2 8*:1467 (1970).

58. (a) P. H. Geil, *J. Appl. Phys. 33*:642 (1962). (b) S. V. Meille, T. Konishi, and P. H. Geil, *Polymer 25*:773 (1984).

59. D. L. Dorset, M. P. McCourt, S. Kopp, J. C. Wittman, and B. Lotz, *Acta Crystallogr. B50*:201 (1994).

60. R. G. Scott, *J. Polym. Sci. 57*:405 (1962).

61. R. M. Gohil and J. Petermann, *J. Macromol. Sci., Phys. B18*:217 (1980).

62. F. Rybnikar, B. L. Yuan, and P. H. Geil, *Polymer 35*:1831 (1994).

63. D. Dean and P. H. Geil, paper presented at American Physical Society Meeting, Pittsburgh, March 1994, in press.

64. F. Rybnikar, J. Liu, and P. H. Geil, *Makromol. Chem. Phys. 195*:81 (1994).

65. P. H. Geil, paper presented at Europhysics Conf. on Macromolecular Physics, Prague, July 1995, in press.

66. J. Liu, T. C. Long, B. -L. Yuan, and P. H. Geil, paper presented at Europhys. Conf. Macromol. Phys., Prague, July 1995, in press.

67. A. J. C. Wilson, *Nature 150*:151 (1942).

68. P. H. Geil, *J. Macromol. Sci., Chem. A1:*325 (1967).

69. (a) J. Liu, F. Rybnikar and P. H. Geil, *J. Polym. Sci., Polym. Phys. Ed. B30*:1469 (1992). (b) J. Liu and P. H. Geil, in press.

70. P. Iannelli, D. Y. Yoon, and W. Parrish, *Macromolecules 27*:3295 (1994).

71. J. Liu, F. Rybnikar, and P. H. Geil, *J. Macromol. Sci., Phys. B32:*395 (1993).

72. D. Y. Yoon, N. Masciocchi, L. E. Depero, C. Viney, and W. Parrish, *Macromolecules 23*:1793 (1990).

73. T. Thomsen, H. G. Zachmann, and H. R. Kricheldorf, *J. Macromol. Sci., Phys. B30*:87 (1992).

74. M. Tsuji, Electron microscopy, *Comprehensive Polymer Science*, Vol. 1 (C. Booth and C. Price, eds.), Pergamon Press, Oxford, 1992, Chap. 34.

75. M. Tsuji, A. Uemura, M. Ohara, S. Isoda, A. Kaweguchi, and K. Katayama, *Koenshu-Kyoto Daigaku Nippon Kagaku Seni Kenhyrucho 44*:1 (1987).

76. (a) M. Tsuji, S. Isoda, M. Ohara, A. Kawaguchi, and K. Katayama, *Polymer 23*:1568 (1982). (b) S. Isoda, M. Tsugi, M. Ohara, A. Kawaguchi, and K. Katayama, *Polymer*

24:1155 (1983). (c) K. Katayama, S. Isoda, M. Tsuji, M. Ohara, and A. Kawaguchi, *Bull. Inst. Chem. Res. Kyoto Univ.* *62*:198 (1984). (d) M. G. Dobb, D. J. Johnson, and B. P. Saville, *J. Polym. Sci., Polym. Phys. Ed.* *15*:2201 (1977).

77. (a) J. L. Koenig, *Spectroscopy of Polymers*, American Chemical Society, Washington, D.C., 1992, Chap. 10. (b) V. D. Fedotov and H. Schneider, *Structure and Dynamics of Bulk Polymers by NMR Methods*, Springer-Verlag, New York, 1989.

78. (a) Y. Fu, W. R. Busing, Y. Jin, K. A. Affholter, and B. Wunderlich, *Macromolecules* *26*:2187 (1993). (b) Y. Fu, W. R. Busing, Y. Jin, K. A. Affholter, and B. Wunderlich, *Macromol. Chem. Phys.* *195*:803 (1994). (c) Y. Fu, B. Annis, Y. Jin, A. Boller, Y. Jin, and B. Wunderlich, *J. Polym. Sci., Polym. Phys.* *B32*:2289 (1994).

79. (a) A. Immirzi and P. Iannelli, *Gazz. Chim. Ital.* *117*:201 (1987); *Macromolecules* *21*:768 (1988). (b) P. Iannelli and A. Immirzi, *Macromolecules* *22*:196 (1989); *23*:2375 (1990). (c) W. R. Busing, *Macromolecules* *23*:4608 (1990).

80. (a) M. J. Richardson, *Thermal Analysis in Comprehensive Polymer Science*, Vol. 1 (C. Booth and C. Price, eds.), Pergamon Press, Oxford, 1992. (b) T. Hatakeyama and F. X. Quinn, *Thermal Analysis*, Wiley, Chichester, 1994.

81. (a) S. Sauerbrunn, B. Crowe, and M. Reading, *Am. Lab.* *24(12)*:44 (1992). (b) S. Sauerbrunn and P. Gill, *Am. Lab.* *25(14)*:54 (1993). (c) S. Sauerbrunn and L. Thomas, *Am. Lab.* *27(1)*:19 (1995).

82. J. DeChant, *Infrared Spectroscopic Investigations on Polymers*, Akademie-Verlag, Berlin, 1992.

83. J. L. Koenig, personal communication.

84. G. Natta and P. Corradini, *Nuovo Cimento (Suppl.)*, *15*:40 (1960).

85. W. O. Statton and P. H. Geil, *J. Appl. Polym. Sci.* *3*:357 (1960).

86. R. Hosemann and A. H. Hindeleh, *J. Macromol. Sci., Phys.* *B34(4)*:327 (1995).

87. (a) W. Ruland, *Acta Crystallogr.* *14*:1180 (1961). (b) W. Ruland, *Polymer 5*:89 (1964).

88. C. G. Vonk, *J. Appl. Crystallogr.* *6*:148 (1973).

89. P. H. Geil, *Polymer Single Crystals*, Wiley, New York, 1963.

90. J. J. Hermans, P. H. Hermans, D. Vermaas, and A. Weidinger, *Rec. Trau. Chim. Pzys Bas 65*:427 (1946).

91. (a) W. O. Statton and G. M. Godard, *J. Appl. Phys.* *28*, 1111 (1957). (b) H. Tadokoro, K. Tatsuka, and S. Murahashi, *J. Polym. Sci.* *59*:413 (1962).

92. (a) E. S. Clark and M. R. Boone, Conf. Proc. Soc. Plastics Eng., ANTEC 90, 542 (1990). (b) E. S. Clark, personal communication.

93. J. J. Lear, Ph.D. thesis, University of Illinois, Urbana, 1990.

94. International Standard ISO 2557/1 (1976E) and 2557/2 (1979E).

95. A. N. J. Heyn, *Fiber Microscopy, A Textbook and Laboratory Manual*, Interscience, New York, 1954.

96. (a) C. R. Desper, J. L. Mead, M. Sussman and E. Kasch, presented at Matl's Res. Soc. Meeting, Boston, Dec 1992. (b) C. R. Desper, *J. Macromol. Sci., Phys. B7*:105 (1973).

97. (a) R. J. Samuels, *Makromol. Chem. (Suppl)*, *4*:241 (1981). (b) F. M. Mirabella, Jr., *J. Polym. Sci., Polym. Phys.* *B25*:591 (1987).

98. L. J. Fina and J. L. Koenig, *J. Polym. Sci., Polym. Phys.* *B24*:2525 (1986).

99. V. J. McBierty, I. R. MacDonald, and I. M. Ward, *J. Phys. D4*:88 (1971).

100. A. P. Unwin, D. I. Bower, and I. M. Ward, *Polymer 26*:1605 (1985).

101. J. R. Dees and J. E. Spruiell, *J. Appl. Polym. Sci.* *18*:1053 (1974).

102. (a) G. Elsner, C. Riekel, and H. G. Zachmann, *Adv. Polym. Sci.*, *67*:1 (1985). (b) C. Riekel, *Chemical Crystallography with Pulsed Neutrons and Synchrotron Radiation* (H. Carrondo and G. A. Jeffrey, eds.), Raidel., 1988.

103. (a) G. Elsner, M. H. J. Koch, J. Bordas, and H. G. Zachmann, *Makromol. Chem. 182*:126 (1981). (b) W. Wu, C. Riekel, and H. G. Zachmann, *Polym. Comm. 25*:76 (1984).

104. P. Forgaes, B. P. Tolochko, and M. A. Sheromov, *Polym. Bull. 6*:127 (1981).

105. H. G. Zachmann and C. Wutz, *Crystallization of Polymers* (M. Dosiere ed.), Kluwer, Dordrecht, 1993, pp. 403–414.

106. (a) A. Guinier and G. Fournet, *Small Angle Scattering of X-rays*, Wiley, New York, (1995). (b) W. W. Beeman, P. Kaesburg, and J. Anderegg, Size of particles and lattice defects, *Encyclopedia of Physics*, Vol. 32 (S. Flugge, ed.), Springer-Verlag, Berlin, p. 321.

107. W. O. Statton, Small angle X-ray studies of polymers, *Newer Methods of Polymer Characterization*, Interscience, New York, 1963, Chap. 6.

108. R. Bonart and R. Hosemann, *Kolloid Z.-Z. Polym. 186*:16 (1962).

109. W. O. Statton, *J. Polym. Sci. 41*:143 (1959).

110. A. Ferrero, E. Ferracini, and R. Hosemann, *Polymer 25*:1747 (1984).

111. R. G. Scott, *J. Polym. Sci. 57*:405 (1962).

112. (a) L. Sawyer and E. L. Thomas, *Polymer Microscopy*, Chapman and Hall, London, 1987. (b) D. C. Bassett, Etching and microstructural crystalline polymers, *Comprehensive Polymer Science*, Vol. 1 (C. Booth and C. Price, eds.), Pergamon, New York, 1992, Chap. 35.

113. A. Siegmann and P. H. Geil, *J. Macromol. Sci. Phys. B4*:557 (1970).

114. (a) A. Peterlin, The role of chain folding in fibers, *Man Made Fibers*, Vol. 1 (H. F. Mark, S. M. Atlas, and E. Cernia, eds.), Interscience, New York, 1967, p. 283. (b) A. Peterlin and S. Sakaoko, *J. Appl. Phys. 38*:4142 (1967). (c) A. Peterlin and K. Sakaoko, *Clean Surfaces* (G. Goldfinger, ed.), Marcel Dekker, New York, 1970.

115. (a) K. Neki and P. H. Geil, *J. Macromol. Sci., Phys. B9*:71 (1974). (b) A. Peterlin, H. Kiho, and P. H. Geil, *Polym. Let. 3*:151 (1965).

116. F. Rybnikar, unpublished data.

117. A. K. Singhania and P. H. Geil, *Makromol. Chem. 143*:231 (1971).

118. D. M. Sadler, Neutron scattering from solid polymers, *Comprehensive Polymer Science* (C. Booth and C. Price, eds.), Pergamon Press, Oxford, 1992, Chap. 32.

119. G. D. Wignall, unpublished data.

120. (a) D. Y. Yoon and P. J. Flory, *Polymer 18*:509 (1977). (b) D. Y. Yoon, *J. Appl. Crystallogr. 11*:531 (1978). (c) D. Y. Yoon and P. J. Flory, *Disc. Faraday Soc. 68*: Paper 17 (1979).

121. J. D. Hoffman, E. A. DiMarzio, and C. H. Guttmann, *Disc. Faraday Soc., 68*: Paper 17 (1979).

122. B. Wunderlich, *Macromolecular Physics*, Vol. 1, Academic Press, New York, 1973.

123. D. C. Bassett, *Principles of Polymer Morphology*, Cambridge University Press, Cambridge, 1981.

124. A. S. Vaughn and D. C. Bassett, Crystallization and morphology, *Comprehensive Polymer Science*, Vol. 2 (C. Booth and C. Price, eds.), Pergamon Press, Oxford, 1992, Chap. 12.

125. (a) H. D. Keith, *Mater. Res. Soc. Symp. Proc., 321*:511 (1994). (b) A. S. Vaughn, *Sci. Prog. Oxford 76*:1 (1992).

126. P. H. Geil, *C & E News 43*:72 (1965).

127. F. Rybnikar and P. H. Geil, *J. Macromol. Sci. Phys. B71* (1973).

128. B. S. Sprague, *J. Macromol. Sci. Phys. B8*:157 (1973).

129. M. J. Miles, New techniques in microscopy, *Characterization of Solid Polymers* (S. J. Spells, ed.), Chapman & Hall, London, 1994, Chap. 2.

130. G. J. Vansco, R. Nizman, D. Sneting, H. Schonherr, P. Smith, C. Ng, and H. Yang, *Colloids Surf., Physiochem. Eng. Aspects 87*:263 (1994).

131. (a) K. D. Jandt, M. Buhk, M. J. Miles, and J. Petermann, *Polymer 35*:2458 (1994). (b) W. Stocker, H. Schumacher, S. Graff, J. Lang, J. C. Wittman, A. J. Lovinger and B. Lotz, *Macromolecules 27*:6948 (1990).

132. (a) N. G. McCrum, B. E. Read, and G. Williams, *Anelastic and Dielectric Effects on Polymeric Solids*, Dover, New York, 1991 (reprint of Wiley, New York, 1967). (b) P. Gradin, P. G. Howgate, R. Seldin, and R. A. Brown, Dynamic mechanical properties, *Comprehensive Polymer Science*, Vol. 2, Pergamon Press, Oxford, 1992, Chap. 16.

133. K. Wangermann and H. G. Zachmann, *Prog. Colloid Polym. Sci. 57*:236 (1975).

134. J. B. Ibar, P. Denning, T. Thomas, A. Bernes, C. deGoys, J. R. Saffell, P. Jones, and C. Lacabanne, Characterization of polymers by thermally stimulated current analysis and relaxation map analysis spectroscopy, *Polymer Characterization* (C. D. Craver and T. Prouder, eds.), American Chemical Society, Washington, D.C., 1990, Chap. 10.

135. P. DeMont, L. Fourmaud, D. Chatain, and C. Lacabanne, Thermally stimulated creep for the study of copolymers and blends, *Polymer Characterization* (C. D. Craver and T. Prouder, eds.), American Chemical Society, Washington, D.C. 1990, Chap. 11.

136. S. N. Zhurkov, V. I. Vettegren, V. E. Korsukov, and I. I. Novak, *Fracture 1979, Proceedings of the 2nd International Conference on Fracture*, Chapman & Hall, London, 1969, p. 545.

137. R. P. Wool, *Polym. Eng. Sci. 20*:805 (1980).

138. K. K. R. Mocherla, Ph.D. Thesis, University of Utah, Salt Lake City, 1976.

139. H. W. Siesler, *Proceedings of the 5th European Symposium on Polymer Spectroscopy* (D. Hummell, ed.), Verlag Chemie, Weinheim, 1979.

3
Surface Characterization of Textiles Using SEM

WILTON R. GOYNES   Southern Regional Research Center,
Agricultural Research Service, United States Department
of Agriculture, New Orleans, Louisiana

I.   INTRODUCTION

In modern materials science, structural characterization is basic to a thorough understanding of the nature of a material. The microscope has played an important part in formulating structural concepts in polymeric materials. The light microscope (LM) provided information on the external structure of bulk and fibrillar polymers, and the transmission electron microscope (TEM) permitted ultrastructural information to be obtained. However, natural concepts of external and internal surface structures were difficult to envision due to the low resolution and depth of focus in the LM, and due to the necessity of dealing with microareas of the sample at high magnifications in the TEM. The scanning electron microscope (SEM) produces surface images that are realistic, three-dimensional reproductions of the sample, and its depth of focus and range of magnifications fill the gap between the LM and the TEM. It has thus opened a whole new field in fiber structure studies.

Fiber scientists have relied heavily on the microscope for gaining information on their product for many years. In 1836 a microscopical study of shape, diameter, maturity, and variability of properties of the cotton fiber was given at the Privy Council of King William IV in a report on the status of the textile industry of Great Britain [1]. Application of electron microscopy to a study of cotton fiber structure was made in the early 1940s by Ruska and Kretschmer [2], by Eisenhut and Kuhn [3], and by Barnes and Burton [4], and this instrument has continued to be used in the study of the fine structure of fibers and changes in structure due to chemical modifications. This "new" microscope was also used for characterization of synthetic fibers as they were developed and came into commercial production [5]. Because of the limited ability of the TEM to provide gross information

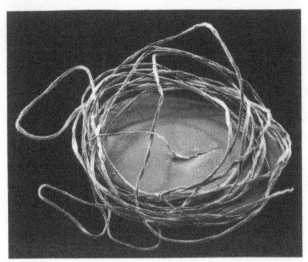

Figure 1   Single coiled cotton fiber showing the extreme length–diameter dimensions.

on fiber structure, fiber scientists were quick to realize the potential for application of the newly developed SEM in this field. Though the instrument did not become commercially available until 1965, it was used to study fibers as early as the mid 1950s. At the conference on Electron Microscopy of Fibers at the University of Leeds in January 1956, Wells described the examination of fibers in the SEM [6]. During developmental work on the SEM at Cambridge University, Smith and Atack of the Pulp and Paper Institute of Canada showed the potential of this instrument in pulp and paper research [7]. When this instrument was commercialized, two of the first three produced were used in textile fiber laboratories. The SEM not only is used in structural studies and basic and developmental research, but also has found great application in industrial troubleshooting and for dramatic illustration.

Most textile fibers, both natural and synthetic, are organic polymers that manifest some characteristic morphological structure with easily discernable features. In natural fibers such as cotton and wool, this structure is genetically determined and may be influenced by variety or growing conditions. Synthetic fibers or filaments are designed and engineered to a predetermined structure. These special structural features may be used for fiber identification, for prediction of fiber characteristics, or for determination of changes that have occurred in the fiber due to chemical reaction and finishing, or to damage by insects, fungus, heat, wear, or chemical attack. Study of deformations or of rupture due to deformation can provide useful information on internal morphology as well as on reaction deformation.

II. SAMPLE PREPARATION

Preparation of fiber samples for SEM is relatively simple, especially as compared to TEM preparation, but it should be carried out with care since even simple preparative techniques, if poorly done, result in poor micrographs or erroneous interpretations. Because of the poor conductivity of most polymers, a layer of conducting material must be applied to their surfaces before SEM examination. Echlin and Hyde [8] list five parameters to be considered in choosing a coating material:

1. Conductivity
2. Mechanical stability
3. Secondary electron emission
4. Backscattered primaries
5. Beam penetration

Poor sample conductivity causes buildup of surface electrons, producing variations in surface potential, or "charging." Charging manifests itself as various image distortions and excessively bright overglow, which prevents true image production. Poor conductivity also reduces thermal dissipation, which may lead to

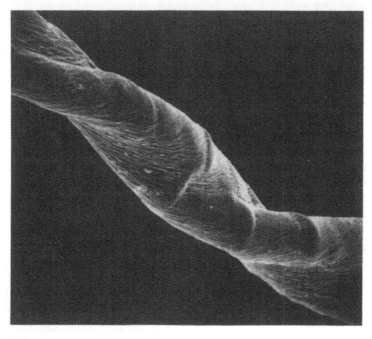

Figure 2 Enlarged fiber segment showing the wrinkled primary wall surface.

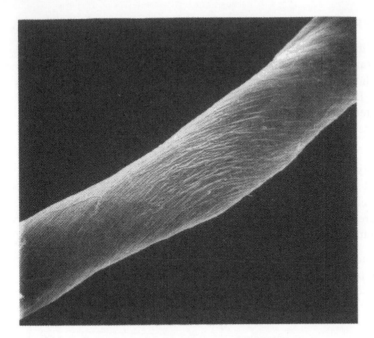

Figure 3 Enlarged fiber segment in which the surface wrinkles indicate a change in the direction of the underlying fibrillar spiral (a reversal).

local sample movement or even heat damage. In addition, polymers, which are usually materials of low atomic density, are readily penetrated by a high-keV (electron voltage) electron beam, causing loss of surface detail. Coating with a microlayer of metal is the most widely used method to suppress charging and decrease beam penetration. Low beam penetration is especially important in polymers since secondary electron emission (those electrons emitted by the sample due to bombardment of the electron beam) is their primary mode of image generation, and deep penetration reduces the number of surface secondaries produced. The amount of metal to be applied depends on the sample and the magnification or resolution to be achieved. Most literature sources specify a 200–300 Å metal layer, which, to be effective in eliminating charging, must be continuous. Continuity may be easily achieved with fibers or even yarns, but the geometry of fabrics presents a problem for obtaining continuous films. However, since most yarn and fabric samples are studied at low magnifications, thicker coatings (500–600 Å) may be applied without obscuring any information available at these magnifications. Such coatings are likely to be continuous over the sample surface as well as from sample to stub. Specimen-to-stub conductivity can be improved by use of conductive tape or paint. Caution should be used in applying conductive paint since it can wick into fibrous samples. As the need for use of higher magnifica-

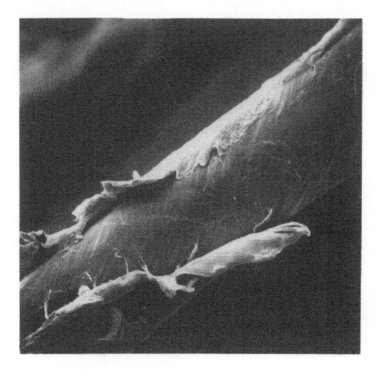

Figure 4 Enlarged fiber segment showing a peeled primary wall, and the underlying fibrillar spiral.

tions arises, sample coating thicknesses should be reduced to improve resolution of sample features.

A. Metal Coating

Of the several methods that have been used to suppress charging, currently the most satisfactory method is metal coating by sputtering. The other method for metal coating that is used in some laboratories is vacuum evaporation, because this equipment is often already available for TEM preparation methods. Procedures other than metal coating have been used but with less success. Pfefferkorn et al. [9] discussed use of metal foil screening, gas discharge etching, and use of "wet" samples. Lane described the use of a humidity chamber to control charging. This method has been incorporated into the design of a scanning electron microscope with a controlled humidity chamber in which no sample coating is used. This instrument is called an environmental electron microscope and is discussed in Section V. Sikorski et al. [10], who was one of the first to use the SEM in fiber science, advocated the use of antistatic agents (usually solutions of organic liquids) rather

Figure 5   Tubular mercerized cotton fiber, showing a smoother surface.

than metal coating. This method has not achieved very great acceptance as a means of charge suppression.

The widely used sputtering method is achieved by bombarding a conductive metal with heavy particles, using inert gas. As the material is sputtered, it condenses on the specimen. Since the paths of the sputtered atoms are not straight in sputtering, theoretically the sample receives a more uniform coating, even in areas out of a direct path to the target material. In addition, the process is carried out at a much higher pressure than thermal evaporation, and is thus valuable for specimens sensitive to high vacuum. Various sputter-coating devices are available from microscopy supply companies.

In vacuum evaporation, the evaporated atoms travel in a straight line, only coating those surfaces directly in their line of travel. Thus, various methods of spinning and rotating the sample must be used to achieve maximum exposure of sample surfaces to the atoms. The difficulty of forming a continuous coating on fibers, yarns, and fabrics due to their multistructured nature cannot be overemphasized, and a method for achieving such coating for each type specimen examined must be worked out before SEM examination can be profitable. Commercially available rotating-tilting devices that fit inside the vacuum bell jar pro-

Figure 6 Wet-abraded fiber showing stringy fibrillation.

vide a very satisfactory means of sample coating. Depending on the heat lability of the polymer, care must be taken to protect it from the radiant heat of the filament. The specimen should be placed as far as possible from the source, and if necessary a shuttered aperture may be placed between the source and the specimen.

B. Coating Materials and Thickness

Choice of coating materials is of prime consideration. Even though it is a relatively poor electrical and thermal conductor, carbon has been used rather extensively as an evaporant [11]. Actually, carbon is best used as an undercoat in conjunction with heavy metal coating, since carbon on vaporizing is reflected inside the vacuum chamber and approaches the sample in directions not possible with metal atoms, thus providing a more uniform film. A metal film can then be overlaid onto the carbon to provide better conductance (unless x-ray analyses are to be done). Although silver and aluminum are the cheapest and easiest metals for coatings, they can be used only when the specimen is to be examined immediately upon coating, since these films deteriorate rapidly on storage. This is of particular concern in preparing fabrics, since fabric construction tends to encourage slight move-

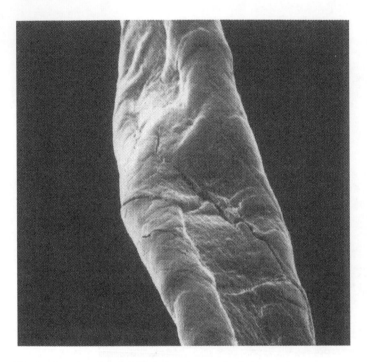

Figure 7 Dry-abraded fiber showing cracks in the surface.

ment ever after the sample is mounted and coated. Movement can cause the film
to crack, making it useless as a conductor. Gold and gold–palladium (60–40%,
w/w) are probably the best choices as coating materials, since they form relatively
stable films, and their higher atomic densities produce greater quantities of secondary elections with thin coating.

Coating thickness may be measured in several ways [8], none of which probably gives an accurate determination for specimens with very irregular surfaces.
For this reason it is difficult to propose that a yarn or fabric surface should have a
uniform coating of any particular thickness. While coating thickness may be more
critical for samples at high magnifications where maximum resolution is needed,
the prime goal for samples photographed at low magnifications is to rid the sample of charging, and with irregular fabrics, thicker coatings than those normally
recommended may be necessary. If it is certain that the sample surface is grounded
to the stub, and charging occurs after a normal coating procedure, the specimen
should be recoated. Such heavy coatings may render the sample useless for studying detailed structures at high magnifications, but do permit examination at the low
magnifications of interest.

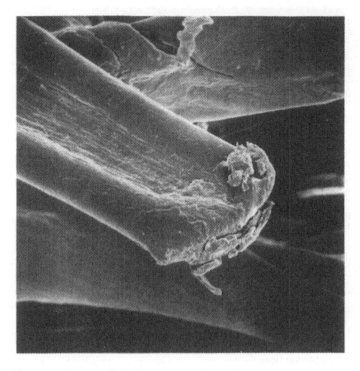

Figure 8   Blunt end of a broken, dry-abraded fiber.

Coating also affects the type of information obtained from the sample. If analyses other than secondary electron collection are to be made on a sample, this should be taken into account during sample preparation, since elemental analyses require coating with a material such as carbon that does not normally produce extraneous, unrelated sample information. Although carbon alone is not considered a good textile coating material for charge suppression, it can be used when elemental analyses are to be made.

C.   Sample-Holder Attachment

In preparation, the samples to be examined are simply cut to a size consistent with the mounting stub and attached to the stub with some type of adhesive. For fabric samples, double-coated adhesive tape is very convenient. This tape is applied to the stub, trimmed so that it does not quite meet the edge, and the fabric is firmly pressed into it. Tape tends to craze in the vacuum and does not provide an esthetic background for fibers, yarns, or particulate polymers at low magnification. A smoother background is usually obtained by the use of a liquid glue. The choice

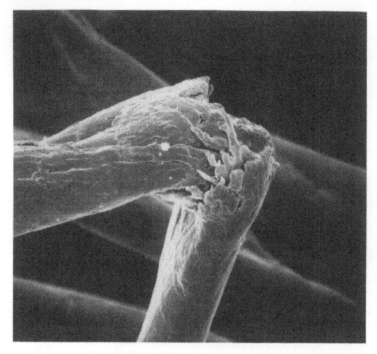

Figure 9   Snapped, dry-abraded fiber.

of glue is not critical, but it should be one that does not seriously outgas in the vacuum, or whose solvent does not affect the sample to be examined. One problem in
using glue is that if the sample is applied while the glue is still liquid, it tends to
infiltrate the sample and presents an interpretation problem. Once most glues have
solidified to the point at which they do not infiltrate, they are usually not tacky
enough to securely hold the specimen to the stub. A glue that remains tacky on drying may be obtained by dissolving the adhesive from cellophane adhesive tape in
an appropriate solvent. A semiliquid coating of this adhesive is placed on the specimen stub and allowed to dry to tackiness. The specimen may then be pressed onto
the stub without infiltration. The liquidity of the adhesive may be adjusted to
choice by use of various amounts of solvent.

Fibrous samples may also be mounted by using adhesive on only the ends of
the fibers. However, this type of mounting does not restrict specimen motion,
which can cause cracking of the conductive film, or specimen drift during examination. Mounting stubs have been devised for examination of fibers under special
conditions.

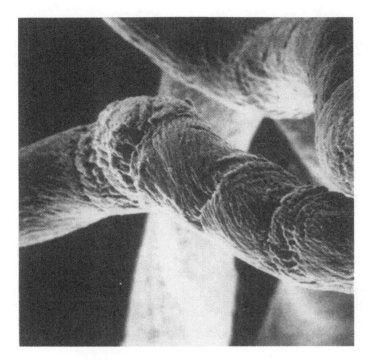

Figure 10 Fungus-damaged fiber after sodium hydroxide swelling.

Special sample preparation aids, such as carbon and metallic tapes, that make sample preparation easier are available from specialty microscopy supply houses. Transfer glues are available in sizes that cover SEM specimen stubs. These tacky glues are attached to a waxy paper and can be transferred to the stub by pressing the sticky surface to the stub and removing the paper. This product provides a tacky, strongly adhesive surface that does not wick into the sample. Silver paints and pastes are also available, but these should be used with caution to prevent infiltration into the sample or interference with elemental analyses.

III. MICROSCOPE OPERATING CONDITIONS

Optimum operating conditions of the SEM for examination of fibers may be somewhat different than for other materials. Parameters that provide good results for samples of high atomic number will not likely provide the same success with most fibers. For that reason it is necessary for the fiber microscopist to have at least a

Figure 11   Cotton fabric surface (LM).

basic understanding of the theory of operation of the SEM in order to set operating conditions that will provide the best results with textile samples.

A.   Beam Voltage

Polymers, generally materials of low density, allow greater penetration of the electron beam than do samples of higher atomic number. If a high-keV beam is used, this penetration produces signal from areas of the specimen below the surface and results in images of lower contrast and a loss of detail. Reduction of the beam voltage to 5 keV gives an image of considerably more detail but reduces the resolving capability of the SEM. However, the increased detail brought about by reducing the beam penetration overbalances this loss in resolution, especially at low magnifications. Use of low voltage is especially important when very thin specimens such as fibrillated or sectioned fibers are studied. Even with most coated specimens, low voltages are recommended as an additional means of reducing charging in fibers. Since resolution is lost by the use of low voltage, care must be taken in SEM operation to set other parameters to give the greatest possible resolution. If chosen operating parameters produce an image with a high signal-to-noise ratio

Figure 12 Cotton fabric surface (SEM).

(a great amount of background noise as compared to the signal generated by the sample), this increased noise may be overcome by using longer exposure times. Exposures of 100 sec are commonly used. Use of a shorter working distance is another means of increasing resolution. With careful operation, acceptable resolutions can be achieved even at very low voltages.

B. Sample Signals

Micrographs usually associated with the SEM are produced by collecting secondary electrons, those emitted by the sample due to bombardment of the electron beam and whose energy is less than 50 eV. They produce the spatially pleasing micrographs that provide surface structural information. Particularly in polymers with low atomic numbers, metallic coatings can control the depth of penetration of the beam and prevent excitation of electrons deep within the sample. They can also help prevent "edge effects," seen as bright areas, on samples that are thinner on the edge than in the center.

Information not available from secondary electrons can also be obtained by collection of other signals emitted from the sample. Backscattered electrons are

Figure 13 Unburned (left) and charred (right) cotton twill fabric, showing shrinkage that occurs during burning.

those electrons from the primary beam that penetrate the sample and are returned through the penetrated surface. Their energy is close to the incident beam energy. These electrons, having energy higher than that of secondary electrons, are less readily affected by charging forces. Therefore, use of the backscattered mode is one way to photograph samples that continue to charge even after coating. However, backscattered electrons contain different information than do secondary electrons, and the micrograph attained will not have the exact same surface image as those attained with secondary electrons alone. Backscattered electrons do contain Z (atomic number) information, and areas of changes in elemental composition may be noted by changes in contrast in the sample image.

A more definitive method for gaining elemental information from the sample is to collect x-rays that are produced in the sample due to rearrangements of the electrons in the atomic structure after emission of secondary electrons. These x-rays are characteristic of the element from which they are emitted, and can be identified by measuring their energies (energy-dispersive analysis) or their wavelengths (wavelength analysis). Spectral analysis of energy-dispersive x-rays (EDS) has been a useful tool in textile finishing studies. This method not only can be used for quantitative and qualitative analyses of the sample, but is spatially

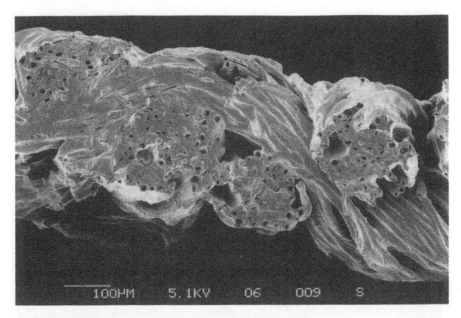

Figure 14   Fabric cross section of a cotton–polyester char.

consistent with secondary images, allowing locations of various elements in the sample to be shown. Qualitative energy-dispersive analysis of textile products is relatively simple, while quantitative analyses are more difficult.

IV.   APPLICATIONS OF SEM TO TEXTILES

Natural fibers may be either of plant or animal origin. Cotton and wood fibers are the two plant fibers of greatest commercial importance. Others include flax, ramie, jute, and sisal. The natural animal fibers are wool, other animal hair, and silk. Until the production of synthetic fibers, cotton, flax (linen), wool, and silk were the principal textile fibers. A survey of textile literature shows that the SEM has been used to characterize many different textile fibers. Jute was studied by A. K. Mukhopadhyay et al. [12, 13] and P. K. Ray et al. [14]. Mukherjee and Satyanarayana studied the structure and properties of vegetable fibers such as sisal [15]. Cheek and Roussel compared mercerization of ramie, flax, and cotton [16], and Cheek and Strusxczyk showed effects of liquid ammonia and sodium hydroxide on viscose [17]. Structure and properties of coir fibers (coconut palm) were investigated by Satyanarayana et al. [18]. SEM of wool fibers has been carried out by Brady et al. [19], Hafey and Watt [20], and Ito et al. [21]. One of the first commercial SEMs was used by Billica to study synthetic fibers. A report on techniques and general applications of SEM for synthetic fibers was published by Billica and

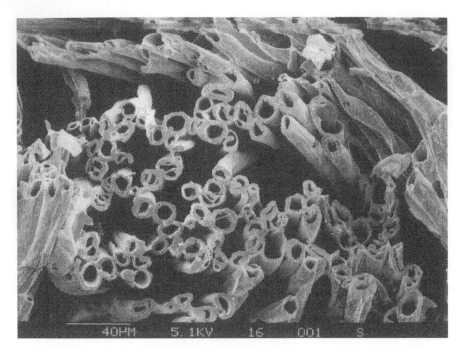

Figure 15 Char of a cotton yarn from a flame-retardant–finished fabric, showing fibers with compressed walls and enlarged lumens.

Van Veld [22]. More recent examples of synthetic fiber work can be found in publications by Stowell and Card [23] and Subramanian et al. [24] (nylon), Ansell [25] (polyester), Kulshreshtha et al. [26] (acrylic), and Schwartz et al. [27] (polyethylene). Structural characteristics can often be used for fiber identification. Sawbridge and Ford have shown longitudinal and cross-sectional views of many natural and manmade fibers [28]. Sich [29] compared SEM and light microscopy for identifying animal fibers. Problems of abrasion and pilling were studied by Paek [30], and damage to a wide range of fibers was reported by Zeronian et al. [31]. Cotton fibers and fabrics have been studied extensively using the SEM. General methods for microscopy of cellulosic textiles have been reviewed by Rollins et al. [32]. One of the earliest applications of SEM to the cotton fiber was the various studies of fiber fracture by Hearle and Lomas [33]. SEM has been used often to examine cotton fabric abrasion. An example of this is shown in the work of Dweltz et al. [34]. Raheel and co-workers have published a series of papers on modifying wear life of cotton fabrics [35]. Obendorf et al. have extensively studied fabric soiling [36]. The effect of chemical modification on cotton was shown by Ghosh and Dalal [37]. References are made in this chapter to SEM applications for many types of fibers, but the major emphasis is on the cotton fiber because that is the area that the author has most extensively studied.

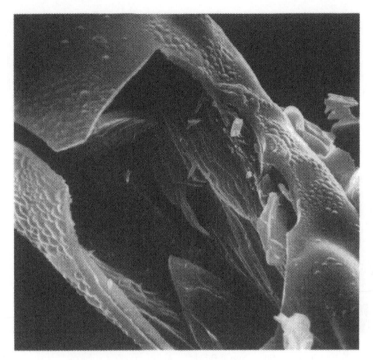

Figure 16 Charred fiber from a flame-retardant–finished fabric. The break in the wall shows the collapsed internal structure.

A. Cotton Fiber Morphology

Cotton is a seed hair that grows in tubular form, developing thickness by depositing layers of cellulose fibrils inside the cell wall (primary wall) and enclosing the central canal, or lumen. These fibrillar layers spiral around the axis of the fiber at an angle of approximately 45 degrees. Periodically the direction of this angle reverses. This point is called a reversal, and the change in spiral angle affects fiber strength at that position. When the boll opens, the fibers dry out and collapse, shrinking and forming twists or convolutions. The spacing of the convolutions is not constant, and their direction may alternate from clockwise to counterclockwise. The size of the fiber is dependent on genetic variety, growth conditions, and degree of maturity, but the average length of spinnable cottons is from 0.5 to 1.5 in. The diameter of the fibers varies from 12 to 22 μm. As a textile fiber, cotton is a "dead" biological substance; thus fixation and other preservation procedures are not necessary for microscopy unless the undried, "living" fiber is to be studied. The general morphological structure of the cotton fiber has been determined by LM, and detailed surface features have been studied by replication procedures in the TEM. However, SEM provides a more natural view of fiber gross struc-

Figure 17   Surface of a blended cotton–wool fabric.

tures. The extreme of the length–diameter dimensions of the cotton fiber may
be seen by viewing the fiber in coiled form (Fig. 1). The coiled fiber was prepared by rolling a single fiber around a needle, setting the point of the needle
down on the tacky sample stub, and pushing the fiber down the needle to the
stub. The shrunken, twisted morphology of the fiber is evident at this magnification. Higher magnifications of portions of the fiber show the wrinkled surface
of the coated primary wall (Fig. 2) and a reversal, made evident by a change in
the direction of the wrinkles (Fig. 3). In many areas of the fiber, ridges or compression marks are evident on the fiber surface. If the primary wall is broken, the
main body of the fiber is exposed, showing the spiral of the fibrils around the
fiber axis (Fig. 4).

B.   Nature of Changes in Fibers

Changes in characteristic structural features are often evident in the SEM. When
chemical finishing causes visual changes in fibers or fabrics, it is possible to use
SEM to determine whether a specific finish has been given, as well as to see the
extent that the finish has changed the nature of the sample.

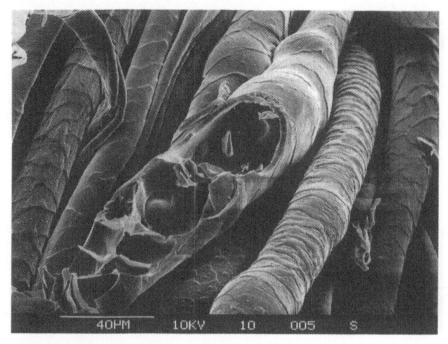

Figure 18   Swelled and ruptured wool fiber caused by heat.

1.   Mercerization

Mercerization, a process of swelling the fiber in a caustic solution (usually NaOH), causes the wrinkled surface to become more smooth (Fig. 5) and the fiber to become more cylindrical. This treatment changes the crystalline structure of the cellulose, and improves textile properties of the fiber. Mercerization under laboratory conditions brings about more thorough conversion of the fiber cellulose, and thus of the fiber appearance, than do commercial processes. In many commercial mercerizations not all of the fibers in a yarn are swelled to the rounded structure seen in Fig. 5. Processes are usually developed that produce the desired increase in absorption, dyeability, hand or sheen. These do not always require full mercerization.

2.   Fiber Damage

The study of damaged cotton fibers in the SEM not only helps in understanding the nature and causes of the damage, but also indicates the nature of the inner structure of the fiber. Fibers damaged by washing and drying processes were shown by Goynes et al. [38]. When cotton fibers are abraded wet in repetitive rubbing ac-

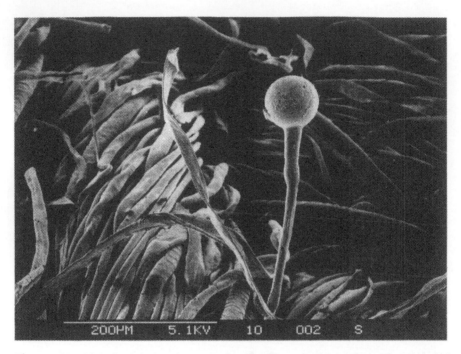

Figure 19 Melted globular tip on polyester fiber on the surface of a blended cotton–wool–polyester fabric that had been heated.

tions, such as can happen in a washing machine, the fibers separate into small, stringy bundles of fibrils (Fig. 6). When dry abrasion damage occurs, such as that from machine drying, instead of fibrillating, the fibers become brittle and usually crack in the direction of the spiral of the secondary layers around the fiber axis (Fig. 7). Cracks may also occur at flat fiber edges. Fiber breaks caused by dry abrasion leave blunt ends with little fibrillation (Fig. 8.). These types of damage illustrate the effect of moisture on the fibrillate structure of the fiber. Wet fibers swell, separating the fibrils and allowing them to be separated by frictional forces. As fibers are dried, the fibrils are drawn closer together, and hydrogen-bonding forces prevent fibrillation. Overdrying causes the fibers to become very brittle, and in some instances they have the appearance of being snapped by abrasion (Fig. 9). Fiber damage from other causes such as fungus and heat can also be identified using the SEM. Figure 10 illustrates damage to the cotton fiber due to fungal attack. The damage has been made more evident by swelling the damaged fiber in NaOH.

C.   Fabric Structure

The SEM is a particularly valuable tool for studying textile fabrics because it is difficult to completely examine the surface of a fabric in the light microscope due

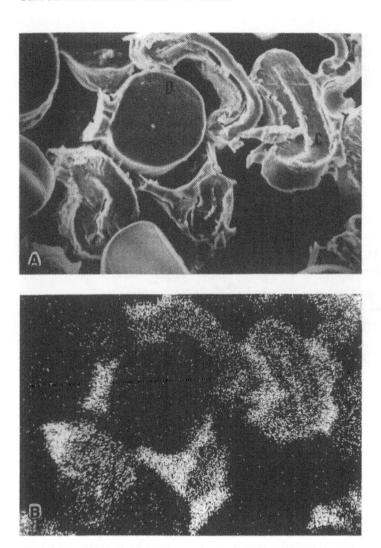

Figure 20 Cross sections from a flame-retardant–finished cotton–polyester fabric (top) showing finishing chemical between fiber sections, and a corresponding energy-dispersive x-ray map (bottom) showing location of phosphorus in the deposited chemical and the cotton fibers but not in the polyester fibers.

to its low depth of focus. The depth of focus in the SEM is great enough to study such parameters as fabric construction, damage, or surface deposition of chemical finishing agents. A comparison of LM and SEM fabric surfaces is illustrated in Figs. 11 (LM) and 12 (SEM), which show the increased information available from the SEM. Although light microscopy has provided important information in

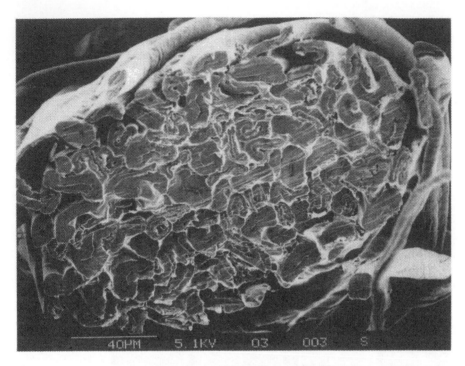

Figure 21 Yarn cross section from a cotton fabric treated for thermal adaptability, showing treatment polymer filling spaces between fibers.

the understanding of textiles, the SEM is complementary technique that greatly increases information and provides images that are more easily interpreted.

1. Chemical Finishing

Chemical finishes intended to provide new or improved properties often change the appearance of the fabric surface. The ability to locate these finishing agents is very important in understanding mechanisms and evaluating treatment methods. Resin-type polymer finishes are often used to coat fabrics to provide weather- and wear-resistant surfaces. In such fabrics, good coating to fabric adhesion is essential. With the SEM it is possible to study (1) amount of coating present, (2) uniformity of treatment, and (3) depth of penetration of the material into the fabric.

a. Flame Resistance. Finishes that are intended to make cotton fabrics flame resistant can be achieved either through reaction (chemical reaction with the cellulose of the fiber) or deposition (polymer deposited on the fiber surface or within the fabric structure). Of these, reactive processes are more desirable since they are more durable, especially in fabrics that are to be laundered. Changes in fiber mor-

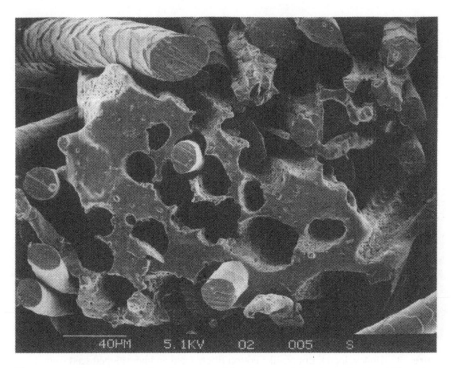

Figure 22   Yarn cross section from a wool fabric treated for thermal adaptability, showing poor adhesion of the fibers to the treatment polymer.

phology due to chemicals deposited on fiber surfaces can be seen by use of secondary electron microscopy. However, chemicals that penetrate the fiber surface are difficult to see by use of the secondary image even when internal surfaces of the fiber are studied. These internally deposited or reacted chemicals are more easily studied by use of one of the elemental analysis procedures (EDS, for example).

Treating cotton textiles with phosphorus-containing flame-retardant (FR) finishes reduces fabric flammability. When an untreated cotton fabric is burned under atmospheric conditions, it produces a wispy, fragile char. Figure 13 compares an unburned and burned cotton twill and illustrates the shrinkage that occurs on burning. Because of the melting and flow character of polyester fibers, fabrics blended of cotton and polyester form a more solid char, as shown in the fabric cross section in Fig. 14. Proper finishing with a flame-retardant agent may not produce noticeable differences in the appearance of the fabric surface but does change the nature of the char greatly. Figure 15 is a cross section of an FR-treated yarn char, showing the enlarged lumen area of the fibers that occurs when the wall of the fiber compresses. In Fig. 16, detailed structure of the internal charred, finished cotton

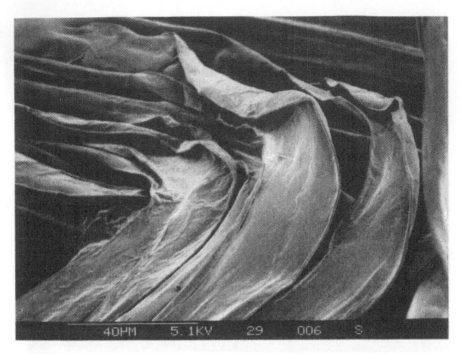

Figure 23 Undeveloped, ribbon-like cotton fibers from an undeveloped seed (mote).

fiber can be seen. The resultant char is stronger than that of an untreated fabric. The presence of fibers other than cotton in blended fabrics changes the burning rates and the resultant char length of these fabrics, as well as the structure of the resultant chars. A blended cotton/wool fabric surface is shown in Fig. 17. When this fabric was burned, the approaching heat edge first visually affected the wool fibers. Swelling and rupture of wool fibers is shown in Fig. 18. When polyester fibers are also included in the blend, those fiber ends that are raised above the fabric surface show the first reaction to heat by melting and forming globules. These melted fiber ends are illustrated in Fig. 19. The effect of heat on cotton polyester and wool fibers has been shown by Goynes et al. [39, 40].

Chemicals within the fiber cannot be detected by surface SEM. Energy-dispersive x-ray microanalysis provides a means of determining locations of specific elements within the fiber, and can show selective interactions with different fibers in blended fabrics. For this procedure, yarns from treated fabrics were embedded in a polymerizable liquid, and the hardened block was sectioned with a microtome [32]. Sections were collected on a carbon disk and the embedment polymer was removed using a solvent. The disk was then attached to a sample stub and prepared for examination by coating with carbon. A secondary electron image

Figure 24 "White speck" defect on the surface of a dyed cotton fabric.

was generated and recorded (Fig. 20a), and then an elemental spectra was generated by EDS analysis. Peaks were identified, the element of interest (phosphorus) was selected, and an elemental map corresponding to the secondary image was recorded (Fig. 20b). Comparison of the two images shows that the phosphorus flame-retardant chemical is deposited heavily on fiber surfaces and in spaces between the fibers, and that it is also within cotton fibers but not within polyester sections. Such analyses provide valuable information in formulating more efficient finishes [41].

b. Thermal Adaptability. Another type of textile finish in which the SEM has provided useful information is the production of thermally adaptable fabrics by use of polyethylene glycol solutions [42]. These fabrics were designed to increase wearer comfort by liberating heat, thus providing a warming effect, when they are cooled, and by absorbing heat, producing a cooling effect, when they are heated. Figure 21, a yarn cross section from a treated cotton fabric, shows that the treating polymer had filled the spaces between the fibers. This fiber/polymer binding produces a yarn that is essentially a low-temperature composite structure, a polymer matrix of embedded fibers. This type of bonding was found to be necessary to a good finish. All fibers did not bind well in the formulation used, as

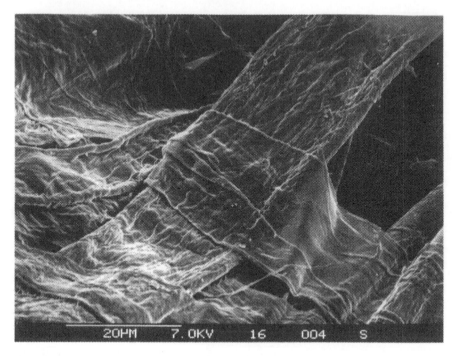

Figure 25   Enlarged area of "white speck" defect, showing that it is made-up of flat, un-developed fibers.

is shown by the treated wool yarn section shown in Fig. 22. To achieve a well-bonded fiber matrix, it was necessary to use a finishing procedure specifically designed for wool fibers.

2. Fabric Defects

The SEM is also valuable in textile problem solving in areas other than finishing. Imperfections on fabric surfaces can produce fabrics that are not acceptable for marketing. In dark-dyed fabrics imperfections can be seen as small specks of undyed materials that appear white on the dark surface. An SEM study of such "white specks" showed them to be clumps of undeveloped, flat, ribbon-like fibers. These undeveloped fibers are produced by undeveloped seed (motes). Their growth was interrupted before the internal layers of cellulose fibrils were deposited. Thus they became tubes that flattened into ribbons on drying. These fibers, which show up as white specks, do not contain enough cellulose to dye in the process used for cotton fabrics. Figure 23 shows a bundle of such flat fibers from a mote. Knots of these compressed fibers can be carried through textile processing and entangled into fibers on the fabric surface. An example of such a defect on a fabric surface is shown in Fig. 24. Higher magnification of the defect more clearly shows the extremely thin fiber walls that form them (Fig. 25). Judicious use of SEM operation parameters was required to obtain this image of a very thin fiber overlying another fiber. A keV level high enough to resolve the details of the structure was necessary, but a keV level whose energy penetrated the thin fiber without generating secondary electrons would not have defined the fiber.

V. NEW INSTRUMENTAL TECHNIQUES

The SEM is a relatively new tool for studying textiles. Many of the techniques developed for application of this microscience were intended to be used with nonbiological materials that have higher atomic numbers than do most textile materials. Because of the great variance in properties between biological and nonbiological materials, these techniques have had to be adapted for use with textiles. Even with these limitations, a casual literature survey indicates that SEM has provided a valuable means of developing textile information that was otherwise unavailable. Instrumental developments continue. One of the greatest obstacles to use of the SEM with biological materials is the necessity for the sample to be in the high-vacuum, low-moisture microscope chamber for examination. These conditions are not ideal for textile study. To overcome these disadvantages, a microscope that operates at a lower chamber pressure, and in which moisture can be increased, is currently on the market. It is called an environmental scanning electron microscope (ESEM) because conditions in the microscope chamber are more closely related to normal environmental conditions than are those commonly used in the SEM [43]. While the theory of operation of this microscope is promising, in actual practice it does not provide optimal operating conditions for textiles. Because of the moisture in the chamber, samples need not be coated before examination. However, coated fibers provide a better image than do uncoated due to reduction of beam penetration; ESEM fiber images often have bright edges with poor contrast, and do not relate well to images obtained from the SEM. Care also has to be taken that surfaces do not have water condensed on them, or details of the fiber surface can be obscured. The microscope does, however, offer the possibility of performing dynamic experiments within the chamber, such as stretching and breaking fibers, and observing effects of wetting and drying on the sample. Tao and Collier have shown results of some textile applications of the ESEM [44].

VI. FUTURE OUTLOOK

The many interactions that occur between the primary electron beam and the atoms in the sample provide a great avenue for further development of SEM applications to textiles. Cathodoluminisence, production of visible light by interaction with the electron beam, has already found some limited application. Techniques such as Auger analysis and electron energy loss analysis (EELS) will eventually be available for textile applications as progress is made in methodology. Energy-dispersive analyses using conventional detector windows are limited to elements of atomic number sodium or above. Windowless detectors are now available for detection of light-element electrons. Even these, however, are not easily applied to textile materials, since many textiles are composed of carbon, hydrogen, and oxygen. Detection of light elements in such a matrix has not yet been highly successful. Progress in detector improvement and in the limits of both qualitative and quantitative elemental detectability is being made by instrument manufacturers. This offers the possibility of better EDS analyses in the future. New methods for electron beam generation, such as field emission, are now available. The field emission gun greatly increases resolution in the microscope and provides more detailed views of surfaces, especially at low keV.

VII. SUMMARY

SEM and related techniques have provided a means of better visual understanding of textile fibers and fabrics and has improved concepts of the role of unit interactions in fibrous matrices. The methods are valuable in both basic research and in process evaluation. Limitations are present in microscope capabilities, such as resolution, and in sample preparation. As improvements are made in these areas more information on textile structures will be available. Opportunities for gaining even greater understanding of fiber and fabric structures will continue to develop as innovations in associated microscopical technology become commercially available and are adapted to use with special materials such as textiles.

ACKNOWLEDGMENT

The author acknowledges the work of the Fiber Microscopy Group developed at the Southern Regional Research Center by the late Mary L. Rollins. Without the pioneering work of this group the material in this chapter from SRRC would have had no basis. Special thanks are expressed to Jarrell H. Carra and Bruce Ingber, who continue to work with the author in this unit.

REFERENCES

1. A. Ure, *The Cotton Manufacturers of Great Britain Systematically Investigated*, Vol. 1, Charles Knight, London, 1836, p. 56.
2. E. Valko and G. Tesoro, *Encyclopedia of Polymer Science and Technology* (H. Mark, N. Gaylord, and N. Bikales, eds.), Vol. 2, Interscience, New York, 1965, pp. 204–229.
3. O. Eisenhut and E. Kuhn, *Angew. Chem. 55*:198 (1942).
4. R. Barnes and C. Burton, *Ind. Eng. Chem. 35*:120 (1943).

5. M. von Ardenne and D. Beischer, *Z. Phys. Chem. B45*:465 (1940).
6. C. Challice and J. Sikorski, eds., Summarized proceedings of a conference on electron microscopy of fibers, *Br. J. Appl. Phys. 8*:1–26 (1956).
7. D. Atack and K. Smith, A new tool in fiber technology, *Pulp Pap. Mag. Can. 57*:245–251 (1956).
8. P. Echlin and P. Hyde, *Scanning Electron Microsc. 5*:137–146 (1972).
9. G. Pfefferkorn, H. Gruter, and M. Pfautsch, *Scanning Electron Microsc. 5*:147–152 (1972).
10. J. Sikorski, J. Moss, P. Neuman, and T. Buckley, A new preparation technique for examination of polymers in the SEM, *J. Phys. E: Sci. Instrum. 1*:29–31 (1968).
11. S. Chatterji, J. Moore, and J. Jeffery, *J. Phys. E: Sci. Instrum. 5*:118 (1972).
12. A. K. Mukhopadhyay, S. K. Bandyopadhyay, and U. Mukhopadayay, Jute fibers under scanning electron microscopy, *Text. Res. J. 55*:733 (1985).
13. T. K. Guha Roy, A. K. Mukhopadhyay, and A. K. Mukherjee, Surface features of jute fibers using scanning electron microscopy, *Text. Res. J. 54*:874 (1984).
14. P. K. Ray, P. Sengupta, and B. K. Das, Scanning electron microscopic Studies on jute fibre: Effect of processing on surface structure, *Ind. J. Text. Res. 11*:31 (1986).
15. P. S. Mukherjee and K. G. Satyanarayana, Structure and properties of some vegetable fibers, Part 1, Sisal fibre, *J. Mater. Sci. 19*:3925 (1985).
16. L. Cheek and L. Roussel, Mercerization of ramie: Comparisons with flax and cotton, *Text. Res. J. 59*:478 (1989).
17. L. Cheek and H. Struszczyk, the effect of anhydrous liquid ammonia and sodium hydroxide on viscose fabric, *Cellulose Chem Technol. 14*:893 (1980).
18. K. G. Satyanarayana, A. G. Kulkarni, and P. K. Rohatgi, Structure and properties of coir fibres, *Proc. Indian Acad. Sci. 4*:419 (1981).
19. D. Brady, J. Duncan, R. Cross, and A. Russell, Scanning electron microscopy of wool fiber degradation by *Streptomyces* bacteria, *S. Afr. J. Anim. Sci. 20*:136 (1990).
20. M. J. Hafey and I. C. Watt, The interaction of polymer latex particles with wool, *J. Colloid Interface Sci. 109*:181 (1986).
21. H. Ito, H. Sakabe, T. Miyamoto, and H. Inagaki, Fibrillation of the cortex of merino wool fibers by freezing-thawing treatments, *Text. Res. J. 54*:397 (1984).
22. H. R. Billica, and R. D. Van Veld, Scanning Electron Microscopy of Synthetic Fibers, *Surface Characteristics of Fibers and Textiles Part I*, (M. J. Schick, ed.). Marcel Dekker, Inc., New York, 1975, pp. 295–327.
23. L. I. Stowell and K. A. Card, Use of SEM to identify cuts and tears in nylon fabric, *J. Forens. Sci. 35*:947 (1990).
24. D. R. Subramanian, A. Venkataraman, and N. V. Bhat, SEM studies on nylon 6 filaments subjected to drawing and swelling treatment, *Text. Res. J. 54*:331 (1984).
25. M. P. Ansell, The degradation of polyester fibres in a PVC-coated fabric exposed to boiling water, *J. Text. Inst. 74*:263 (1983).
26. A. K. Kulshreshtha, V. N. Garg, and Y. N. Sharma, Plastic deformation, crazing, and fracture morphology of acrylic fibers, *Text, Res. J. 56*:484 (1986).
27. P. Schwartz, A. Netravali, and S. Sembach, Effects of strain rate and gauge length on the failure of ultra-high strength polyethylene fibers, *Text. Res. J. 56*:502 (1986).
28. M. Sawbridge and J. E. Ford, Textile fibers under the microscope, Shirley Institute Publication S50, Didsbury, Manchester, U.K., 1987.

29. J. Sich, Which is best for identifying animal fibers: SEM or light microscopy? *Text. Chem. And Col.* 22:23 (1990).
30. S. L. Paek, Pilling, abrasion, and tensile properties of fabrics from open-end and ring spun yarns, *Text. Res. J.* 59:577 (1989).
31. S. H. Zeronian, K. W. Alger, M. S. Ellison, and S. M. Al-Khayatt, Studying the cause and type of fiber damage in textile materials by SEM, *Historic Textile and Paper Materials, Conservation and Characterization* (H. Needles and S. Zeronian, eds.), American Chemical Society, Washington, D.C., 1986, p. 77.
32. M. L. Rollins, A. M. Cannizzaro, and W. R. Goynes, Electron microscopy of cellulose and cellulose derivatives, *Instrumental Analysis of Cotton Cellulose and Modified Cotton Cellulose* (R. J. O'Connor, ed.), Marcel Dekker, New York, 1972, pp. 261–271.
33. J. W. S. Hearle and B. Lomas, The study of fiber fracture, *Appl. Polymer Symp.* 23:147 (1974).
34. N. E. Dweltz, J. W. S. Hearle, G. E. Cusick, and B. Lomas, The surface abrasion of cotton fabrics as seen in the SEM, *J. Text. Inst.* 69:250 (1978).
35. M. Raheel and M. Lien, Modifying wear life of all-cotton fabrics, Part I: Liquid ammonia treatment and durable press finish, *Text. Res. J.* 52:493 (1982).
36. S. K. Obendorf and J. J. Webb, Detergency study: Distribution of natural soils on shirt collars, *Text. Res. J.* 57:557 (1987).
37. P. Ghosh and J. C. Dalal, Effect of chemical modification of cotton cellulose on its surface as studied by SEM, *Ind. J. Tech.* 27:189 (1989).
38. W. R. Goynes and M. L. Rollins, A SEM study of washer-dryer abrasion in cotton fibers, *Text. Res. J.* 41:226 (1971).
39. W. R. Goynes, E. K. Boylston, L. L. Muller, and B. J. Trask, A microscopical survey of flame-resistant cotton fabrics, Part I: Treatments based on THPOH-NH$_3$, *Text. Res. J.* 44:197 (1974).
40. W. R. Goynes and B. J. Trask, Effects of heat on cotton, polyester, and wool fibers in blended fabrics—A SEM study, *Text. Res. J.* 55:402 (1985).
41. W. R. Goynes and J. H. Carra, Microscopical analysis of chemically modified textile fibers, *Characterization of Metal and Polymer Surfaces*, Vol. 2 (L. H. Lee, ed.), Academic Press, New York, 1977, p. 251.
42. W. R. Goynes, T. L. Vigo, and J. S. Bruno, Microstructure of fabrics chemically finished for thermal adaptability, *Text. Res. J.* 60:277 (1990).
43. R. E. Cameron, Environmental SEM: Principles and applications, *Microsc. Anal.* 6:17 (1994).
44. W. Tao and B. Collier, The environmental scanning electron microscope: A new tool for textile studies, *Text. Chem. Color.* 26:29–31 (1994).

4
Investigation of Textiles by Analytical Pyrolysis

IAN R. HARDIN   University of Georgia, Athens, Georgia

I. INTRODUCTION

There are numerous methods that are used to investigate and identify polymers, fibers, and the textiles they make up. These include reaction to burning, solubility, optical and electron microscopy, infrared spectrophotometry, and thermal analysis. Each of these methods gives information that is valuable in identification of the fiber and any modifications to the fiber that may have been done. The application of analytical pyrolysis is another method that can provide a powerful approach to polymer, fiber and textile investigations.

Two generally agreed on definitions in the field of pyrolysis techniques are [1]:

Pyrolysis: a chemical degradation reaction that is induced by thermal energy alone.

Analytical pyrolysis: the characterization of a material or a chemical process by the instrumental analysis of its pyrolysate.

Pyrolysis is the chemical process where heat energy breaks down large molecules into smaller fragment molecules. It is done by the direct application of high temperature energy to the substrates. The molecular bonds in the large molecules are broken by thermolysis, and characteristic smaller molecules are created. The terms thermolysis and thermal reaction are often used in place of pyrolysis [2]. Examples of pyrolysis to produce information about the nature of organic materials include the production of isoprene from natural rubber, the destructive distillation of wood to yield methanol, and the range of chemicals derived from coal tar, produced by the pyrolysis of coal. It is important to understand that reactions caused by heating with the involvement of catalysts, or the simultaneous or subsequent reaction with ambient oxygen, do not constitute pyrolysis.

The types of reactions that are a part of pyrolysis are dissociation into free radicals, elimination of small molecules, and carbon–carbon fission. Because of the size of the polymer molecules, you may have random scission at many points along the polymer backbone, or scission of the side groups in either a random or systematic fashion. There may also be reversion of the polymer back to monomer, dimer, or other oligomers. Given the large sizes of many of the molecules that are part of the substances examined, it is not surprising that pyrolysis gives rise to complex mixtures of products resulting from the many possible reactions. When pyrolysis is coupled with separation of the volatile fragments by gas chromatography (PY-GC), one has a simple analytical method that can be applied to all manner of substances and materials to gain information regarding their nature and identity. The result of such an analysis is referred to as a pyrogram. An example for cotton is given in Fig. 1. The addition of mass spectrometry to identify the individual fragments (Py-GC-MS) gives one a very powerful tool that can be applied to a wide range of chemical problems. One very significant advantage to this method is that only a few milligrams of the specimen are necessary to give clear identification.

Analytical pyrolysis is particularly suited to analysis of polymers, whether they are natural proteins or polysaccharides, or manufactured plastics, adhesives, paints, or fibers. The literature on analytical pyrolysis is voluminous and covers a

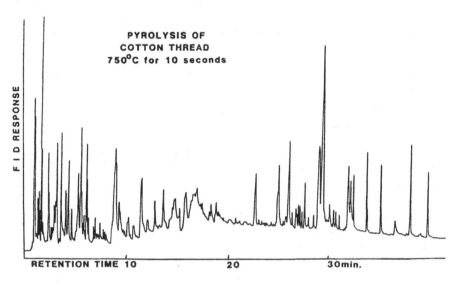

Figure 1   Pyrogram of cotton fiber, 750°C (From Ref. 41.)

broad range of areas, including biomass conversion [3], plant cell walls [4], microorganisms [5,6], wood lignin [7], characterization of suspended matter in sea water [8], smoke aerosols [9,10], food sciences [11], tobacco smoke [12], coffee flavors [13], and forensics. In the following sections, the method of analytical pyrolysis and its use are described, and applications to polymers, fibers, finishes, and dyes are surveyed.

II. INSTRUMENTATION AND ANALYSIS

Analytical pyrolysis is normally done by using a pulse of heat to quickly heat the sample and break it down into the individual components. The sample size is usually 10 to 100 μ. Pyrolyses can be run at heating rates ranging from 0.01°C per second to 20,000°C per second. Very slow rates result in pyrograms that resemble those produced by pulse pyrolyses at lower end-point temperatures. Lower rates may permit one to see more of the primary pyrolysis products, which may be formed at lower temperatures and then removed before secondary reactions occur [14]. The process of pyrolysis can be linked to gas chromatography (GC), mass spectrometry (MS) or Fourier-transform infrared spectroscopy (FTIR), either individually (PY-GC, PY-MS, PY-FTIR) or in combination (PY-GC-MS). Each of these has particular advantages. Certainly, PY-GC is the most widely used and the most widely reported in the literature. Pyrolysis is dynamic, with a flow of carrier gas sweeping volatile products away from the pyrolysis zone to the gas chromatograph, mass spectrometer, or the FTIR.

A. Pyrolyzers

The instruments employed for the pyrolysis step are usually either resistance-heated filaments, inductively heated filaments or wires, microfurnaces, or radiative laser heating. Most of these pyrolyzers are small units that can be inserted into a gas chromatograph or mass spectrometry port, or linked with an interface. While each type of pyrolyzer has certain advantages and disadvantages, the most widely used type is the resistance-heated filament or coil.

1. Resistance Heating

In the resistance-heated pyrolyzer the sample is placed on platinum filament or inside a platinum coil that is part of the pyrolysis probe. The temperature rise time (TRT) is dependent on the probe type. For the instrument cited most often in the literature [15] a temperature of 600°C can be reached in 8 msec with the ribbon filament; or 600 msec for the coil with a quartz tube included within the coil. The temperatures reached can be varied from ambient to 1400°C. With models now available, the temperatures are exact, and are reproducible. The quartz tube is used to help hold certain kinds of samples in place and to minimize possible catalytic effects from the platinum.

The resistance-heated pyrolyzer has the general advantages of a fast temperature rise time, a good temperature range, and the ability to program the heating in a flexible way. It has the additional advantage of being widely used with numerous publications citing applications. Disadvantages include the possibility of catalytic effects (although few problems with this are cited in the literature), and the fact that the temperature actually experienced by the sample may be difficult to determine exactly.

2. Inductive Heating

The inductively heated filament pyrolyzers are usually referred to as the Curie-point method. This refers to the property of ferromagnetic metals or alloys (FE, NI, Co) whereby when subjected to radiofrequency radiation they heat up rapidly by the absorption of the energy. At their "Curie point," which is specific for each material, the energy absorption ceases and the material stabilizes at the characteristic temperature. The temperature–time characteristics of the filament are determined by the ferromagnetic properties of the sample holder, its dimensions, and the power of the radiofrequency field applied to it [16]. The end-point temperature occurs at the temperature where the wire loses its ferromagnetic properties. This temperature is very reproducible and is determined by the composition of the ferromagnetic material chosen. The temperature can be varied between 160 and 1040°C [17]. The heating rates are normally in the range of 1000–10,000°C per second.

The primary advantage of this type of pyrolysis is the ability to have exact control of the temperature because of the Curie-point principle [18]. A number of wires are available with Curie points between 300 and 981°C. The temperature rise time (TRT) is dependent on the composition and geometry of the wire and power of the oscillator.

3. Microfurnaces

This type of instrument can have several designs. One incorporates a very small quartz tube maintained in a furnace assembly at the pyrolysis temperature by a thermocouple feedback controller [19]. The pyrolysis is isothermal, and liquid or solid samples can be injected directly into the furnace. Short lengths of sample are injected directly into the quartz furnace with a "solids injecting syringe." Another design, available in Japan, utilizes a small sample holder made of platinum, which is then dropped into the reactor zone [20]. The key to getting a fast TRT is to eliminate dead space in the furnace. Oguri [17] modified the maintenance oven so that the heated area of the transfer needle was increased to improve the analysis of the high-molecular-weight pyrolysates. Improved programs resulted, particularly for high-molecular-weight products. A heated aluminum block prevented the condensation of the pyrolysates in the sample and transfer tubes.

In both designs the advantages seem to be the control of the fragmentation temperature and sure reproducibility of that temperature, thus improving the replication of the programs. A disadvantage would be less flexibility in programming the thermal conditions.

4. Radiative Heating

This is usually done with a carbon dioxide laser source. The advantage for this method is that the sample can be heated in remote locations or on any material. The disadvantages are the lack of reasonably priced instruments and the difficulty of controlling accurately both the TRT and the final temperature. A new instrument has recently been described that utilizes an Nd–Cr–GGG laser [21].

B. Techniques for Characterization

1. Pyrolysis–Gas Chromatography

Pyrolysis–gas chromatography is a relatively simple and inexpensive experimental technique. A pyrolysis unit is linked directly or by an interface to the gas chromatograph. The pyrolysis products are formed and then moved away by carrier gas

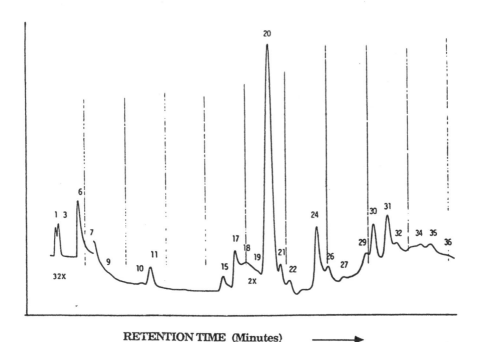

RETENTION TIME (Minutes) ⟶

Figure 2   Pyrogram of cotton fiber, 805°C. (From Ref. 32.)

to the gas chromatograph. The result is a pyrogram, as shown in Fig. 2. This method uses very small samples, is very sensitive, and the chromatograms obtained are much less complicated than the mass spectra in PY-MS. In addition, the power of the method can be increased by using mass spectrometry. The mass spectrometry can be combined with PY-GC by a suitable interface so that each of the GC peaks can be mass analyzed. Table 1 presents the results of the analysis of the Fig. 2 pyrogram. This method can be most useful for initial investigations of a product so that PY-GC can be used on the material with much smaller sample sizes [22]. The drawbacks include the need to take particular care in reproducing all conditions, the very large amount of data that a program gives, and the length of time involved (30 min to 1 hr) in running one sample.

Table 1   Identification of Volatile Pyrolysis Products for Cotton Fabrics

GC peak	Compounds identified	Molecular weight	Identification by GC/MS[a]
1	Carbon monoxide	28	P
3	Carbon dioxide	44	P
6	Water	18	P
7	Propene	46	HP
11	Oxirane	44	HP
12	Butene	56	HP
15	Furan	68	HP
16	Dimethyl ether	46	HP
17	Acetone	58	HP
20	Glycolaldehyde	60	HP
21	Acetic acid	60	HP
23	2,3-Butanedione	86	HP
24	1-Hydroxy-2-propanone	74	HP
26	3-Methyl-3-butene-2-one	84	LP
27	2,5-Dimethylfuran	96	HP
28	3-Hydroxy-2-butanone	88	HP
30	Acetic anhydride	102	HP
31	Fufural	96	HP
32	2-Furan methanol	98	LP
35	1-(2-Furanyl)ethanone	110	LP
36	5-Methyl-2-Furaldehyde	110	HP
38	2-Hydroxy-2-cyclopentene-1-one	112	HP

[a]P = positively identified by the mass spectrum and retention time on GC, HP = high-purity fit with the standard spectrum in the MS library (800–850), LP = relatively low-purity fit with the standard spectrum of the MS library (800–850).
Source: Ref. 32.

The conditions used for gas chromatography can have an important effect on the pyrogram obtained. Certain GC instruments may require a different interface between the pyrolyzer and the GC. One can use either packed or capillary columns. Capillary columns of chemically inert fused silica have high numbers of theoretical plates and can give better resolution, and therefore more details in the pyrogram, than packed columns [23]. See, for comparison, Fig. 1–capillary column versus Fig. 2–packed column for the same material. The smaller peak width means a lower detection limit of an individual component is possible [24]. On the other hand, capillary columns generally take a longer time to produce a pyrogram and usually require smaller samples and splitting of the carrier gas stream. The use of a cryogenic focusing step can allow very small samples, splitless injection, and prevention of the loss of low-molecular-weight compound resolution [25]. The smaller sample size can present some problems in handling and can raise questions regarding the representative nature of the microgram sample used. A packed column will allow larger samples to be used, but resolution is lost and thermal gradients within the sample are likely.

If properly controlled pyrolysis is coupled with gas chromatographic separation, one can obtain pyrograms that are repeatable. The pyrograms of cotton in Figs. 1 and 2 are each replicable, even though they are significantly different in detail. These pyrograms both were of cotton, but done with different gas chromatographic columns and conditions. When the same instrument and conditions are used, and care is taken with the condition of the pyrolyzer and sample preparation, then the pyrograms obtained on successive specimens of the same sample will be virtually identical. Both of the pyrograms in Figs. 1 and 2 will be useful for many purposes. The extra detail in Fig. 1 will aid in mechanistic studies and will be very useful in detecting small differences in the materials. Figure 2 is probably just as useful as Fig. 1 for fingerprint identification purposes.

2. Pyrolysis–Mass Spectrometry

When mass spectrometry is used directly in conjunction with pyrolysis the sample is introduced directly into the ion source and the mass spectrometer itself provides the pyrolytic system. The mass spectral profiles produced by pyrolysis–mass spectrometry (PY-MS) are highly characteristic and reproducible and can be used as fingerprints [26]. The data produced are automatically available to the computer for storage and manipulation. The runs can be conducted in a few minutes, and analysis time is short. The drawbacks are that PY-MS is less sensitive than PY-GC and the results can be complicated with the MS cracking patterns superimposed. The equipment is more expensive than that for PY-GC, although the relatively low-cost GC-MS units now on the market may well change this in the future.

Two important considerations for both of the methods presented are that high temperature rise time (TRT) and small sample size are critical to obtain reproducible results. If these are achieved, reproducibility will be good. If TRT is slower and sample size larger, then much more attention to the actual heating program will be required to obtain useable results. In large samples there may be diffusion control of volatile product evolution, which will restrict the rate of escape of the pyrolysis products. Increased residence time may lead to increased probability of forming secondary products, thus obscuring the true pyrolysis profile and leading to false conclusions [22].

3. Pyrolysis–Infrared Spectroscopy

Pyrolysis can be combined with infrared spectroscopy to aid in the identification of polymers and fibers. The first applications of this method used a pyrolysis chamber where the resulting vapors were trapped and the infrared spectrum of the resulting mixture was taken. A typical study was that done on a variety of flame-retardant–treated fabrics where the fabrics were pyrolyzed at intervals of 200°C from 200 to 1200°C [27].

Pyrolysis–Fourier transform infrared spectroscopy (PY-FTIR) of polymers provides a reproducible pyrogram that is used as a fingerprint for polymer identification [28] or for polymer quality control [29]. This technique can be useful in quick determinations of polymer identify. Pyrolysis of the sample directly into the infrared beam using a unique interface enables spectra to be obtained before side reactions and condensation occur. For example, the direct PY-FTIR of polystyrene gave a spectrum almost identical to the styrene monomer. The nylons were easily distinguished through the relative strength of the C-H stretching bands near 3000 cm^{-1} compared to the amide stretching band at 1650 cm^{-1}. The analysis can be run in a fraction of the 20–60 min that a typical pyrolysis–gas chromatography run requires. In addition, PY-FTIR lends itself directly to studies of evolved gas during thermal treatment and processes.

C. Analysis of Results

1. Fingerprints

The product of pyrolysis–gas chromatography is a pyrogram. These plots of compounds in the pyrolyzate by retention time are truly "fingerprints." To be useful, the experimenter must pay attention to details such as sample size, preparation, and the long-term reproducibility of the GC column performance. Fingerprints can be used to identify a polymer or a fiber and to detect small changes made in a given material. The use of fingerprints does not require knowledge of the composition of the pyrolysis products, nor how they relate to the original molecular structure. It should be understood that when using fingerprints that interlaboratory reproducibility is difficult, since instruments and techniques will inevitably vary. Certainly, a key to the success of fingerprints is their use in conjunction with a site-established library of pyrograms. On the other hand, fingerprints cannot be used to identify polymers in commercial mixtures, or with complicated compositions in the presence of other polymers where the individual constituent programs are superimposed [30]. The absence of compositional information prevents identifying polymers that are similar in nature, with slight differences in macromolecular structure, but having different properties.

2. Classification System

Another example of an identification approach is that used by Milina et al. [31] to quickly identify low-molecular-weight substances from their pyrograms. There are some similarities to the empirical approach used by Hardin and Wang [32] and it may be worthwhile in application to fibers. The system does not incorporate data from the entire pyrogram but only uses three characteristics to ensure a unique recognition of a pyrogram. Pyrograms are presented to the computer by giving relative retention times and areas of individual peaks. Calculations were done by using a maximum of 10 peaks, with the requirement that any peak used was larger than one percent of the total area subtended by the pyrogram. The characteristic values used were:

1. $PI = \Sigma A_i T_i$, pyrolysis index, where A is the area and T is the relative retention time of the individual peaks
2. $C = A_i T_i$, the ratio of two adjacent peaks
3. T_{max}, the relative retention time of the most intense peak

No one of the three characteristics permitted the complete recognition of the pyrograms, but the simultaneous use of all three resulted in 100% recognition of all compounds.

3. Compositional Analysis

Comparisons between pyrograms that are similar and complex can pose seemingly intractable analysis problems. Fifty or more individual components in a pyrogram are not unusual, and small changes can be difficult to differentiate. Compositional analysis relies on the identification of individual pyrolysis products. This method uses preliminary chromatographic investigation of the polymers to establish a relationship between pyrolysis products and the polymer composition [30]. Once the presence of repeating units, cyclic oligomers, mass markers, fragments, and series is correlated to known polymer structures, then unknown samples can be run and their patterns can be used to establish molecular structure. For those samples clearly different, there are no problems. For those close in structure, one must pay close attention to pyrogram quality. There may be differences in quantitative distribution rather than each possessing unique fragments. Composition analysis of copolymers or polymer blends (or copolymer fibers or fiber blends) can be done by pyrolyzing pure components to establish the chromatographic peaks characteristic of each component. The composition of

true copolymers and blends is then established by calibration charts created from
the pure components [33].

4. Computational Techniques

In order to deal with the complexity of many pyrograms and to use pyrograms to
detect small changes in the polymers or fibers, it is necessary to use methods that
try to rationalize data and make them amenable to statistical analysis. Truly complex pyrograms can have up to 200 components. Only more complex methods of
pattern recognition techniques can give the differentiation needed.

The general term applied to this approach is chemometrics. This approach has
been greatly enhanced by the spectacular rise of the capabilities of small computers. One definition of chemometrics is "the application of numerical techniques to
the identification of, or discrimination between, chemical substances" [34]. This
is a particularly important approach for analytical pyrolysis in trying to use the
method to give the maximum amount of information to detect small changes in
materials as a result of treatments.

One approach was applied to untreated cotton and cotton treated with inorganic salts. Small changes in the pyrograms were evident but a systematic analysis was needed. The analysis involved classifications of the volatiles into groups,
with partial correlation coefficients for each volatile. Sign matrix analysis was
used to group the compounds to determine the mechanisms of the reactions involved [35]. A number of other studies can be looked at for more sophisticated examples. These include the work by Jones [16], where statistical procedures were
coupled with discriminate analysis, the probability method developed by
Ryabikova et al. [36], and the compositional analysis of carbon fiber composites
by Bradna and Zima [37]. There are applications of chemometrics to pyrolysis-
mass spectrometry of rubber triblends by Lattimer et al. [38], and to polyesters and
polyethers by Georgakopoulos [39]. Chapters on chemometrics and their application to pyrolysis can be found in Irwin's book on analytical pyrolysis [40] and
Maddock and Ottley's Chapter 3 in Wampler's handbook [41].

D. Reproducibility

The question of reproducibility with analytical pyrolysis methods was reviewed
by Wampler and Levy [42]. Sample size and geometry both affect the transfer of
heat. The smaller the sample size, the less the thermal gradient will be and the more
likely it is that pyrolysis will be uniform. Generally, a sample size in the low micrograms range will produce better and more reproducible results than larger samples. The best geometry for the sample is a thin film, preferably melted onto the
filament. When melting is not possible the reproducibility of the sample shape becomes crucial. Homogeneity should be sought by grinding, powdering, and mixing of the sample when possible.

Contamination from several sources has to be acknowledged as possible and avoided. Successive pyrograms of the same substance that show erratic behavior may simply be the result of introducing finger oil to certain samples. In addition, the pyrolyzer itself could be a source of reproducibility problems. Quartz linings to minifurnaces, or quartz tubes for resistance heated pyrolyzers, can obviate problems if metal catalysis is suspected. Sample transfer as a source of erratic results will depend on the tightness of connections and the maintenance of an evenly heated transfer system.

If sampling and handling techniques are carried out with proper attention, then reproducibility can be quite good. The instruments currently available, when used properly, are capable of giving replications with relative standard deviations of less than 3%, and sometimes much better than this [42].

III. POLYMERS

Analytical pyrolysis is particularly well suited to characterize polymers and fibers and to detect small changes in the chemical composition of these materials. It is a rapid and very sensitive method that can provide structural information for polymers and fibers. Pyrolysis does not depend on polymers or fibers being solvent soluble, and can readily handle cross-linked polymers with three-dimensional networks. It works by thermal nonoxidative breakdown of the long polymer chains into a series of structure-specific smaller volatile compounds. Three kinds of reactions may result from the applications of high heat to polymers:

1. Crosslinking within the polymer
2. Chain scission leading to a decrease in the molecular weight
3. Large yields of small molecules, including monomeric and oligomeric moieties

The latter effect is typically seen if heat is applied quickly and a high temperature is reached in a very short time. Various quantitative studies have shown that pyrolysis processes in polymers give little or no evidence for recombination reactions that create new or different molecules [43]. Therefore, if pyrolysis conditions are kept constant, the sample should always degrade into the same constituent molecules. When the degradation products are swept into a chromatograph under particular given conditions, the resulting chromatogram will always be the same and a fingerprint is obtained.

The limitations of pyrolysis–gas chromatography in its early days included difficulty in getting sufficient separation of degradation products and problems with identification of the individual products. The first reference to the use of pyrolysis in examining polymeric products was that by Davison et al. in 1954 [44], when they suggested pyrolyzing the samples and then separately separating the products by gas chromatography. They gave the results for eight polymers. Later, other workers [45,46] combined the pyrolysis apparatus with the gas chromatograph and showed that it could be a general tool for the identification of polymers and quantification of groups within the polymers. There were also studies that applied the beginning method to groups of polymers such as the acrylics and the polyurethanes [47–51].

Nelson et al. [52] examined a large group of plastics and showed that the method could be extended to various classes of polymers and that milligrams and submilligram quantities could be used. They concluded that reproducibility was adequate for identification purposes and that the method could have general application. Since the early 1960s, when this work was done, the method of analytical pyrolysis has become a widely used method for polymers, and hundreds of papers have been published. The bibliography by Wampler [41] contains 140 references on synthetic polymers for the years 1980–1989 alone.

IV. FIBERS

The amount of information about polymers and analytical pyrolysis is voluminous. The articles and book chapters that deal specifically with fibers are much less so, though certainly significant. In general, information is contained in articles concerned with forensic investigations or aimed towards a textile audience. Early studies on fibers included those by Haase and Rau [53], Janiak and Damerau [54] and Gunther et al. [55], where natural and synthetic fibers were pyrolyzed to produce fingerprint programs. In other early work [56], blends of fibers were also included. Forensic applications involving fibers are discussed by Wheals and Noble [57], Saferstein and Manura [58], Hughes et al. [59], Wheals [60,61], and Challinor [62]. There are several references that are concerned with the application of analytical pyrolysis to fibers in general. These include the early work by Janiak [63], the publication on identification of textile materials by the Textile Institute [64], and articles by Chrighton [65] and Focher et al. [66].

Hardin and Wang [32] examined a wide range of examples of textile applications and created pyrograms for cotton and treated cottons, nylon, acetate, acrylic, polyester, and aramid. See Figs. 3–7 for examples of polypropylene, acrylic, Nomex aramid, acetate, and polyester fibers. Perlstein, in a comprehensive study [67], examined the pyrograms from polyester, nylon 66, aramids, acrylic, polypropylene, cellulose, and protein fibers. Blends of polyester and wool, polyester and cotton, polyester and polypropylene, and others were also examined. The pyrograms were interpreted with the aid of diagnostic peaks for degradation products, rather than by the pattern recognition method. Diagnostic peaks proved especially useful in identification of fiber blends. Wright et al. [68] used a furnace-type pyrolyzer to examine PY-GC profiles of nylon 6, polypropylene, polyethylene, and polyester fibers. They noted that these pyrograms are typically very complex and used heart-cutting, cold-trapping, and back-flushing techniques to

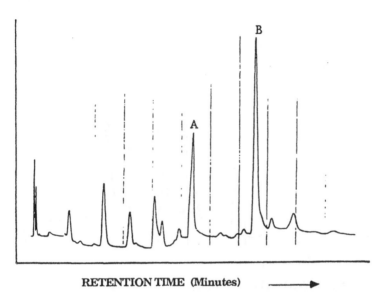

Figure 3   Pyrogram of acetate fiber. (From Ref. 32.)

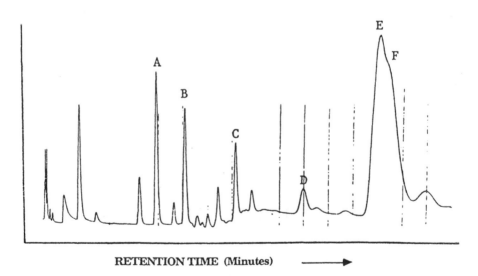

Figure 4   Pyrogram of acrylic fiber. (From Ref. 32.)

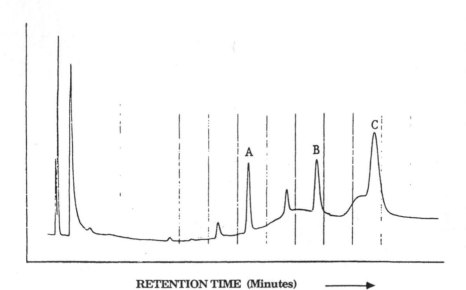

Figure 5 Pyrogram of Nomex aramid fiber. (From Ref. 32.)

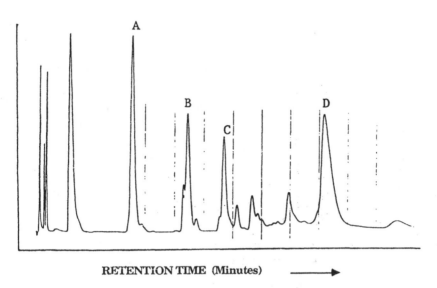

Figure 6 Pyrogram of polypropylene fiber. (From Ref. 32.)

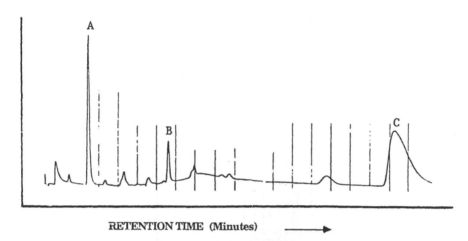

Figure 7 Pyrogram of polyester fiber. (From Ref. 32.)

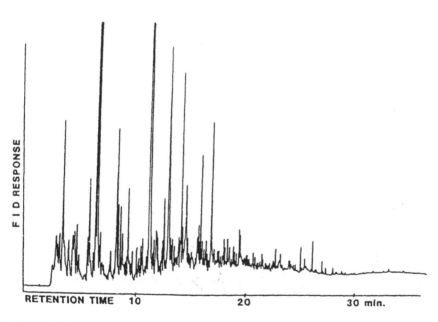

Figure 8 Pyrogram of raw wool. (From Ref. 41.)

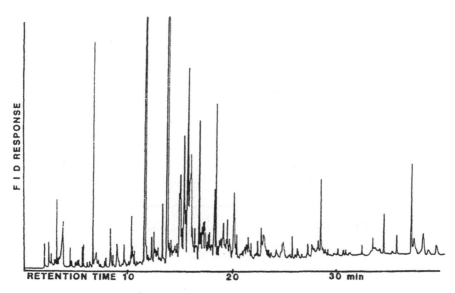

Figure 9 Pyrogram of silk fiber. (From Ref. 41.)

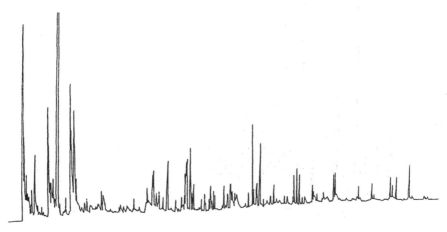

Figure 10 Pyrogram of nylon 6/6. (From Ref. 41.)

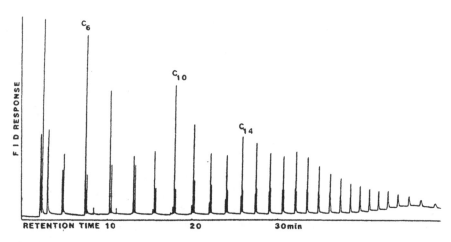

Figure 11 Pyrogram of polyethylene. (From Ref. 41.)

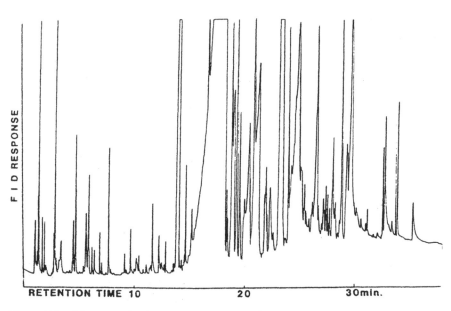

Figure 12 Pyrogram of polyester fiber. (From Ref. 41.)

eliminate less valuable information from the programs and shorten the scan time. In his handbook on pyrolysis [41], Wampler includes an index of sample pyrograms, including several fibers. Figures 8–12 are from that source.

A. Polyesters

Polyethylene terephthalate (PET) is by far the most common polymer used for polyester fiber. When pyrolyzed, it typically yields compounds such as benzene, benzoic acid, biphenyl, and vinyl terephthalate [69]. Hardin and Wang [32], Perlstein [67], and Wright et al. [68] all show pyrograms of polyester, each characteristic of the particular conditions and setup. Day and Wiles [70] examined PET untreated and treated with TRIS, the now-banned flame retardant. Ordeyaar and Rowan [71] studied the interaction of polyester and cotton blends during pyrolysis. Yang and Hardin used a different approach to gain quantitative determinations of cotton and polyester in blends of unknown composition [72]. Instead of separating the individual compounds after pyrolysis, the pyrolyzates were moved quickly through the column and the total area of the reaction products peaks was measured. By using calibration charts created by known blend composition samples, the blend ratios in unknowns could be determined accurately and quickly.

B. Nylons

Several studies have dealt with specifically with nylons. In early work Krull examined stabilized nylon 6 [73]. A detailed study of nylons was done by Ohtani et al. [74]. They examined 12 different nylons (5 aminocarboxylic and 7 dicarboxylic types). These were pyrolyzed in a microfurnace with peaks separated by glass capillary GC and then identification of the peaks by mass spectrometry. Characteristic degradation products were identified and general mechanisms of degradation formulated. Challinor [62] showed how nonpolar capillary columns can be used to clearly differentiate nylon 6 and nylon 66 (Fig. 13).

C. Acrylics

Differentiation between microgram quantities of acrylic and modacrylic fibers was demonstrated by Bortnish et al. [75]. Further studies of polyacrylonitrile were done by Ferguaor and Mahapotra [76]. Urbas and Kullik [77] analyzed the production of toxic compounds from nitrogen-containing fibers like the acrylics. The copolymer constituents of acrylic fibers were qualitatively and quantitatively determined by Saglam [78]. Methyl acrylate and methyl methacrylate were determined by dissolving the fibers in dimethyl formamide, depositing the solution on a platinum spiral, pyrolyzing, and then separating the products by gas chromatography. The compositions arrived at agreed with other methods.

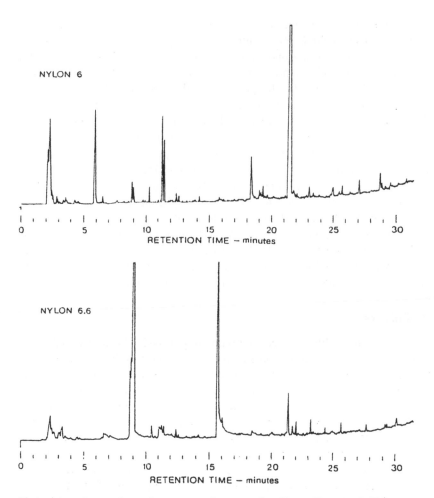

Figure 13   Comparison of pyrograms for two nylon fibers. (From Ref. 62.)

D.   Proteins

Analytical pyrolysis work on protein fibers has interesting links with animal and human hair and other proteins. Some of the earliest studies in PY-GC of wool dealt with analysis of volatile gases, particularly those that might be toxic. Kerrit and Kullik [79] looked at the evolution of some toxic gases (CO_2, CH_4, H_2S, and CS_2) from amino acids and at differences between wools from various sources and silk. Urbas and Kullik took the work further into other nitrogen-containing fibers such polyacrylic and polyamide fibers [80].

Several studies relate directly to wool. Danielson and Rogers examined tryptophan in proteins by PY-GC [81]. This done by on pyrolysis followed by GC separation of skatole, a unique pyrolysis product of tryptophan. Urbas and Kullik [82] analyzed untreated and flame-proofed wools. A series of detailed papers on wool from Marmer, Magidman and colleagues dealt with detection and characterization of commercially applied chemical agents, tryptophan and tyrosine in wool [83–85]. Characteristic peaks from benzoylated wool, from wool treated with the insect resistant agent Mitin FF, and from wool treated with Lisoamin yellow 26 were examined, all of which gave characteristic GC peaks. The latter two studies examined compositional changes induced by various kinds of chemical treatment normally done in textile processing. Py-GC was used to quantitate tryptophan in wool using product ratios (indole to *para*-cresol + phenol) and by detecting tyrosyl residues caused by chemical processing of wool.

Perlstein also examined wool and silk for diagnostic peaks in their pyrograms [67]. Although both are proteins, the difference in their mix of amino acids allows easy identification. Specific studies concerning human hair that linked PY-GC to MS were those by Munson and Vick [86] and Ishizawa and Misawa [87].

E. Cellulosics

There is a voluminous literature dealing with the pyrolysis of cellulose and its relationship to such interests as biomass conversion, flame retardancy, and flammability in general. Analytical pyrolyses of cellulosic fibers are exemplified by the PY-GC work on rayon fibers by Heinsoo et al. [88], Perlstein's comprehensive fiber identification paper [67], the series of papers by Hardin and co-workers [32,72,89–92], the examination of microcrystalline cellulose by Pouwels et al. [93], the PY-MS work on polysaccharides by Lomax et al. [94], and the review paper on the application of PY-GC to the paper and textile industries by Ferrari [95]. An ancillary study was the thermal characterization of milkweed fibers by Gu et al. [96]; which included differential thermal analysis (DTA), FTIR, and pyrolysis. The data were compared with those from cellulose, hemicellulose, and lignin.

V. FINISHES AND DYES

There is a limited amount of literature dealing with the application of analytical pyrolysis to finishes and dyes. Cope [97] used a small furnace at 400°C to pyrolyze two commercial fire resistant finishes, tetrakis(hydroxymethyl)phosponium hydroxide (THPOH) and N-methylol dimethylphosphonium propionamide (Pyrovatex CP), and cotton fabrics after they had been treated with these agents. The THPOH was also applied to a 80/20 cotton/Nomex blend fabric. The now-banned flame retardant TRIS—tris(2,3-dibromopropyl) phosphate—was pyrolyzed alone and after being applied to a polyester (PET) fabric. Day and Wiles [70] also examined this flame retardant on polyester. Another study by Thoma and Hutzinger examined a series of brominated flame retardants by PY-GC-MS [98].

The THPOH and Pyrovatex CP finishes both react with fibers. The pyrograms of the neat finish and the pyrograms of the finished fabric showed little similarity. The TRIS, on the other hand, did not react with the PET but was absorbed physically within the fiber structure. The principal peaks of TRIS pyrolyzed alone were present when the treated fabric was pyrolyzed, although the latter pyrogram was much less complex.

Hardin and Wang [32] compared the pyrograms from unfinished cotton and THPOH-NH$_3$ and Pyrovatex CP–treated cotton (Figs. 14–16). The pyrograms for all three contained the same compound peaks, although the relative amounts of the lower molecular weight compounds and the intense peak for glycolaldehyde were distinctive. With both finishes the amount of glycolaldehyde was greatly suppressed. In other studies by Hardin and colleagues analytical pyrolysis was used to investigate the effect of organophosphorus flame retardants on the flammability of cotton cellulose [91,92] and the effect of deposited salts on the pyrolysis of untreated cotton cellulose [89,90].

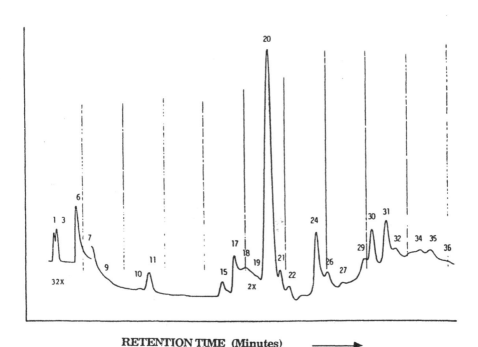

RETENTION TIME (Minutes) ⟶

Figure 14   Pyrogram of unfinished cotton. (From Ref. 32.)

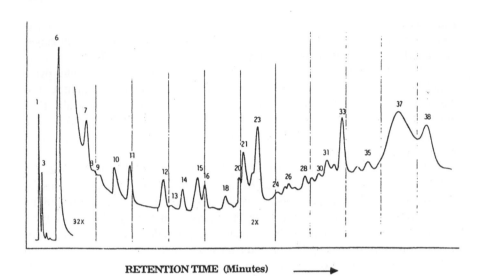

Figure 15   Pyrogram of THPOH-NH₃ finished cotton. (From Ref. 32.)

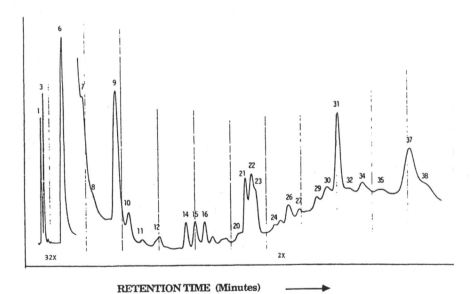

Figure 16   Pyrogram of Pyrovatex CP finished cotton. (From Ref. 32.)

Hercosett cationic resin is a diethyltriamine condensed with adipic acid and crosslinked with epichlorohydrin. It is the most widely used finish for shrink-proofing wool. Cutler et al. developed a sensitive and accurate method for measuring the level of the agent on fabrics using PY-GC [99]. The method is based on pyrolysis–gas chromatography. A small sample was pyrolyzed at 764°C and conveyed to a capillary column for the detection of cyclopentanone, the primary pyrolyzate of adipic acid. A linear correlation was shown to exist between the Hercosett content of wool and the ratio of the cyclopentanone peak to a nearby reference peak arising from wool protein. The method, which takes about 2 hr, is faster and more accurate than other methods that might be used, and can give the amount of Hercosett by setting up a standard curve.

In other work on textile finishes the emphasis was on using analytical pyrolysis to identify finish materials. Casanovas and Rovira [100] used PY-GC to identify acrylic copolymers and their mixtures with formaldehyde and melamine resins used in textiles. The pyrolysis products formed could be used to identify the copolymers and the mixtures. Hardin and Wang [32] presented pyrograms for sizing materials such as corn starches, modified potato starches, hydroxypropyl methylcellulose, and two different polyvinyl alcohols (Figs. 17–22).

The identification of aniline dyes by PY-GC-MS was demonstrated by Abbey et al. [101]. Four dyes were obtained from commercial sources and five others were synthesized. The samples used were about 2 μg pellets. The pyrolysis products were identified using computer database searches. The products obtained were complete subunits of the original dyes, making it easy to get information about the original structure. It was also possible to quantify the amount of dye

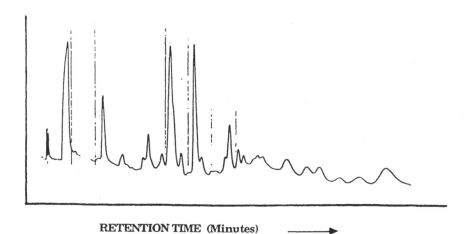

RETENTION TIME (Minutes) ⟶

Figure 17 Pyrogram of potato starch. (From Ref. 32.)

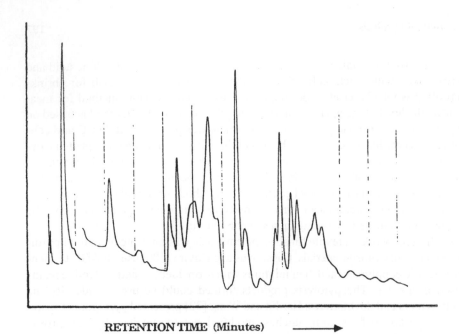

RETENTION TIME (Minutes) ⟶

Figure 18 Pyrogram of corn starch. (From Ref. 32.)

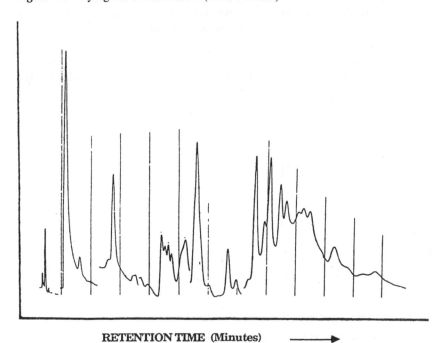

RETENTION TIME (Minutes) ⟶

Figure 19 Pyrogram of hydroxypropyl potato starch. (From Ref. 32.)

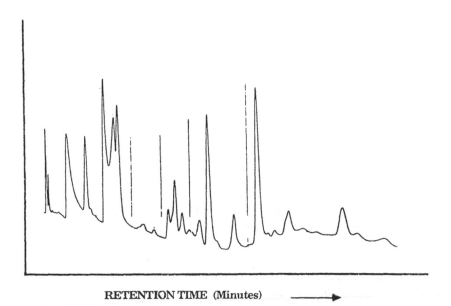

Figure 20 Pyrogram of hydroxypropyl methyl cellulose. (From Ref. 32.)

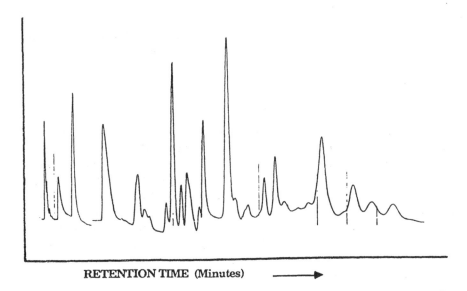

Figure 21 Pyrogram of polyvinyl alcohol, 88% hydrolyzed. (From Ref. 32.)

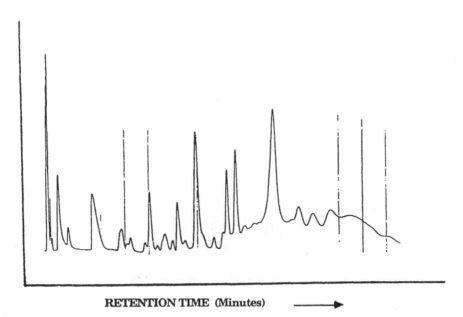

RETENTION TIME (Minutes)   ⟶

Figure 22   Pyrogram of polyvinyl alcohol copolymer. (From Ref. 32.)

pyrolyzed. The method was applied to several environmental waste samples and was shown to be useful in the rapid identification of aniline-based dyes in the environment.

VI.   REVIEWS ON ANALYTICAL PYROLYSIS

There are a number of good sources of information available on analytical pyrolysis. These contain much information on polymers and some specific information on the application of the method to fibers. Journal articles include reviews by Wolf et al. [102], Irwin [103], Liebman and Levy [104], Jones [16], and Shedrinsky et al. [105]. There is also an extensive bibliography by Wampler that covers analytical pyrolysis applications from 1980 to 1989 [106]. The section on instrumentation in the bibliography contains 34 references, that on biopolymers has 34 references, and there are more than 140 references on synthetic polymers. Articles focused on PY-MS include Haverkamp and Kistemaker [107], Gale et al. [108], Pausch et al. [109], Israel et al. [110], and Garozzo and Montaudo [26].

Books on analytical pyrolysis include the monograph on pyrolysis–gas chromatography by May et al. [111], the compendium and atlas by Meuzelaar et al. [112], comprehensive reviews by Irwin [40] and Voorhees [113], and the book devoted to analytical pyrolysis and polymers by Liebman and Levy [114]. In addition, there is a recently published analytical pyrolysis handbook edited by Wampler that contains chapters giving an overview of the method, instrumentation and analysis, and various sections on polymers and fibers [41]. Other sources include the chapter on polymer identification using PY-GC by Brown [115] and the section on pyrolysis techniques in Crompton [14].

REFERENCES

1. I. Ericsson and R. P. Lattimer, Pyrolysis nomenclature, *J. Appl. Anal. Pyrol. 13:*219 (1989).
2. J. A. Moore, *Encyclopedia of Chemistry, 2nd ed.* (S.B. Parker, ed.), McGraw-Hill, New York, 1993, p. 891.
3. H. R. Schulten, Relevance of analytical pyrolysis studies to biomass conversion, *J. Anal. Appl. Pyrol. 6:*2561 (1984).
4. R. D. Hartley and J. Haverkamp, Pyrolysis mass spectrometry of the phenolic constituents of plant cell walls, *J. Sci. Food Agric. 35:*14 (1984).
5. C. Gutteridge, Characterization of microorganisms by pyrolysis mass spectrometry, *Methods Microbiol. 19:*227 (1987).
6. W. Windig, H. L. C. Meuzelaar, and J. Haverkamp, Pyrolysis-mass spectrometry of virus-infected plants and fungi, *Phytopathology 3:*437 (1983).
7. W. Genuit, J. J. Boon, and O. Faix, Characterization of beech milled wood lignin by pyrolysis gas chromatography-photoionization mass spectrometry, *Anal. Chem. 59:*508 (1987).
8. A. Saliot, A. Ulloa-Guevara, T. C. Viets, J. W. De Leeuw, P. A. Schenck, and J. J. Boon, The application of pyrolysis-mass spectrometry and pyrolysis-gas chromatography/mass spectrometry to the chemical characterization of suspended matter in the ocean, *Adv. Org. Geochem. 6:*295 (1983).
9. R. Tsao and K. J. Voorhees, Analysis of smoke aerosols from nonflaming combustion by pyrolysis/mass spectrometry pattern recognition, *Anal. Chem. 56:*368 (1984).
10. K. J. Voorhees and R. Tsao, Smoke aerosol analysis by pyrolysis mass spectrometry/pattern recognition for assessment of fuels involved in flaming combustion, *Anal. Chem. 57:*1630 (1985).
11. R. Aries and C. Gutteridge, Applications of pyrolysis mass spectrometry to food science, in *Applied Mass Spectrometry in Food Science* (J. Gilbert, ed.), Elsevier Applied Science, New York, 1985, p. 377.
12. R. E. Fresenius, Analysis of tobacco smoke condensate, *J. Anal. Appl. Pyrol. 8:*561 (1985).
13. K. Harada, O. Nishimura, and S. Mihara, Rapid analysis of coffee flavor by gas chromatography using a pyrolyzer, *J. Chromatog. 391:*457 (1987).
14. T. R. Crompton, *Practical Polymer Analysis*, Plenum Press, New York, 1993, p. 73.
15. CDS Analytical Inc., P.O. Box 277, Oxford, PA 19363-0277.
16. S. Jones, Applications of pyrolysis gas chromatography in an industrial research laboratory, *Analyst 109:*823 (1984).

17. N. Oguri and A. Onishi, Curie-point pyrolysis with a modified maintenance oven for capillary column GC, *Liquid Chromatogr.-Gas Chromatogr.* *12(7)*:541 (1994).
18. R. L. Levy, D. L. Fanter, and C. J. Wolf, *Anal. Chem.* *44*:38 (1972).
19. Scientific Glass Engineering, Inc., 2007 Kramer Lane, Austin, TX 78758-4095.
20. A. Ohnishi, K. Kata, and E. Takagi, Curie-point pyrolysis of cellulose, *Polymer* *7*:431 (1975).
21. N. Cecchetti and R. Polloni, A new pyrolyzer for laser pyrolysis–gas chromatography–mass spectrometry experiments, *J. Anal. Appl. Pyrol.* *23*:165 (1992).
22. R. S. Lehrle, Polymer pyrolysis mechanisms: Experimental approaches for investigating them, *J. Appl. Anal. Pyrol.* *11*:55 (1987).
23. S. Tsuge, Characterization of polymers by pyrolysis/high resolution gas chromatography with fused-silica capillary columns, *Chromatog. Forum* Nov.-Dec.:44 (1986).
24. A. Venema, Potential of capillary gas chromatography in industrial research laboratories, *J. Chromatogr.* *279*:103 (1983).
25. T. P. Wampler and E. J. Levy, Cryogenic focusing of pyrolysis products for direct (splitless) capillary chromatography, *J. Anal. Appl. Pyrol.* *8*:65 (1985).
26. D. Garozzo and G. Montaudo, Identification of polymers by library search of pyrolysis mass spectra and pattern recognition analysis, *J. Appl. Anal. Pyrol.* *9*:1 (1985).
27. E. R. McCall, N. M. Morris, and R. J. Berni, Infrared analysis of vapor-phase pyrolysis products of flame-retardant fabrics, *Text. Res. J.* *49(5)*:288 (1979).
28. J. W. Washall and T. P. Wampler, Direct pyrolysis–Fourier transform infrared spectroscopy for polymer analysis, *Spectroscopy* *6*:38 (1991).
29. J. W. Washall and T. P. Wampler, A pyrolysis/Fourier transform–infrared system for polymer quality control: A feasibility study, *J. Appl. Polym. Sci. Appl. Polym. Symp.* *45*:379 (1990).
30. K. V. Alekseeva, Gas chromatographic identification of polymers using individual pyrolysis products, *J. Anal. Appl. Pyrol.* *2*:19 (1980).
31. R. Milina, N. Dimov, and M. Dimitrova, Classification system for recognition of low-molecular-weight substances from their program, *Chromatographia* *17*:29 (1983).
32. I. R. Hardin and X. Q. Wang, The use of pyrolysis–gas chromatography in textiles as an identification method, *Text. Chem. Color.* *21(1)*: 29 (1989).
33. R. S. Lehrle, Pyrolysis-gas chromatography for polymer analysis and characterisation and for studying thermal degradation mechanisms, *Anal. Proc.* *20*:574 (1983).
34. J. Maddock and T. W. Ottley, *Applied Pyrolysis Handbook* (T. P. Wampler, ed.), Marcel Dekker, New York, 1995, pp. 57–76.
35. I. R. Hardin, X. Q. Wang and Y. P. Qiu, Application of multiple correlation techniques to the analysis of cellulose pyrolysis products, *AATCC Book of Papers* 106 (1989).
36. V. M. Ryabikova, S. D. Tumina, and A. N. Zigel, Probability method for pyrolysis–gas chromatographic data treatment and elaboration of a computer system for the identification of polymers, *J. Chromatogr.* *520*:121 (1990).
37. P. Bradna and J. Zima, Compositional analysis of epoxy matrices of carbon-fibre composites by pyrolysis–gas chromatography–mass spectrometry, *J. Anal. Appl. Pyrol.* *24(1)*:75 (1992).

38. R. P. Lattimer, K. M. Schier, W. Windig, and H. L. C. Meuzelaar, Quantitative analysis of rubber triblends by pyrolysis–mass spectrometry, *J. Anal. Appl. Pyrol. 8*:95 (1985).
39. C. G. Georgakopoulos, M. Statheropoulos, J. Korloyannakos, and G. Parrisakis, Hephestus: An expert system in Prolog for the interpretation of pyrolysis mass spectra of polyesters and polyethers, *Chemometric Intelligent Lab. Syst. 19*:75 (1993).
40. W. J. Irwin, *Analytical Pyrolysis—A Comprehensive Guide*, Marcel Dekker, New York, 1982.
41. T. P. Wampler, ed., *Applied Pyrolysis Handbook*, Marcel Dekker, New York, 1995.
42. T. P. Wampler and E. J. Levy, Reproducibility in pyrolysis—Recent developments, *J. Appl. Anal. Pyrol. 12*:75 (1987).
43. H. L. C. Meuzelaar, W. Windig, A. M. Harper, S. M. Huff, W. H. McClennen, and J. M. Richards, Pyrolysis mass spectrometry of complex organic materials, *Science 226*:268 (1984).
44. W. H. T. Davison, S. Slaney, and A. L. Wragg, A novel method of identification of polymers, *Chem. Ind. (Lond.)* 1356 (1954).
45. R. S. Lehrle and J. C. Robb, Direct examination of the degradation of high polymers by gas chromatography, *Nature 183*:1671 (1959).
46. W. H. Pariss and P. D. Holland, *B. Plastics 33*:372 (1960).
47. J. Haslam and A. R. Jeffs, Gas-liquid chromatography in a plastics analytical laboratory, *J. Appl. Chem. (Lond.) 7*:24 (1957).
48. E. A. Radell and H. C. Strutz, Identification of acrylate and methacrylate polymers by gas chromatography, *Anal. Chem. 31*:1890 (1959).
49. J. Strasburger, G. M. Brauer, M. Tryon, and A. F. Forziati, Analysis of methyl methacrylate copolymers by gas chromatography, *Anal. Chem. 32*:454 (1960).
50. C. E. R. Jones and A. F. Moyles, Rapid identification of high polymers using a simple pyrolysis unit with a gas-liquid chromatograph, *Nature 189*:222 (1961).
51. H. Szymanski, C. Salinas, and P. Kwitkowski, A technique for pyrolysing or vaporizing samples for gas chromatographic analysis, *Nature 188*:403 (1960).
52. D. F. Nelson, J. L. Yee, and P. L. Kirk, The identification of plastics by pyrolysis and gas chromatography, *Microchem. J. VI*:225 (1962).
53. H. Haase and J. Rau, *Melliand Textilberichte 47*:434 (1966).
54. R. A. Janiak and K. A. Damerau, The application of pyrolysis and programmed temperature gas chromatography to the identification of textile fibers, *J. Criminal Law, Criminol. Police Sci. 59*:434 (1968).
55. W. Gunther, K. Koukoudimas, and F. Schlegelmilch, Characterization of textile fibrous materials by pyrolysis and capillary gas chromatography, *Melliand Texilberichte 60*:501 (1979).
56. U. Gokcen and D. M. Cates, *Appl. Polym. Sci. Symp.2*:15 (1966).
57. B. B. Wheals and W. Noble, Forensic applications of pyrolysis–gas chromatography, *Chromatographia 5*:553 (1972).
58. R. Saferstein and J. J. Manura, Pyrolysis mass spectrometry—A new forensic science technique, *J. Foren. Sci. 22*:748 (1977).
59. J. C. Hughes, B. B. Wheals, and M. J. Whitehouse, Pyrolysis–mass spectrometry of textile fibers, *Analyst 103*:482 (1978).

60. B. B. Wheals, Analytical pyrolysis techniques in forensic science, *J. Anal. Appl. Pyrol. 2*:277 (1980).
61. B. B. Wheals, Forensic application of analytical pyrolysis technique, *Analytical Pyrolysis* (C. E. R. Jones and C. A. Cramers, eds.), Elsevier, Amsterdam, 1985, p. 89.
62. J. M. Challinor, Forensic applications of pyrolysis capillary gas chromatography, *Foren. Sci. Int. 21*:269 (1983).
63. R. A. Janiak and K. A. Damerau, The application of pyrolysis and programmed temperature gas chromatography to the identification of textile fibers, *J. Criminal Law, Criminol. Police Sci. 59*:434 (1968).
64. The Textile Institute, *Identification of Textile Materials*, 7th ed., The Textile Institute, Manchester, England, 1975.
65. J. S. Chrighton, Characterization of textile materials by thermal degradation: A critique by pyrolysis GC and thermogravimetry, *Analytical Pyrolysis* (C. E. R. Jones and C. A. Cramers, eds.), Elsevier, Amsterdam, 1985, p. 337.
66. B. Focher, A. Sewa, and M. Ballini, PY-GC of textile materials, *Tinctoria 69*:411 (1972).
67. P. Perlstein, Identification of fibers and fiber blends by pyrolysis gas chromatography, *Anal. Chim. Acta 155*:173 (1983).
68. D. W. Wright, K. O. Mahler, and L. B. Ballard, The application of multidimensional techniques to the rapid pyrolysis–GC profiling of synthetic polymers, *J. Chromatogr. Sci. 24*:13 (1986).
69. J. M. Challinor, Examination of forensic evidence *Applied Pyrolysis Handbook* (T. P. Wampler, ed.), Marcel Dekker, New York, 1995, pp. 207–241.
70. M. Day and D. M. Wiles, Influence of temperature and environment on the thermal decomposition of poly(ethylene terephthalate) fibres with and without the flame retardant tris (2,3-dibromopropyl)phosphate, *J. Anal. Appl. Pyrol. 7*:65 (1984).
71. N. F. Ordeyar and S. M. Rowan, Interaction observed during pyrolysis of binary mixtures of textile polymers, *Thermochim. Acta 23*:371 (1978).
72. X. N. Yang and I. R. Hardin, Analytical pyrolysis as a method to determine blend levels in cotton/polyester yarns, *Text. Chem. Color. 23(4)*:15 (1991).
73. M. Krull, A. Kogerman, O. Kirret, L. Kutyina, D. Zapalski, Py-GC of Capron (nylon 6) fiber stabilized with ethers of 4-oxydiphenylamine, *J. Chromatogr. 135*:212 (1977).
74. H. Ohtani, T. Nagaya, Y. Sugimura, and S. Tsuge, Studies on thermal degradation of aliphatic polyamides by pyrolysis–glass capillary gas chromatography, *J. Anal. Appl. Pyrol. 4*:117 (1982).
75. J. P. Bortnish, S. E. Brown, and E. H. Sild, Differentiation of microgram quantities of acrylic and modacrylic fibers using pyrolysis gas-liquid chromatography, *J. Foren. Sci. 14*:380 (1971).
76. J. Ferguaor and B. Mahapotra, Pyrolysis studies on polyacrylonitrile fiber, *Fiber Sci. Technol. 11*:55 (1978).
77. E. Urbas and E. Kullik, PY-GC analysis of some toxic compounds from N-containing fibers, *J. Chromatogr. 137*:210 (1977).
78. M. Saglam, Qualitative and quantitative analysis of methyl acrylate or methyl methacrylate of acrylonitrile fibers by pyrolysis–gas chromatography, *J. Appl. Polym. Sci. 32*:5719 (1986).

79. O. Kirret and E. Kullik, Pyrolysis–gas chromatography of amino acids and protein fibers (Ger.), *Proceedings of the Fifth International Wool Textile Research Conference, Aachen 2*:523–537 (1975).

80. E. Urbas and E. Kullik, Pyrolysis–gas chromatographic analysis of some toxic compounds from nitrogen-containing fibres, *J. Chromatogr. 137*:210 (1977).

81. N. D. Danielson and L. B. Rogers, Determination of Tryptophan in proteins by pyrolysis gas chromatography, *Anal. Chem. 50(12)*:1680 (1978).

82. E. Urbas and E. Kullik, PY-GS analysis of untreated and flame-proofed wools, *Fire Mater. 2*:25 (1978).

83. W. N. Marmer, P. Magidman, and C. M. Carr, Pyrolysis gas chromatography of wool. Part I: Detection and Quantitation of commercially applied agents, *Text. Res. J. 57*:681 (1987).

84. W. M. Marmer, P. Magidman, and H. M. Farell, Pyrolysis gas chromatography of wool. Part II: Detection and quantitation of tryptophan in wool and simple protein, *Text. Res. J. 59*:616 (1989).

85. W. M. Marmer and P. Magidman, Pyrolysis gas chromatography of wool. Part III: Detection and quantitation of tyrosine, *Text. Res. J. 60*:417 (1990).

86. T. O. Munson and J. Vick, Comparison of human hair by pyrolysis–gas chromatography and gas chromatography–mass spectrometry, *J. Anal. Appl. Pyrol. 8*:493 (1985).

87. F. Ishizawa and S. Misawa, Capillary column pyrolysis–gas chromatography of hair: A short study in personal identification, *J. Forens. Sci. 30(4)*:201 (1990).

88. H. Heinsoo, A. Kogerman, O. Kirret, E. Coupek, and S. Vilkova, Stepwise pyrolysis–gas chromatography of viscose fibers, *J. Anal. Appl. Pyrol. 2*:131 (1980).

89. I. R. Hardin and B. L. Slaten, Effect of sodium, magnesium, and calcium salts on the flammability and thermal decomposition of fire-retardant cotton fabrics, *ACS Organic Coatings and Plastics Preprints 36*:462 (1975).

90. I. R. Hardin, L. J. Martin, and Y. P. Qiu, The effects of mode of calcium carbonate deposition on thermal decomposition of cellulose and flame retardant treated cellulose, *Cellulose and Wood—Chemistry and Technology*, (C. Schuerch, ed.), Wiley, New York, 1989, p. 885.

91. Y. L. Hsieh and I. R. Hardin, Effects of inorganic salts on cotton flammability, *Text. Res. J. 54(3)*:171 (1984).

92. Y. L. Hsieh and I. R. Hardin, Thermal conditions and pyrolysis products in flammability tests, *Text. Chem. Color. 17(3)*:41 (1985).

93. A. D. Pouwels, G. B. Eijkel, and J. J. Boon, Curie-point pyrolysis–capillary GC–high resolution MS of microcrystalline cellulose, *J. Anal. Appl. Pyrol. 14*:237 (1989).

94. J. A. Lomax, J. M. Commandeer, and J. J. Boon, Pyrolysis mass spectrometry of polysaccharides, *Trans. Biochem. Soc. Lond. 19*:935 (1991).

95. G. Ferrari, Gas chromatography in the textile and paper sector, *Tinctoria 87(4)*:58 (1990).

96. P. Gu, R.K. Hessley, and W. P. Pan, Thermal characteristics analysis of milkweed fibers, *J. Anal. Appl. Pyrol. 24(2)*:147 (1992).

97. J. F. Cope, Identification of flame retardant textile finishes by pyrolysis–gas chromatography, *Anal. Chem. 45(3)*:562 (1973).

206   **Hardin**

98. H. Thoma and O. Hutzinger, Pyrolysis and GC-MS analysis of brominated flame retardants in online operation, *Chemosphere 18*(1–6):1047 (1989).
99. E. T. Cutler, P. Magidman, and W. N. Marmer, Pyrolysis–gas chromatography of Hercosett 125 finish on wool, *Text. Chem. Color. 25*(6):27 (1993).
100. A. M. Casanovas and X. Rovira, Determination of reactive comonomers and/or amino-resins in acrylic copolymers used in the textile industry, by pyrolysis gas chromatography–mass spectrometry, *J. Anal. Appl. Pyrol. 11*:227 (1987).
101. L. E. Abbey, J. P. Gould, and T. F. Moran, Identification of aniline dyes by pyrolysis–GC–mass spectrometry, *J. Water Pollut. Control Fed. 54*:474 (1982).
102. C. J. Wolf, M. A. Grayson, and D. L. Fanter, PY-GC of polymers, *Anal. Chem. 52*:348A (1980).
103. W. J. Irwin, Analytical pyrolysis—An overview, *J. Anal. Appl. Pyrol. 1*:3 (1979).
104. S. A. Liebman and E. J. Levy, Advances in pyrolysis GC systems: Applications to modern trace organic analysis, *J. Chromatogr. Sci. 21*:1 (1983).
105. A. M. Shedrinsky, T. A. Wampler, N. Indictor, and N. S. Baer, Application of analytical pyrolysis to problems in art and archaeology: A review, *J. Anal. Appl. Pyrol. 15*:393 (1989).
106. T. P. Wampler, A selected bibliography of analytical pyrolysis applications 1980–1989, *J. Anal. Appl. Pyrol. 15*:291 (1989).
107. J. Haverkamp and P. G. Kistemaker, Recent developments in pyrolysis mass spectrometry, *Int. J. Mass Spectrom. Ion Phys. 45*:275 (1982).
108. P. J. Gale, B. L. Bentz, and W. L. Harrington, Characterization of polymers by pyrolysis mass spectrometry, *RCA Rev. 47*:380.
109. J. B. Pausch, R. P. Lattimer, and H. L. C. Meuzelaar, A new look at direct compound analysis using pyrolysis mass spectrometry, *Rubber Chem., Technol. 56*(5):1031 (1983).
110. S. C. Israel, W. C. Yang, and M. Bechard, Characterization of polymers by direct pyrolysis/chemical ionization mass spectrometry, *J. Macromol. Sci. Chem. A22*:779 (1985).
111. R. W. May, E. F. Pearson, and D. Scothern, *Pyrolysis–gas Chromatography*, The Chemical Society, London, 1977.
112. H. L. C. Meuzelaar, J. Haverkamp, and F. D. Hileman, *Pyrolysis Mass Spectrometry of Recent and Fossil Biomaterials*, Elsevier, New York, 1982.
113. K. J. Voorhees, *Analytical Pyrolysis—Techniques and Applications*, Butterworths, London, 1984.
114. S. A. Liebman and E. J. Levy, *Pyrolysis and GC in Polymer Analysis*, Marcel Dekker, New York, 1985.
115. G. M. Brown, PY-GC techniques of polymer identification, *Thermal Characterization Techniques* (P. E. Slade and L. T. Jenkins, eds.), Marcel Dekker, New York, 1970, p. 41.

5
Liquid Chromatographic Technique in Textile Analysis

YIQI YANG   The Institute of Textile Technology, Charlottesville, Virginia

I. INTRODUCTION

Liquid chromatographic (LC) technique has been more and more widely and quickly accepted in the textile industry during recent years. As reported in a survey [1], LC ranked fourth, after ultraviolet-visible (UV-visible) spectrophotometer, optical microscope, and gas chromatograph, as the most widely used instrument in the textile industry. As a matter of fact, the invention of chromatography was due to the separation of normal colorants [2].

The advantages of using LC for textile analyses can be summarized as follows.

1. Strong ability in separation and purification
2. Suitability for molecules that won't evaporate without destroying the structure
3. Sensitivity to small amount of chemicals
4. Simulation of the wet processing environment for the study of textile properties
5. Convenience for analysis

LC analysis related to textiles could be roughly separated into three areas: using conventional stationary phases, such as, C18, for textile analysis; packing textile materials into LC column for textile property and wet processing studies; and using fibrous materials as stationary phases for other operations such as bioseparation and purification.

Conventional LC has been used for dye identification, separation, and purification in textile, food, drug, and cosmetic industries and in forensic examinations [3–10]. It has also been used as a powerful technique for the qualitative and quantitative analysis of textile finishing processes such as the study of flame re- tardants, stain-resist compounds, durable press (DP) finishes, formaldehyde, DP finishing mechanism, and contaminants on fiber [11–23]. Packing textile material into the column as stationary phase for the investigation of properties such as pore structures and related dyeing and finishing behaviors, dyeability, and dyeing mechanism has been successful [24–36]. Using supercritical fluids as mobile phase, LC also has been reported as an efficient tool for chemical extraction from textile fibers [37] and dye analysis [38].

In addition to the textile analysis, LC using textile materials as stationary phase has also been applied in other areas such as bioseparation and purification. Fibers—natural, synthetic, and hollow—and even a single piece of fabric were all reported as LC stationary phases with advantages of low pressure drop, high chem- ical and physical stabilities (excellent stability to acid and alkali, high tempera- ture), high liquid flow, and low price [39–45]. This chapter focuses on applications of LC in textile analysis and characterization.

II.   BACKGROUND

A.   Composition of a Liquid Chromatograph

As shown in Figure 1, a liquid chromatograph is basically composed of a reservoir of liquid (I) that serves as the mobile phase; an LC pump (II) that moves the mo- bile phase through the system; a pressure gauge (III) indicating the differential pressure between the LC system and its atmosphere; a relief valve (IV) to avoid damage due to unexpected high pressure of the system; an injector (V) that quan- titatively controls the sample injection; a sample syringe (VI) that introduces sam- ples for analysis; an LC column (VII) as the key component for the system; a detector (VIII) measuring the concentration of solutes out of the column; a recorder (IX) to record the signals from the detector with calculations such as peak area, slope, and peak location; and an eluate collector (X) to collect the eluate. For dye, finish, and other chemical analysis, samples can be introduced through the in- jector (V) into the LC system by a sample syringe (VI). The eluent from reservoir (I) could be a single-component solution with a fixed concentration, such as 20 g/L NaCl aqueous solution, or a gradient that is a multicomponent solution with changing concentrations of some or all of the components, such as an aqueous mo- bile phase with sodium lauryl sulfate concentration changing linearly from zero to 0.1 M in 1 hr. A gradient programmer is usually used together with LC pumps (II) to control the amount of different components to be mixed at different time. An LC column (VII) used for the analysis of textile chemicals or fibrous materials, such as fibers or fabrics, can be packed with conventional spherical or particulate supports as stationary phases or with the fibers or fabrics being tested. Choosing

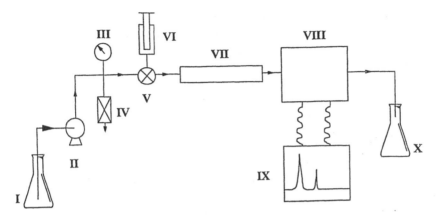

Figure 1 Basic composition of liquid chromatograph: I, a reservoir of eluent served as mobile phase; II, LC pump; III, pressure gauge; IV, relief valve; V, sample injector; VI, sample syringe; VII, LC column; VIII, detector; and X, eluate collector.

the right stationary and mobile phases is most important toward the success of LC analysis. To measure the concentration of solutes elute out of the column, the most commonly used detector is a UV-visible light detector, which can be continuously selected to any wavelength in the ultraviolet and visible light range or be selected within specific wavelengths in the UV-visible range using the right filter. Determination of textile colorants or chemicals with aromatic structures often requires such a detector. For other textile chemicals, such as mineral electrolytes, aliphatic surfactants, or for D_2O, sugars, dextrum, polyethylene glycol (PEG) used for pore structure studies, and any other chemicals that do not have strong absorption in the UV-visible range, a differential refractometer is often effective. The differential refractometer is also called an RI detector (RI stands for reflective index). The concentration of chemicals determined by the detector is interpreted and recorded by a chart recorder as a chromatogram or by an integrator as a chromatogram with peak position, peak area, and other calculated parameters.

B. Basic Concepts

There are some basic terms and concepts of the LC application in textiles that need to be introduced for better understanding of further discussions. Only those mentioned in this chapter are discussed. A thorough discussion of basic LC theory could be obtained from one of the many LC books.

1. Solute Elution

When a solute is introduced into the LC column, it moves with the mobile phase at the same speed in the spaces between materials of stationary phases. These spaces are occupied by mobile phase. Solute moves at a slower speed when it contacts the stationary phase if it has some kind of attractive interaction with the stationary phase. The solvent will move the solute out of the column because of the solubility of the solute in the solvent. But due to the dispersion of solute in the voids between solid materials of stationary phase and the different adsorption energies on the surface of stationary phase, the solute does not elute out at the same time. Therefore a peak is observed on chromatogram as shown in Figure 2. The flat portion of a chromatogram where there is no solute being eluted is called the baseline. To describe the position of a peak in the chromatogram, the place that has most of the solute eluted out, that is, the top of the peak, is used.

As shown in Figure 2, time zero is the start of solute movement, that is, the injection time. The term t_r is called retention time, which is the time between injection and the peak maximum. The magnitude of t_r depends on the extent of attraction between the solute and the stationary phase; t_r decreases with decreasing attraction. The elution time of a solute that can not penetrate into any of the pores of the stationary phase and has no attraction to the stationary phase only depends on the empty spaces, that is, the magnitude of voids in the stationary phase. The elution time of this excluded component is called the dead time (t_o) (cf. Fig. 2). In addition to the property of mobile and stationary phases, the elution time also de-

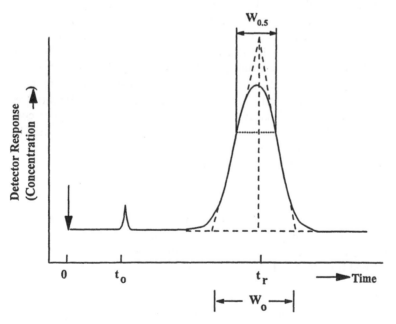

Figure 2   Schematic diagram of a chromatogram.

pends on the flow rate (q) of the mobile phase. Therefore, in addition to dead time and retention time, the terms dead volume (V_o) and retention (V_r) are also used.

$$V_r = qt_r \qquad V_o = qt_o \tag{1}$$

The fundamental equation which relates retention volume to the interaction between solute, solvent and stationary phase is given in Eq. (2),

$$V_r = V_m + KV_s \tag{2}$$

where V_m and V_s are the total volumes of mobile phase and that of stationary phase of the column, respectively; K is the distribution coefficient of the solute between two phases.

2. Plate Concept

Plate count (N), that is, the number of plates in an LC column, and plate height (H), the size of a single plate, are often used to describe the quality of a column. The terms are borrowed from the distillation theory. Because of the continuous exchange of solute between the mobile phase and stationary phase as the solute moves through the column, equilibrium between the phases is never achieved. Therefore, a column is considered to be divided into a number of theoretical plates. The size of the plate is such that the solute is assumed to have enough time to achieve equilibrium with both mobile and stationary phases. So, if the column performance is good, that is, the sorption and desorption of solute onto stationary phase are fast and efficient, then the size of the plate (H) will be small and there will be a large number of plates (N) in the column. Assuming the peak is Gaussian-shaped, plate count is a function of the standard deviation (σ) of the peak and the effective length (L) of the column,

$$N = \frac{L^2}{\sigma^2} \tag{3}$$

where

$$L = \rho \times t_r \tag{4}$$

with ρ as the linear velocity of mobile phase and t_r the retention volume of the peak. For the Gaussian-shaped peaks, σ can be derived from the width of the peak at the baseline (W_o), the width of the peak at half-height ($W_{0.50}$) or the width of 0.6065 of the peak height ($W_{0.6065}$) [Eq. (5) and Fig. 2].

$$\sigma = 0.250W_o = 0.426W_{0.50} = 0.500W_{0.6065} \tag{5}$$

Plate height is also used to describe the column efficiency. It is simply calculated by Eq. (6).

$$H = \frac{L}{N} \tag{6}$$

From Eq. (3), (4), and (5), we can also achieve:

$$N = 16(\frac{t_r}{t_{w_0}})2 = 5.51(\frac{t_r}{t_{w_{0.500}}}) = 4.00(\frac{t_r}{t_{w_{0.6065}}})^2 \qquad (7)$$

where t_w is the peak width in units of time.

3. Resolution

A good chromatogram for multicomponent analysis requires not only a relatively high plate count but also a good resolution factor (R_s), a measure of how well two bands are resolved [77]. It is defined as

$$R_s = \frac{t_{r_2} - t_{r_1}}{\frac{1}{2}(t_{w_1} + t_{w_2})} \qquad (8)$$

where 1 and 2 represent two peaks.

4. Capacity Factor

The capacity factor, k, for a solute is defined as:

$$\frac{\text{Moles of the solute in stationary phase}}{\text{Moles of the solute in mobile phase}} = K\frac{V_s}{V_m} = \frac{X_s V_s}{X_m V_m} \qquad (9)$$

where X_s and X_m are the concentrations of the solute in the stationary and mobile phases, respectively. From Eq. (2) and (9),

$$k = \frac{V_r - V_m}{V_m} \qquad (10)$$

where V_m is equivalent to the column void volume V_o) measured by t_o of an excluded component. Hence, Eq. (10) can be rewritten as

$$k = \frac{t_r - t_o}{t_o} \qquad (11)$$

C. Column Packing

In addition to the selection of the right solvent, choosing the suitable column is the key factor for a successful LC study. For conventional stationary phases, a packed LC column can be bought directly. For other particular stationary phases, the packing techniques can be easily found from many reference books. But the packing of fibrous materials and a whole piece of fabric is worthy of discussion.

1. Configuration of Fibrous Stationary Phases

The currently available packing technology of fibrous stationary phases could be roughly divided into the five following categories.

1. Yarn or fiber is wound in a spiral manner around a rod and the yarn or fiber wound rod is packed into a column [34,46].
2. Yarn or fiber in a parallel alignment is packed into a column longitudinally [41,42,47].
3. Randomly oriented fiber, yarn, fabric chunk and powder [29,34,43,48].
4. Ordered packing of disks of batting and fabric [29,30,31,34].
5. A piece of fabric rolled tightly and pulled into a column in which the yarns are oriented both parallel and perpendicular to the flow direction [32–36,44,45].

2. Three Successful Packing Techniques for Textile Studies

Although there are at least five different packing configurations for textile stationary phases, the ones most successfully used in textile studies are methods 3, 4, and 5 discussed in Section II.C.1. The common problem these textile stationary phases have is the large dead volume (V_o) and channeling due to the high resiliency of textile fibers. Because of that, lower plate count (N) and poor resolution (R_s) are often the disadvantages. Tight and even packing is most frequently used for the improvement of column quality [31,43,45]. Decreasing the size of the column after package [48] and using inactive materials to fill the voids between fiber and yarn [41] also were reported for the improvement of column quality. After packing, the textile column, no matter what packing method is used, needs to be stabilized through swelling and reorganization of the fibrous stationary phase by a continuous flow of the solvent through the column for 12 hr or up to several days [31,34].

a. Randomly Oriented Fiber, Yarn, Fabric Chunk, and Powder   Large pieces of textile materials are first cut into small ones using a mill. The size of textiles is suitable for LC column if the material cut could pass through an 80-mesh screen. To decrease the variation of particle size for better N value of the column, the textile material is successively passed through a series of screens with different sizes of holes, from larger to smaller. Screens of 20, 40, 60, and 80 meshes were used by Bertoniere and King [29] for stationary-phase preparation. The prepared material can be packed dry by the tap-and-fill method [49] or, more often, packed wet as an aqueous suspension [29,31].

For the wet packing, the ground textile is suspended in water and the slurry is allowed to settle down into the column by filling the column and an extension with the fiber suspension at the top of the column. A vacuum is applied to the bottom of the column to improve the packing density [43].

b. Disks of Fabric and Batting   Stacks of pieces of fabrics [30,31,34], such as four to seven layers [31], are cut into disks with a die. The diameter of the cut fabric disks should match the column's inner diameter exactly. The disks are filled into the column dry by pressing gently into the column with a suitable piston. After closing the column, the mobile phase later required for the analysis is intro-

duced to stabilize the column. More disks can be added into the column after wetting if empty space becomes available during the stabilization process [34]. The
advantage of disk packing compared to the ground fabric packing is that the property of the whole fiber can be studied. Some textile properties such as pore volume and its distribution may be affected by cutting the fibers.

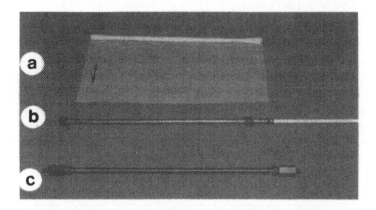

Figure 3 Column packing technique for a whole fabric rolled stationary phase: (a) fabric
rolling, (b) rolled fabric partially inserted into metal column, (c) a packed LC column [33].

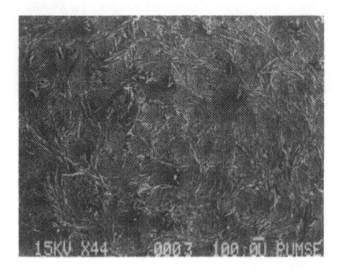

Figure 4 Scanning electron micrograph of the cross section of a cotton fabric roll. The
center of the roll is in the middle of the picture [33].

c. Whole Fabric Rolled Stationary Phase   The advantages of the whole-fabric stationary phase over the other methods are the retention of the complete structure of a fabric, easy packing as a whole-fiber method compared to the disk method, and the production of a tightly packed column flow.

As indicated in Figure 3, a fabric is rolled tightly along the warp direction (a) and then pulled into an LC column (b). Sometimes wetting helps the tight rolling. Fabric roll can be packed into the column dry or wet. Wetting the fabric roll makes the packing easier, but a dry packed column may have less dead volume due to the successive swelling in the column when water is used to stabilize the column. This packing method gives a column with very tightly rolled fabric as shown by the scanning electron micrograph of the cross section of the rolled fabric (Figure 4). The tight packing ensures the low V_o and good quality of the column.

III.   LC APPLICATIONS IN TEXTILE ANALYSIS

LC applications can be classified according to the properties of stationary phases or the application in different textile areas. The LC analyses discussed in this chapter are mainly those related to textile wet processing. They will be analyzed in the order of processing sequence. We first discuss the determination of pore structure (and surface area) and their distribution in textiles, their relation to dyeing and finishing, and then the LC application in dye identification, separation, and purification. Following that, the LC study on dyeing is discussed and finish evaluation also is included.

A.   Pore Structure and Surface Area

The importance of pore structure and surface area to the sorption rate, capacity, and uniformity of chemical reactions involved in dyeing is well known. Wet processing treatments such as scouring, bleaching, and mercerization improve the accessibility of cellulosic fibers. Bertoniere and King [29], Rowland et al. [34], Grunwald et al. [31], and Bredereck and Bluher [30] related this improvement to the fibers' pore volume, which increases upon treatment. Pore volume also affects properties of some synthetic fibers, which become more hydrophilic by increasing the microvids in the fiber [50]. Characterizing the pore structure in textile fibers is also very important when studying dyeing mechanisms [51,52], the dyeability, and the effect of durable press finishing on textile properties [26,27,29,30,31].

The pore structure of fibers can be characterized by mercury intrusion, water adsorption, or size exclusion techniques. Mercury intrusion is limited to determining the pore structure of dried fibers [50], but such a structure changes greatly in water. For example, the surface area determined by vapor sorption of N_2 is only 1.2% of that from water vapor sorption [53]. Since the pore structure of wetted fibers is what really influences properties related to wet processing of textiles, the water adsorption technique is a more appropriate method to determine this structure. The size exclusion method gives pore volume distribution based on the volume needed for molecular probes of known sizes to enter the accessible pores in a fiber. Again, since water is used as eluent, fibers swell in a manner similar to that observed in textile wet processing. Aggebrandt and Samuelson [54] were the first to use this method for textile fibers. Using size exclusion chromatography (SEC), pore structure, including pore size and surface area distribution, could be obtained from V_r of molecular probes of known sizes. The requirement for the molecular probes is that they have no interactions, either attractive or repulsive, with fibers being studied. The retention of these molecules in the column is only due to the time required for these molecules to travel through all the accessible pores in the column at the same speed as the mobile phase.

Although different models and calculations were reported for the data treatment [30,31,33,34], the basic principle of obtaining pore structure information through SEC is the same. Using a molecule that is totally excluded from the stationary phase—that is, it is too large to penetrate into any pore of the stationary phase and have no interactions with the surface of that stationary phase—dead volume, V_o, could be obtained from the retention volume of that molecule. Large molecules such as Dextran T-40, average molecular weight (MW) 40,000 [34], or the larger one, Dextran T-2000 (MW = 2,000,000) [30,31], are often used. If the molecule has smaller size, its retention volume (V_r) will increase because it moves into some pores of the stationary phase. The pores that the molecule could move into are called accessible pores. If the total weight of the stationary phase is W, the specific accessible pore volume for i molecule (V_t) is

$$V_i = \frac{V_{r,i} - V_o}{W} \tag{12}$$

The maximum accessible pore volume is the one that solvent molecules can enter into. For the study of textile behavior during wet processing, the solvent is water. To measure the water-accessible pore volume, molecules with similar size should be used. D_2O seems to be a good candidate because of its similar size to water and sensitivity to detectors such as RI detector. But it was found in many different laboratories while cellulosics were tested that V_{r,D_2O} is much larger than the retention volumes of molecules with similar size [30,31,33,34]. This is probably because of the interaction between D_2O and cellulose, and because of OH/OD interchange reactions [34].

An alternative approach is to calculate the water accessible pore volume from the regression equation

$$V_i = f(MW_i) \tag{13}$$

by substituting 18 (MW of H_2O) into the equation, or simply from the extrapolation of V_i vs. MW_i or $log(MW_i)$ plot to $MW_i = 18$. Increasing the size of molecular probe decreases the accessible pore volume. The pore volume difference (ΔV)

between two probes ($i - 1$ and i, $i > i - 1$) is the pore volume accessible to molecules with the size smaller than i but larger than $i - 1$:

$$\Delta V = V_{i-1} - V_i \qquad (14)$$

The commonly used molecular probes are a series of polyethylene glycols (PEG), sugars, and dextrans. Their diameters could be found from papers of Nelson and Oliver [51], Stone and Scallan [56], Ladisch et al. [33], and Bredereck and Bluher [30].

Strictly, ΔV calculated from Eq. (14) is the accessible pore volume for molecular probes larger than $i - 1$ but smaller than or equal to i. To obtain the real pore structure of the stationary phase, further calculation is necessary. There are basically three different treatment to obtain pore structure from V_1 vs. MW_i data.

1. Using ΔV_i directly as the total volume of pores with diameters (D):

   $$D_{i-1} \leq D \leq D_i$$

   This approach assumes that molecular probe could penetrate into the pores with same and larger diameters. But because of the uncircular shaped pore openings, the spherical probe could only enter into the pores larger than itself.
2. Using ΔV_i as the volume of pores with diameters (D):

   $$3D_{1-1} \leq D \leq 3D_1$$

   This is the commonly used adjustment to calculate the pore structure. The assumption is that molecular probes could penetrate into pores with diameter three times its size or larger. This threefold adjustment was often used for cellulosics [33,55].
3. Instead of using an unchanged threefold adjustment, Knox and Scott [57] developed an equation, Eq. (15), to convert pore volume accessible to molecular probes to pore volume related to the sizes of the pores of the stationary phase.

   $$g = K(\ln r) - 1.5 \frac{dK}{d(\ln r)} + 0.5 \frac{d^2K}{d(\ln r)^2} \qquad (15)$$

   where r is the radius of a molecular probe, K is the ratio of accessible pore volume of the molecular probe to the volume of total accessible pores, and g is the ratio of volumes of pores with radii equal to or larger than R to the total pore volumes of the stationary phase. R is the radius of the smallest pores that the molecular probe with the radius of r can reach. As a function of r, K could be obtained from:

   $$K(\ln r) = a \times \exp[b \times (\ln r + c)^2] \qquad (16)$$

   where a, b, and c are constants that could be obtained from the regression of experimental K vs. r data by Eq. (16).

If V(R) represents the volume of pores with radii equal to and larger than R, dV/dR vs. R gives a pore volume distribution curve. Furthermore, if the shape of the pores is known, the surface area distribution of the pores can also be calculated. Ladisch et al. [33] gave the detailed calculation with the assumption that the shape of the pores is elliptical.

Bertoniere and colleagues have made important contributions in this area relating pore structure of cellulosics to their preparation, dyeing, and finishing properties. Their research on accessible internal volume of cotton fiber is discussed in another chapter in this volume.

B.   Colorant Identification, Separation, and Purification

To separate a specific dye from other dyes, impurities, and other chemicals in the test system, LC can allow the dye to elute out at specific time and can thus achieve dye identification, separation, and/or purification. The right stationary phase, eluent, and detector are the three key factors to ensure a high-quality LC analysis of dyes.

The commonly used stationary phases are reverse phases of C8, C18, and ODS, silica gel, and polymeric materials such as polystyrene and divinyl benzene beads. Eluents often used are polar solvents such as water, alcohol, and acetonitrile, aqueous solutions of acids and salts, such as citric acid, acetic acid, perchloric acid, sodium sulfate, and ammonium acetate, and solutions of amines such as triethanol amine and t-butylammonium hydroxide. Because dyes have large conjugated systems they all have strong absorption in ultraviolet (UV) and visible wavelengths. Thus the most sensitive detectors are UV and/or visible wavelength spectrophotometers. Choosing the specific wavelength with maximum absorbance (λ_{MAX}) due to the maximum sensitivity, the concentration of dye eluted out of the column can be detected and recorded by a chart recorder or an integrator. For a mixture of two or more dyes, which is usually the case for dye analysis, this single-wavelength detector is not enough. The dyes in the mixture usually have different color. Therefore their λ_{MAX} values are different. When these dyes are separated in the column, they come out at slightly different time. If a fixed wavelength is used, it may only be sensitive to one dye. For other dyes, there might be a very small peak or even no peak at all. To overcome this, several detectors set at different wavelengths must be used. However, for the analysis of unknowns, it is impossible to use this preset wavelength method. The development of fast scanning UV/visible light and diode array detectors is an important movement for colorants analysis. Scanning the eluate with the complete UV/visible lights and repeating such scanning quickly can obtain the absorbance for each wavelength. Saving and treating the saved signals can obtain chromatographs in which all peaks are obtained from their λ_{MAX} values. However, no matter how fast the scanning is, the measurement of a specific wavelength by these fast scanning UV/visible light detectors is discontinuous. This may cause a loss of information, especially for trace and fast LC analyses. A diode array detector [2,4,58,59] measures the entire region light absorbance continuously. The equipment covers the light region up to 600 nm at least. A broad-emission lamp generates light of all wavelengths passing through the flow cell. The light, after absorption by the colorants in the cell and dispersion by a holographic grating, falls onto an array of diodes. The array contains hundreds of diodes, each for a specific wavelength. The output from the diodes is stored and analyzed by a computer.

Using multiwavelength detectors not only improves the sensitivity of the detector by selecting the signals from λ_{MAX} of a specific colorant, but also allows some unique analysis [2]. An important application is to verify the purity of a solute after separation. Assume there is a dye mixture that has two dyes, a and b, with λ_{MAX} of 430 nm and 575 nm, respectively. For a pure dye (e.g., a), at no matter what concentration, the ratio of absorbance (e.g., A_{430}/A_{575}) should be constant. If the eluate is a mixture, the plot of the absorbance ratio versus retention time will not be a rectangle. An irregular peak top is usually an indication of impurities. This could be easily proved by Lambert–Beer's law:

$$A = kC \tag{17}$$

where A is absorbance, C is concentration, and k is a constant.

When color a elutes out of the column, the concentration changes because of the dispersion. Assuming at time 1, the concentration is $C_1(a)$; at time 2, it is $C_2(a)$. Using $\lambda = 430$ nm, it could have

$$A_{430}[C_1(a)] = K_{430}(a) \times C_1(a) \tag{18}$$

$$A_{430}[C_2(a)] = k_{430}(a) \times C_2(a) \tag{19}$$

Using $\lambda = 575$ nm, we have

$$A_{575}[C_1(a)] = k_{575}(a) \times C_1(a) \tag{20}$$

$$A_{575}[C_2(a)] = k_{575}(a) \times C_2(a) \tag{21}$$

For color b,

$$A_{430}[C_1(b)] = K_{430}(b) \times C_1(b) \tag{22}$$

$$A_{430}[C_2(b)] = k_{430}(b) \times C_2(b) \tag{23}$$

$$A_{575}[C_1(b)] = K_{575}(b) \times C_1(b) \tag{24}$$

$$A_{575}[C_2(b)] = k_{575}(b) \times C_2(b) \tag{25}$$

Absorbance at a specific wavelength (i) and time is the summation of other A_i values. Using this example of 430 nm and time 1, if the eluate contains both colors,

$$A_{430} = A_{430}[C_1(a)] + A_{430}[C_1(b)] \tag{26}$$

If the eluate has only color a, $A_i[C_i(b)]$ equals to zero. At time 1,

$$A_{430} = A_{430}[C_1(a)] = k_{430}(a) \times C_1(a) \tag{27}$$

$$A_{575} = A_{575}[C_1(a)] = k_{575}(a) \times C_1(a) \tag{28}$$

$$A_{430}/A_{575} = k_{430}/k_{575} \tag{29}$$

The same reasoning can be applied at time 2:

$$A_{430}/A_{575} = k_{430}/k_{575} \tag{30}$$

The ratio is a constant, and it will not change with changing concentrations of the color. If the color is not pure, at time 1 we have

$$
\begin{aligned}
A_{430} &= A_{430}[C_1(a)] + A_{430}[C_1(b)] \\
&= K_{430}(a) \times C_1(a) + k_{430}(b) \times C_1(b)
\end{aligned} \tag{31}
$$

$$
\begin{aligned}
A_{575} &= A_{575}[C_1(a)] + A_{575}[C_1(b)] \\
&= K_{575}(a) \times C_1(a) + k_{575}(b) \times C_1(b)
\end{aligned} \tag{32}
$$

At time 2,

$$A_{430} = k_{430}(a) \times C_2(a) + k_{430}(b) \times C_2(b) \tag{33}$$

$$A_{575} = k_{575}(a) \times C_2(a) + k_{575}(b) \times C_2(b) \tag{34}$$

Obviously,

$$A_{430}/A_{575}(\text{time } 1) \neq A_{430}/A_{575}(\text{time } 2) \tag{35}$$

The dyes studied by LC method include direct, acid, basic, disperse, vat, and dyes used for cosmetics and food. Table 1 summarizes some of the methods for dye identification, separation, and purification. The detailed techniques can be found from the references in the table.

C. Dyeing Behavior

Using textile materials as a stationary phase, the dyeing behavior of a specific textile material can be obtained from retention of dye introduced into the column through an injector (VI in Figure 1), and frontal analysis of breakthrough curves can be obtained from running dye solutions with certain concentrations as mobile phases. Dye affinity, dyeing enthalphy, dye compatibility, dye sorption isotherm, and dye–fiber interactions all can be studied through LC.

1. Dyeing Thermodynamics

If we change the form of Eq. (2), we have

$$K = \frac{V_r - V_m}{V_s} \tag{36}$$

Table 1 LC Method for Dye Analysis

Dye class	Stationary phase	Eluent	Reference
Basic dye	Silica	Methanol/water buffered to pH 9.7 with ammonium acetate	[72]
Disperse dye	ODS	Acetonitrile/water (4/1) buffered to pH 3.2 with citric acid	[9]
	C18	Acetonitrile/water (70/30)	[5]
Direct dye	C18	Methanol/aqueous triethanolamine ion-pairing gradient	[73]
Vat dye	C18	0.05 M sodium acetate/0.05 M acetic acid in water (19%) and methanol (81%)	[74]
Acid dye	C18	Tetrabutylammonium ion pair	[58]
	Polystyrene/ divinyl benzene	Acetonitrile/water/citric acid/ t-butylammonium hydroxide, etc.	[59/77]
Cosmetic dye	C8	Acetonitrile/perchloric acid (pH 3) ion pair	[78]
	C18	Methanol/water/acetic acid (89/10/1); methanol/acetic acid/0.01 M tetrabutyl-ammonium hydroxide, pH 3.5 with phosphoric acid	[6]
Food dye	ODS	Methanol/aqueous sodium sulfate	[3]
	C18	Isopropanol aqueous solutions	[10]

where $V_m = V_o$, assuming the volume from injection to column inlet is negligible, and $V_s = V_t - V_o$, where V_t is the total solvent volume in the column.

The expression of affinity of a dye is

$$-\Delta\mu° = RT \ln K \tag{37}$$

where $-\Delta\mu°$ is the affinity, R is the gas constant, T is absolute temperature, and K is the distribution coefficient. Combining Eqs. (36) and (37), we can write:

$$-\Delta\mu° = RT \ln \frac{V_r - V_m}{V_s} \tag{38}$$

Thus, the dye affinity can be obtained from elution study by an LC system.

It is well known that

$$\frac{d \ln K}{dT} = \frac{\Delta H°}{RT^2} \tag{39}$$

where $\Delta H°$ is the heat of dyeing. Substituting $(V_r - V_m)/V_s$ for K,

$$\frac{d \ln [(V_r - V_m)/V_s]}{dT} = \frac{\Delta H°}{RT^2} \tag{40}$$

From integration we have:

$$\Delta H^\circ = \frac{RT_1T_2}{T_2 - T_1} \ln\left[\frac{(V_{r2} - V_{m2})/V_{s2}}{(V_{r_1} - V_{m_1})/V_{s_1}}\right] \tag{41}$$

where T_2 and T_1 are two different temperatures and V_{r_2}, V_{r_1}, V_{s_1}, V_{m_2}, and V_{m_1} are retention volumes, solvent volumes in the stationary phase(s), and solvent volumes in the mobile phase (m) at temperatures T_2 and T_1, respectively.

Assuming the change of V_m and V_s with temperature is negligible, Eq. (41) could be rewritten as:

$$\Delta H^\circ = \frac{RT_1T_2}{T_2 - T_1} \ln(\frac{V_{r_2} - V_m}{V_{r_1} - V_m}) \tag{42}$$

The successful application of textile LC to obtain dyeing thermodynamic parameters is reported by Grunwald et al. [31] and Achwal [24].

2. Dye Sorption Isotherm

Adsorption isotherms of dyes on textiles give thermodynamic parameters from which optimum dyeing conditions can be specified. The primary methods currently used to determine a dye adsorption isotherm [60,61] are based on (a) determining the concentration of the batch solution before and after dyeing equilibrium, (b) extracting and quantifying the adsorbed dye from the fiber after dyeing equilibrium has been established, and (c) immersing the dyed material in a dyebath of zero or small dye concentration, reaching the desorption equilibrium, and then determining the dye adsorbed by method a or b.

During the 1970s, Sharma and Fort [46] introduced the use of LC to study adsorption on fibers; most of the LC work published since that time has not addressed dye adsorption on textiles. Most of the adsorbates studied are much smaller in size than dyes [42,62–67]. The LC method is more accurate and sensitive than the classical batch method of shaking the adsorbent with the solution, measuring the change in solution concentration due to adsorption, and calculating the amount of solute adsorbed by the concentration difference [46,67]. Ladisch and Yang [32] developed a method of using whole fabric as the stationary phase to determine direct dye adsorption isotherms.

Two major approaches for determining adsorption isotherm measurements from liquid chromatography measurements are the minor disturbance method and the frontal analysis method [63].

a. Minor Disturbance Method. DeVault [68] gave the adsorption isotherm f(C) in terms of ideal equilibrium chromatography. The major assumption is that the equilibrium between solution and adsorbent is instantaneously established and that the effect of diffusion is negligible. The equation that gives x, the distance a band of dye has traveled from the column inlet, is

$$x = \frac{V}{V_0 / L + Mf'(C)} \tag{43}$$

where L is the column length, V is the volume of solution that has passed any given point since the initial time, V_0/L is the effective void volume per unit column length, M is the amount of adsorbent per unit length of the column, and f(C) is the function that gives the amount of solute adsorbed per unit weight of adsorbent, that is, the adsorption isotherm of the solute on the adsorbent; $f'(C)$ is the first-order derivative of f(C).

The linear velocity of the moving concentrated band along the column is given by u, while S is the flow rate of the eluent:

$$u = \frac{S}{V_0/L + Mf'(C)} \tag{44}$$

For a column of length L, V_0 is the total dead volume.

The retention volume V_r is LS/u, and the total amount of adsorbent in the column W is L $\times$ M. Therefore, Eq. (44) can be transformed into Eq. (45):

$$f'(C) = \frac{V_r - V_0}{W} \tag{45}$$

Integration of Eq. (45) gives

$$f(C_i) = \frac{V_{r,i} - V_0}{W}(C_i - C_{i-1}) + f(C_{i-1}) \tag{46}$$

where $f(C_i)$ is the amount of adsorbate adsorbed per unit weight of adsorbent at an equilibrium concentration of C_i. In the following discussion, $f(C_i)$ is replaced by the commonly used symbol $Q(C_i)$.

Experiments consist first of equilibrating the column with a solution of concentration C_i. A pulse of a concentration close to C_i is then injected, and the retention time of the pulse is used to obtain $V_{r,i}$ [63]. If $C_i = C_1$, and $C_{i-1} = 0$, Eq. (46) becomes

$$Q(C_1) = f(C_1) = \frac{V_r - V_0}{W}C_1 \tag{47}$$

For a linear isotherm, or at low dye concentration $(V_r - V_0)/W$ is a constant, which can be obtained by injecting a very dilute solution into the column. This method is applicable only when the assumptions of DeVault's equation are satisfied.

It was reported [32] that the minor disturbance method was unsuitable for studying direct dye sorption on cotton. The sorption data obtained by this method through Eq. (47) was too low compared to what was obtained from conventional batch study. This is probably because of the insufficient time the dyes have in the column for diffusion and obtaining equilibrium sorption.

b. *Frontal Analysis Method.*   Solute will be adsorbed when a solution of concentration C_i passes through a bed equilibrated with the same solution of $C_{i-1}(C_{i-1} < C_i)$. The concentration of the solute C eluting from the column initially has a lower concentration than C_i and gradually approaches C_i as the new equilibrium is attained. An example is given in Figure 5, which shows a breakthrough curve of CI direct green 26 using the cotton column described by Ladisch et al. [33].

From Figure 5, the total solute coming out of the column (q_0) is

$$q_0 = \int_0^{V_e} g(V) \, dV \tag{48}$$

where V_e is the elution volume at the point where the outlet concentration is equal to the inlet concentration, and $g(V)$ represents the solute concentration as a function of the volume of eluent. The total solute coming into the column is

$$q_T = C_i V_e \tag{49}$$

and the solute retained in the column (q_R) is the difference:

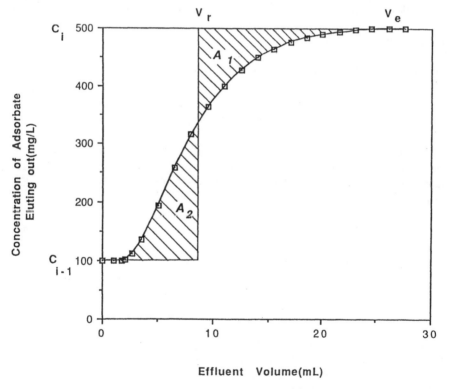

Figure 5 Breakthrough curve of CI direct green 26 with a concentration of 500 mg/L at 30°C. The cotton column was previously equilibrated with a 100 mg/L dye solution. V_r is the eluent volume at which the shadowed area $A_1 = A_2$. [32]

$$q_R = q_T - q_0 = C_i V_e - \int_0^{V_e} g(V) \, dV \tag{50}$$

The solute adsorbed q is the difference between the total retained solute q_R and the net solute retained in the void volume of the column, q_v:

$$
\begin{aligned}
q &= q_r - q_v \\
&= C_i V_e - (C_i - C_{i-1}) V_0 - \int_0^{V_e} g(V) \, dV
\end{aligned} \tag{51}
$$

where

$$q_v = (C_1 - C_{i-1}) V_0 \tag{52}$$

$C_{i-1} V_0$ represents the solute retained in the column void volume from the previous run done at an inlet concentration of C_{i-1}. If the total weight of the adsorbent in the bed is equal to W, the solute adsorbed per unit weight of the adsorbent is q/W. If the unit adsorption from the previous study ($C = C_{i-1}$) is Q_{i-1}, then the total adsorption at C_i is

$$
\begin{aligned}
Q_i &= q/W + Q_{i-1} \\
&= 1/Wt[C_i V_e - (C_i - C_{i-1}) V_0 - \int_0^{V_e} g(v) \, dV] + Q_{i-1}
\end{aligned} \tag{53}
$$

Assuming at V_e the adsorption in the column was equilibrated at C_i, then Q_i expressed in Eq. (53) gives the equilibrium adsorption of the solute at concentration C_i.

Equilibrium conditions may only be assured if the lowest possible flow rate is used or the eluent flow is periodically stopped once V_e is obtained. Equation (48) would then be rewritten as

$$q_0 = \sum_{j=1}^{n} \int_{V_{ej-1}}^{V_{ej}} g_j(V) \, dV \tag{54}$$

where N is the number of runs made until true equilibrium is attained. The total amount of dye adsorbed is additive; the volume that elutes for each run is represented by the difference $V_{ej} - V_{ej-1}$, where $V_{e0} = 0$. To calculate q_0 from either Eq. (48) or (54), the area above the breakthrough curve g(V) is usually integrated directly. Alternately, the elution volume V_r, often called the retention volume, is the volume at which the areas A_1 and A_2 are equal and can be used to obtain q_R (see Fig. 5),

$$q_R = \sum_{j=1}^{n} V_{rj} \, \Delta C_j \tag{55}$$

where

$$\Delta C_j = C_i - C_{ij} = C_i - g_j(V_{ej-1}) \tag{56}$$

V_{rj} is illustrated in Figure 5. If n = 1 (i.e., only one run is done at a given concentration), then this results in

$$V_{e0} = 0 \qquad g(0) = C_{i-1} \qquad \text{and} \qquad \Delta C = C_i - C_{i-1}$$
$$q_r = V_r(C_i - C_{i-1}) \tag{57}$$

and

$$q = (V_r - V_0)(C_i - C_{i-1}) \tag{58}$$

Thus

$$Q_i = \frac{V_r - V_0}{W}(C_i - C_{i-1}) + Q_{i-1} \tag{59}$$

Using frontal analysis, Ladisch and Yang [32] obtained equilibrium sorption isotherms for direct red 81 and direct green 26 at both 30 and 60°C on cotton. As presented in Figure 6, LC results were very close to that from conventional batch adsorption determination.

3.   Dye Compatibility

Combining several dyes in the same bath for textile dyeing is very common, so knowledge of the behavior, or compatibility, of all the dyes in the mixture is necessary to obtain the expected hue on the fabric. The definition of compatibility is "the propensity of individual dye components in a combination shade to exhaust at similar rates resulting in a buildup of shade that is constant, or nearly constant, in hue throughout the dyeing process" [69].

Compatibility is a problem for almost all dye classes, such as basic, acid, direct, disperse, reactive, and sulfur [70]. This problem is most serious in the basic dyeing of acrylic fibers, since basic dyes show virtually no migration in acrylic fibers under normal dyeing conditions [71]. Therefore, using compatible dyes, especially for the basic dyeing of acrylic, is very important. The most commonly used constant to evaluate dye compatibility is the so-called "compatibility value." In general, there are five compatibility values assigned to dyes, 1 to 5. For a more precise evaluation, the compatibility value of a dye can lie between two adjacent standards, such as 1.5, 2.5, 3.5, or 4.5. The higher the compatibility value of the dye, the lower the affinity of the dye for the fiber. This rating system provides a guideline for selecting dyes to be used in dyebath mixtures. Ideally, dyes used in mixtures should have the same compatibility values.

The standard method used to evaluate the compatibility of basic dyes entails determining the dyeing behavior of the dye in question when combined with each of five standard dyes having predetermined compatibility values of 1, 2, 3, 4, and 5. The experiment with each standard dye/unknown dye mixture involves dyeing four to six fabrics one after another in the same dyebath at prescribed time intervals. A total of five dye baths, one for each level of compatibility, and 20 to 30 pieces of fabric make up one evaluation for one unknown dye. The compatibility

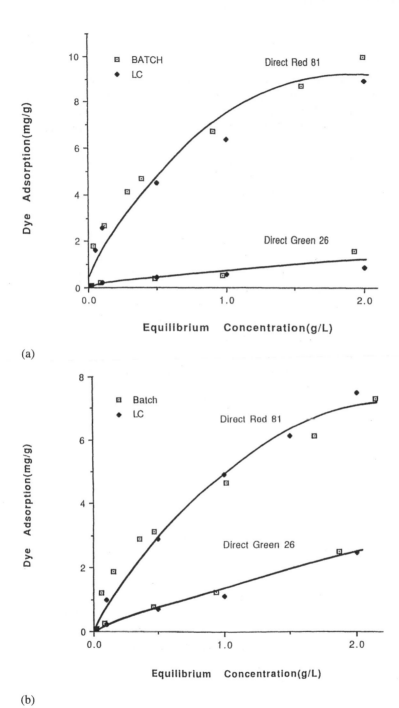

(a)

(b)

Figure 6 Comparison of adsorption isotherms obtained from LC and batch adsorption methods on cotton fabric at (a) 30°C and (b) 60°C for CI direct red 81 and direct green 26. Curves represent the polynomial regression of both batch and LC data [32].

value of the dye in question is that of the standard dye with which it gives on-tone dyeings throughout the sequence. This method has been adopted by both the American Association of Textile Chemists and Colorists (AATCC) and the Society of Dyers and Colourists (SDC) [69,70] and is used in both laboratories and industries throughout the world.

A method of using fabric LC to determine the compatibility value based on the theory of frontal analysis was developed by Yang and Ladisch [36]. Instead of preparing five dye baths and dyeing 20 to 30 pieces of fabric, one breakthrough study could achieve the compatibility value of a dye.

If two dyes are totally compatible and their initial concentrations are the same, their concentrations in the dye bath will be the same at any time throughout the dyeing. If the same fabric is packed in an LC column and a dye solution is passed through the column, both dyes would again show the same rates of adsorption if they were compatible. The rate of dye sorption would be inversely proportional to the dye eluting from such a column. Thus, the study of compatibility can be related to the frontal analysis of breakthrough curves of LC.

Similar breakthrough curves for two dyes in a mixture indicate compatibility. If the compatibility value C_v of a dye is desired, the breakthrough curve of that dye can be compared with that of each of the five standard dyes described previously. The breakthrough curve of the standard dye that most closely matches the unknown indicates the C_v value of the unknown. This procedure still requires a minimum of five experiments.

The C_v value can also be related to the difference in retention of two dyes in an LC column.

$$R = (V_r^0 - V_r)/\overline{V}_r \tag{60}$$

$$\overline{V}_r = (V_r^0 + V_r)/2 \tag{61}$$

$$\Delta C_v = C_v^0 - C_v \tag{62}$$

R is a factor that reflects the difference in the retention volume of the known and unknown dyes. Since both dyes would be passed through the same column, R is independent of the amount of fabric used for the test, and is also independent of the concentrations of the dyes used if the adsorption isotherms of these two dyes are parallel with each other within the concentration range studied. V_r^0 is the retention volume of the known dye with a compatibility value of C_v^0; V_r is the retention volume of the unknown dye with a compatibility value of C_v, which needs to be determined. Retention volume from a breakthrough curve was calculated the same way as discussed in Section III.C.2.a. (cf. Figure 5). V_r is the average retention volume of the known and unknown.

Obviously, ΔC_v is related to R:

$$\Delta C_v = f(R) \tag{63}$$

where f(R) is a function of R.

Equations (62) and (63) give the compatibility value C_v as

$$C_v = C_v^0 - f(R) \qquad (64)$$

Using an LC system as shown schematically in Figure 7, two breakthrough curves, one for the standard dye (with known C_v^0), the other for the test dye, could be obtained form one test. If the real expression of $f(R)$ in Eq. (64) is known, the compatibility value of the tested dye could be calculated from Eq. (64). From the work of Yang and Ladisch [36],

$$f(R) = 0.108 - 2.484R \qquad (65)$$

Examples of their study on basic dye compatibility are shown in Figure 8. The compatibility values of 10 standard dyes obtained by the LC method were compared with that from AATCC standard values in Table 2. The results of the LC method matched well those of the ATCC Test Method 141-1984 [69].

4. Dye–Fiber Interaction

Textile LC could also be used to study the dye–fiber interactions. The dye to be tested is injected into an LC system using fabric or fiber being studied as stationary phase (cf. Fig. 1). By changing the properties of eluent, and studying the retention volume and shape of the peak of the dye recorded from the outlet of the column, the dye–fiber interactions could be studied.

Yang and Ladisch [35] studied the interactions between cationic dyes and acrylic fiber by this method. Using aqueous solutions of concentrated salt (e.g., 2 M NaCl), organic compounds with different sizes of hydrophobic parts, and a combination of both the salt and organic compounds to study the retention of cationic dyes in acrylic column, they found that both ionic and hydrophobic interactions were important for cationic dyeing of acrylic fibers.

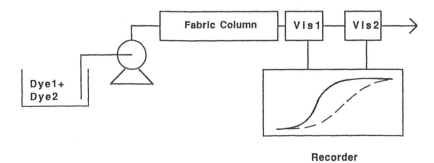

Figure 7 Schematic diagram of LC instrument for the study of basic admixture dyeing. Vis 1, UV-visible light detector with wavelength equal to λ_{max} of dye 1. Vis 2, UV-visible light detector with wavelength equal to λ_{max} of dye 2 [36].

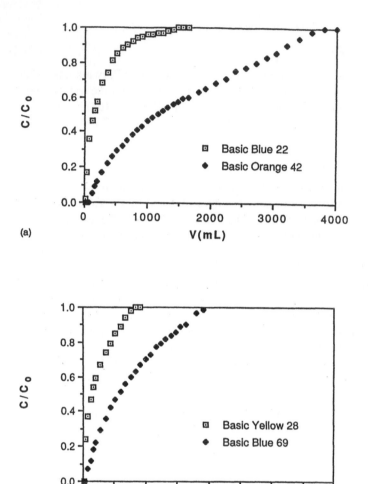

Figure 8 Breakthrough study on basic dye compatibility [36].

D.   Finish Evaluation

LC has also been used for the study of finishes and finishing processes. It was applied to soil resist finishing [13], flame-resist finishing [23], durable press finishing [11,12,14–18], measurement of low level of formaldehyde [20,21], and low-molecular-weight additives and oligomers in synthetic fiber spinning [22].

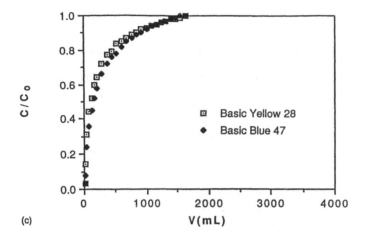

(c)

Table 3 summarized some of the LC systems for the analyses of different finishes and finishing processes.

Further information about LC analysis of textile finishes and finishing processes can be found in the chapter in this volume written by K. R. Beck, a pioneer in applying LC to the study of durable-press finishing.

Table 2   Compatibility Values of Standard Basic Dyes Obtained from AATCC Test Method 141-1987 and the Rolled Fabric Liquid Chromatography Column [69]

CI basic dye	Compatibility value, C_v	
	AATCC	LC
Blue 69	1	0.7
Blue 45	2	1.8
Blue 47	3	2.7
Blue 77	4	4.3
Blue 22	5	4.6
Orange 42	1	1.0
Yellow 29	2	1.7
Yellow 28	3	2.7
Yellow 15	4	3.9
Orange 48	5	4.6

Table 3 LC Method for Finish Analysis

Chemical	Stationary phase	Eluent	Detector	Reference
Formaldehyde[a]	C18	Acetonitrile/ water (60/40)	UV 340 nm	[20,21]
Sulfonated aromatic compounds finish	C18	Water/acetonitrile (90/10 to 10/90 with linear gradient elution)	UV 254 nm	[13]
Tris(2,3-dibromopropyl) phosphate finish	C18	Methanol/water (70/30)	UV 254 nm	[23]
Poly-*m*-phenylene isophthalamide fiber precipitation and plasticizing baths	C18	Tetrahydrofuran/ water (53/47)	UV 260 nm	[22]
M-Methylolpyrolidone finish	C18	Water/methanol (70/130)	RI	[17]
DMDHEU finish[b]	Cationic exchange resin	Water	RI	[14,15, 19]
	C18	Water	RI	[11,12]
DMEU[c]	C18	Water	RI	[16]

[a]Formaldehyde first reacts with 2,4-dimitrophenyl hydrazine to form the corresponding hydrazone, which is more easily detectable than formaldehyde itself.
[b]DMDHEU, dimethyloldihydroxyethyleneurea.
[c]DMEU, dimethylolethyleneurea.

IV. CONCLUSIONS

As an analytical tool, LC has penetrated the textile industry rapidly and is being utilized widely, especially in the chemistry related areas. Although colorant analysis was its original use, it has found many and varied areas of applications from characterization of internal pore structures of fibers to the effects of processing treatments and conditions on fiber/treatment interactions.

REFERENCES

1. K. R. Beck, B. F. North, and C. M. Player, Jr., Analytical instrumentation in the textile industry, *Textile Chem. Color.* 21(10):16–17 (1989).
2. R. P. W. Scott, *Liquid Chromatography for the Analyst*, Marcel Dekker, New York, 1994, pp. 2–3.
3. J. P. Clayton and R. L. Heal, Separation of synthetic dyes by high-performance liquid chromatography on 3-μm columns, *J. Chromatogr.* 368:450–455 (1986).
4. K. P. Evans and N. J. Truslove, Advances in chromatography for dyestuff, *Rev. Prog. Coloration* 23:36–39 (1993).

5. Z. Ma and C.-P. Yen, Identification of disperse dyes on fabrics by reverse phase chromatography, Book of Papers, AATCC Int. Conf. Exhibition, American Association of Textile Chemists and Colorists, Research Triangle Park, NC, 1988, pp.40–46.

6. A. -M. Sjöberg and C. Olkkonen, Determination of synthetic organic colours in lipsticks by thin-layer and high-performance liquid chromatography, *J. Chromatogr.* *318*:149–154 (1985).

7. W. A. Straw, Principles of chromatography and separative techniques—Adsorption and partition chromatography, *J. Soc. Dyers Colour.* *101*(12):409–416 (1985).

8. P. Y. Wang and I. J. Wang, Photolytic behavior of some azo pyridone disperse dyes on polyester substrates, *Textile Res. J.* 62(1):15–20 (1992).

9. B. B. Wheals, P. C. White, and M. D. Paterson, High-performance liquid chromatographic methods utilizing single or multi-wavelength detection for the comparison of disperse dyes extracted from polyester fibres, *J. Chromatogr.* *350*:205–215 (1985).

10. M. L. Young, Rapid identification of color additives, using the C18 cartridge: Collaborative study, *J. Assoc. Off. Anal. Chem.* 71(3):458–461 (1988).

11. D. M. Pasad and K. R. Beck, Quantitative analysis of commercial DP finishing agents, *Textile Chem. Color.* *18*(5):27–32 (1986).

12. D. M. Pasad and K. R. Beck, Influence of reagent residues and catalysts on formaldehyde release from DMDHEU-treated cotton, *J. Appl. Polym. Sci.* *34*:549–558 (1987).

13. M. Bauers, R. W. Keown, and C. P. Malone, Separating and identifying the active ingredient of a stain resist compound, *Textile Res. J.* 63(9):540–544 (1993).

14. K. R. Beck and D. M. Pasad, The effect of pad-bath pH and storage period on the hydrolysis of DMDHEU, *Textile Res. J.* 52(4):269–274 (1982).

15. K. R. Beck and D. M. Pasad, Liquid-chromatographic determination of rate constants for the cellulos-dimethyloldihydroxyethyleneurea reaction, *J. Appl. Polym. Sci.* 27:1131–1138 (1982).

16. K. R. Beck and D. M. Pasad, Reagent residues of DMEU on cotton fabric as a function of pad-bath pH and storage period of the treated fabric, *Textile Res. J.* *53*(9):524–529 (1983)

17. K. R. Beck, D. M. Pasad, and K. S. Springer, High performance liquid chromatographic analysis of durable press finishes, *Textile Chem. Color.* *16*(5):15–18 (1984).

18. K. R. Beck and D. M. Pasad, Regeant residues on *N*-methylolpyrrolidone-treated cotton, *J. Appl. Polym. Sci.* 29:3579–3585 (1984).

19. K. R. Beck, B. J. Leibowitz, and M. R. Ladisch, Separation of methylol derivatives of imidazolidines, urea and carbomates by liquid chromatography, *J. Chromatogr.* *190*:226–232 (1980).

20. D. A. Ernes, An alternative method for formaldehyde determination in aqueous solutions, *Book of Paper, American Association of Textile Chemists and Colorists International Conference and Exhibition*, 1984, pp. 209–211.

21. D. M. Pasad and C. L. Cochran, Optimization of the AATCC sealed jar and HPLC methods for measurement of low levels of formaldehyde, *Textile Chem. Color.* *21*(6): 13–18 (1989).

22. T. I. Podol'skaya, P. V. Smirno, N. I. Kuz'min, K. G. Khabarova, N. M. Kvasha, and A. S. Chegolya, Identification and quantitative determination of low-molecular-weight compounds and precipitation and plasticizing baths, and the effect of monomer- and solvent-contamination on their composition, *Fiber Chem.* 23(5):371–374 (1992).

23. T. L. Smith and B. N. Whelihan, Determination of surface tris (2,3-dibromopropyl) phosphate on FR polyester fabrics, *Textile Chem. Color.* *10*(5):35–37 (1978).
24. W. B. Achwal, Assessment of cellulosic dyeing parameters by a fiber column chromatographic method, *Colourage* *41*(1):21–22 (1994).
25. W. B. Achwal, Assessment of cellulosic dyeing parameters by a fiber column, *Colourage* *40*(4):16–17 (1993).
26. N. R. Bertoniere, W. D. King, and C. M. Welch, Effect of catalyst on the pore structure and performance of cotton cellulose cross-linked with butanetetracarboxylic acid, *Textile Res. J.* *64*(5):247–255 (1994).
27. N. R. Bertoniere and W. D. King, Pore structure of cotton fabrics cross-linked with formaldehyde-free reagents, *Textile Res. J.* *62*(6):349–356 (1992).
28. N. R. Bertoniere, Pore structure analysis of cotton cellulose via gel permeation chromatography, in *Cellulose, Structural and Functional Aspects* (J. F. Kennedy, G. O. Phillips, and P. A. Williams, eds.), Ellis Horwood Limited, Chichester, West Sussex, England, 1989, pp. 99–104.
29. N. R. Bertoniere and W. D. King, Pore structure and dyeability of cotton cross-linked with DMDHEU and with DHDMI, *Textile Res. J.* *59*(10):608–615 (1989).
30. K. Bredereck and A. Bluher, Determination of the pore structure of cellulose fibres by exclusion chromatography, *Melliand Textilberichte* *73*(8):652–662, E297–E302 (1992).
31. M. Grunwald, E. Burtscher, and O. Bobleter, HPLC determination of the pore distribution and chromatography properties of cellulosic textile materials, *J. Appl. Polym. Sci.* *39*(2):301–317 (1990).
32. C. M. Ladisch and Y. Yang, A new approach to the study of textile dyeing properties with liquid chromatography, Part I: Direct dye adsorption on cotton using a rolled fabric stationary phase, *Textile Res. J.* *62*(8):481–486 (1992).
33. C. M. Ladisch, Y. Yang, A. Velayudhan, and M. R. Ladisch, A new approach to the study of textile properties with liquid chromatography, comparison of void volume and surface area of cotton and ramie using a rolled fabric stationary phase, *Textile Res. J.* *62*(6):361–369 (1992).
34. S. P. Rowland, C. P. Wade, and N. R. Bertoniere, Pore structure analysis of purified, sodium hydroxide-treated and liquid ammonia-treated cotton celluloses, *J. Appl. Polym. Sci.* *29*:3349–3357 (1984).
35. Y. Yang and C. M. Ladisch, Hydrophobic interaction and its effect on cationic dyeing of acrylic fabric, *Textile Res. J.* *63*(5):283–289 (1993).
36. Y. Yang and C. M. Ladisch, A new approach to the study of textile dyeing properties with liquid chromatography, Part II: Compatibility of basic dyes for acrylic fabric, *Textile Res. J.* *62*(9):531–535 (1992).
37. Anonymous, Supercritical fluids attracting new interest, *Inform* *1*(9):810–820 (1990).
38. W. P. Jackson and D. W. Later, Analysis of commercial dyes by capillary column supercritical fluid chromatography, *J. High Resolut. Chromatogr.* *9*:175 (1986).
39. M. Czok and G. Guiochon, Aligned fiber columns for size-exclusion chromatography, *J. Chromatogr.* *506*:303–317 (1990).
40. H. Ding and E. L. Cussler, Overloaded hollow-fiber liquid chromatography, *Biotechnol. Prog.* *6*:472–478 (1990).
41. H. Ding, M.-C. Yang, D. Schisla, and E. L. Cussler, Hollow-fiber liquid chromatography, *AICHE J.* *35*(5):814–820 (1989).

42. Y. Kiso, K. Jinno, and T. Nagoshi, Liquid chromatography in a capillary packed with fibrous cellulose acetate, *J. High Resolut. Chromatogr. Commun* 9(12): 763–764 (1986).

43. P. Wikstrom and P.-O. Larsson, Affinity fibre—a new support for rapid enzyme purification by high-performance liquid affinity chromatography, *J. Chromatogr.* 388: 123–134 (1987).

44. Y. Yang, A. Velayudhan, C. M. Ladisch, and M. R. Ladisch, Protein chromatography using a continuous stationary phase, *J. Chromatogr.* 598:169–180 (1992).

45. Y. Yang, A. Velayadhan, C. M. Ladisch, and M. R. Ladisch, Liquid chromatography using cellulosic continuous stationary phases, in *Advances in Biochemical Engineering Biotechnology*, Vol. 49, *Chromatography* (G. T. Tsao, ed.), Springer-Verlag, Berlin, Germany, 1993, pp. 147–160.

46. S. C. Sharma and T. Fort, Jr., Adsorption from solution via continuous flow frontal analysis solid-liquid chromatography, *J. Colloid Interface Sci.* 43:36–42 (1973).

47. R. F. Meyer, P. B. Champlin, and R. A. Hartwick, Theory of multicapillary columns for HPLC, *J. Chromatogr. Sci.* 21(10):433–438 (1983).

48. R. D. Hegedus, The dependence of performance on fiber uniformity in aligned fiber HPLC columns, *J. Chromatogr. Sci.* 26(9):425–431 (1988).

49. J. J. Kirkland, *J.* High performance liquid chromatography with porous silica microsphere, *Chromatogr. Sci.* 10:129 (1972).

50. J. Qin, Z. Lin, M. Yang, P. Mao, and F. Li, Relation between microporous structure of PAC/AS blend fibers and sequence distribution of AS copolymers, *Textile Res. J.* 57:433–439 (1987).

51. S. Chen, Dyeing behaviors of cationic dyes on porous polyacryl-nitrile fibers, *J. East China Inst. Textile Sci. Technol.* 2:47–53 (1982).

52. P. Mao, Z. Liu, B. Zhao, M. Yang, and F. Li, Study of relationship between dyeing properties and structure of acrylic fiber, *Synth. Fibers* 4:17–22 (1985).

53. N. I. Klenkova and G. P. Ivaskin. On the internal surface and capillary structure of natural and mercerized cotton cellulose, *J. Appl. Chem. USSR* 36:378–387 (1963).

54. L. G. Aggebrandt and O. Samuelson, Penetration of water soluble polymers into cellulose fibers, *J. Appl. Polym. Sci* 8:2801–2812 (1964).

55. R. Nelson and D. W. Oliver. Study of cellulose structure and its relation to reactivity, *J. Polym. Sci. Part C* 36:305–320 (1971).

56. J. E. Stone and A. M. Scallan, A structural model for the cell wall of water-swollen wood pulp fibers based on their accessibility to macromolecules, *Cell. Chem. Technol.* 2:343–358 (1968).

57. J. H. Knox and H. P. Scott, Theoretical models for size-exclusion chromatography and calculation of pore size distribution from size-exclusion chromatography data, *J. Chromatogr.* 316:311–332 (1984).

58. K. M. Weaver and M. E. Neale, High-performance liquid chromatographic detection and quantitation of synthetic dyes with a diode array detector, *J. Chromatogr.* 354:486–489 (1986).

59. P. C. White and A. M. Harbin, High performance liquid chromatography of acidic dyes on a dynamically modified polystyrene divinylbenzene packing material with multiwavelength detection in the absorbance ratio characterization, *Analyst (Lond.)* 114:877 (1989).

60. R. H. Peters, *Textile Chemistry, III: The Physical Chemistry of Dyeing*, Elsevier Scientific, Amsterdam, 1975, pp. 61–65.
61. T. Vickerstaff, *The Physical Chemistry of Dyeing*, Oliver and Boyd, London, 1954, pp. 91–96.
62. S. A. Busev, S. I. Zverev, O. G. Larionov, and E. S. Jakubov, Study of adsorption from solutions by column chromatography, *J. Chromatogr. 241*:287–294 (1982).
63. Y. A. Eltekov and Y. V. Kazakevitch, Comparison of various chromatographic methods for the determination of adsorption isotherms in solutions, *J. Chromatogr. 395*: 473–480 (1987).
64. Y. A. Eltekov and Y. V. Kazakevitch, Investigation of adsorption equilibrium in chromatographic columns by the frontal method, *J. Chromatogr. 365*:213–219 (1986).
65. J. Jacobson, J. Frenz, and C. Horváth, Measurement of adsorption isotherms by liquid chromatography, *J. Chromatogr. 316*:53–68 (1984).
66. A. E. Osawa and C. L. Cooney, Abstract of paper MBTD-3, The Use of Fibrious Beds for Chromatographic Separation of Proteins, presented at Miami, Fla., American Chemical Society Meeting, September 11, 1989.
67. H. L. Wang, J. L. Duda, and C. J. Radke, Solutions adsorption from liquid chromatography, *J. Colloid Interface Sci. 66*:153–165 (1978).
68a. D. DeVault, The theory of chromatography, *J. Am. Chem. Soc. 65*:532–540 (1943).
68b. J. N. Wilson, A theory of chromatography, *J. Am. Chem. Soc. 62*:1583–1591 (1940).
69. Technical Manual, American Association of Textile Chemists and Colorists, Research Triangle Park, N.C., *AATCC Test Method 141-1987, Compatibility of Basic Dyes for Acrylic Fibers*, Vol. 64, 248–249, 1989.
70. F. Hoffman, Compatibility of dyes, *Rev. Prog. Color. 18*:56–64 (1988).
71. D. G. Evans and C. J. Ben, Tentative tests for evaluation of the dyeing properties of basic dyes on acrylic fibres, *J. Soc. Dyers Colour. 87*:60, 61 (1971).
72. R. M. E. Griffin, T. G. Kee, and R. W. Adams, High-performance liquid chromatographic system for the separation of basic dyes, *J. Chromatogr. 445*(2):441–448 (1988).
73. A. Shan, D. Harbin, and C. W. Jameson, Analysis of two azo dyes by high-performance liquid chromatography, *J. Chromatogr. Sci. 26*(9):439–442 (1988).
74. J. L. Allen and J. R. Meinertyz, Post-column reaction for simultaneous analysis of chromatic and leuco forms of malachite green and crystal violet by high-performance liquid chromatography and photometric detection, *J. Chromatogr. 536*:217–222 (1991).
75. P. C. White and T. Catterick, Use of color coordinates and peak parity parameters for improving the high performance liquid chromatographic qualitative analysis of dyes, *Analyst 115*:919 (1990).
76. L. Gagliardi, G. Cavazzutti, A. Amato, A. Basili, and E. Tonelli, Identification of cosmetic dyes by ion-pair reversed-phase high-performance liquid chromatography, *J. Chromatogr. 394*:345–352 (1987).
77. M. R. Ladisch, Separation by sorption in *Advanced Biochemical Engineering*, (H. R. Bungay and G. Belfort eds.), John Wiley and Sons, New York, 1987, pp. 219–237.

6
Evaluation of DP Finishes by Chromatographic and Spectroscopic Methods

KEITH R. BECK   College of Textiles, North Carolina State University, Raleigh, North Carolina

I. INTRODUCTION

Chemicals that cross-link cellulose to generate restorative forces for crease retention and smooth drying are called durable press (DP) agents or DP finishes. Analysis of these materials is important for both their production and their application. Chromatography yields both qualitative and quantitative information and provides separation capability for further analyses, such as mass spectrometry. Spectroscopic analysis typically generates information about the nature of the components in the finish. This chapter describes both chromatographic and spectroscopic analyses of DP agents.

II. BACKGROUND INFORMATION ON DP AGENTS

Modern durable press (DP) treatments of cotton involve the application of cross-linking agents to the cellulosic polymer in order to impart smooth drying, wrinkle resistance, and/or crease retention. The very early durable press finishing agents were urea-formaldehyde (UF) and melamine-formaldehyde (MelF) products. These materials formed three-dimensional polymeric networks inside the fiber and were legitimately called resins. When 1,3-dihydroxymethyl-4,5-dihydroxy-2-imidazolidinone (more commonly known as dimethyloldihydroxyethyleneurea or DMDHEU) was introduced in 1964 as a promising new durable-press finish, the term DP resin incorrectly persisted. Unlike the UF and MelF finishes, DMDHEU did not react with itself to form a network polymer, but formed nearly monomeric cross-links between cellulose molecules as shown in Figure 1.

Synthesis of DMDHEU is accomplished [1] by reacting urea, glyoxal, and formaldehyde as shown in Figure 2. Reductions in the amount of formaldehyde released by fabrics finished with DMDHEU have been accomplished through improved control of curing conditions, better catalysts, and by reacting DMDHEU with alcohols or polyols to convert one or more of the *N*-hemiacetal groups to an acetal. The resulting "capped" finishes are the dominant cross-linking finishes in today's textile market.

Because of the concern over formaldehyde released from DP-finished fabrics, a search for formaldehyde-free cross-linkers has been in progress for several years. One solution to this problem has been to utilize glyoxal-based finishes, such as 1,3-dimethyl-4,5-dihydroxy-2-imidazolidinone (DMDHI), that do not contain formaldehyde. Cross-linking of cellulose by DMDHI is shown in Figure 3.

Polycarboxylic acids, such as 1,2,3,4-butanetetracarboxylic acid (BTCA), have also been shown to be effective nonformaldehyde cross-linking agents. Figure 4 shows the cross-linking of cellulose by BTCA.

Information in this chapter emphasizes analysis of DMDHI, DMDHEU and its derivatives, and polycarboxylic acids. Because of its importance in durable-

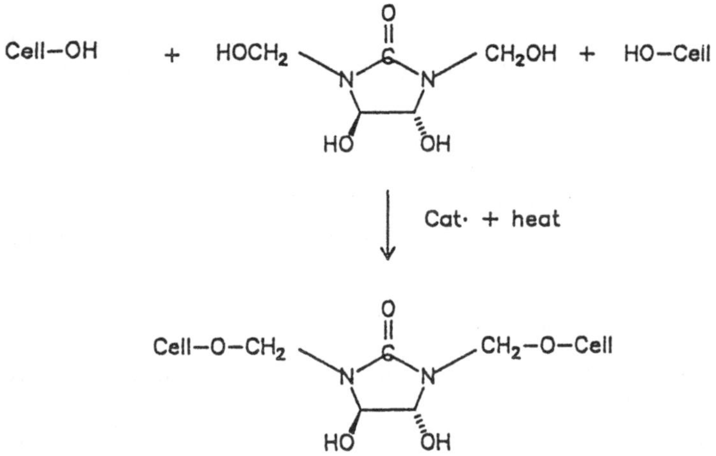

Figure 1   Cross-linking of cellulose with DMDHEU.

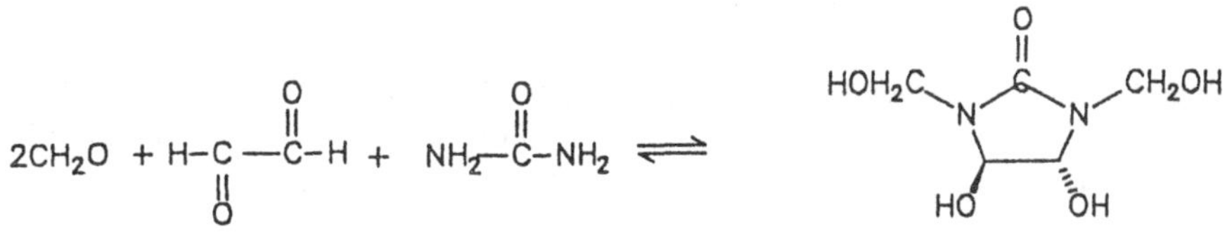

Figure 2   Synthesis of DMDHEU.

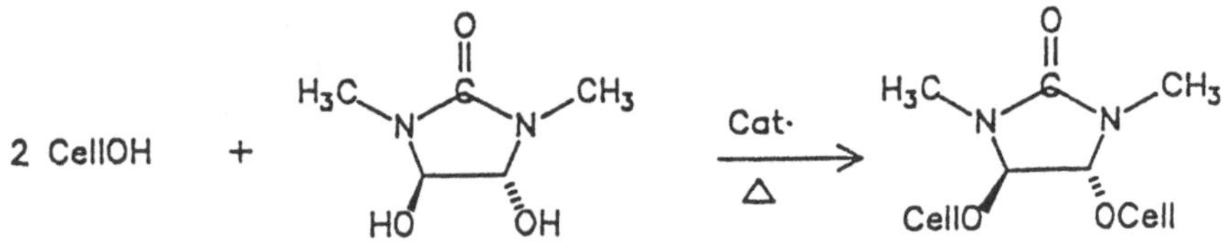

Figure 3 Cross-linking of cellulose by DMDHI.

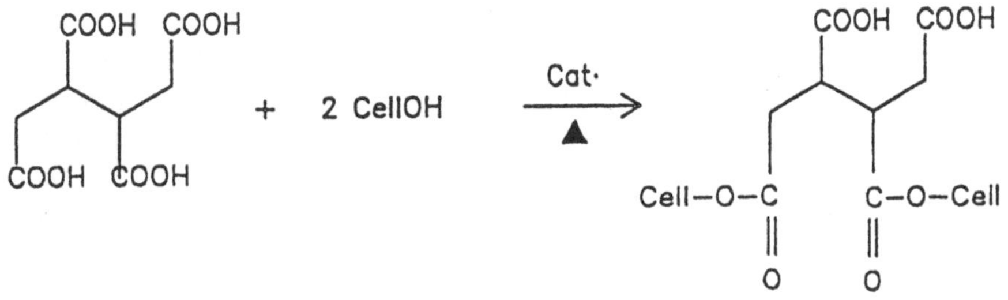

Figure 4 Cross-linking of cellulose by BTCA.

press finishing, appropriate references for formaldehyde analysis are discussed. Analytical techniques include thin-layer chromatography (TLC), gas chromatography (GC), high-performance liquid chromatography (HPLC), ultraviolet-visible spectroscopy (UV-VIS), mid-infrared spectroscopy (IR), near-infrared spectroscopy (NIR), ^{1}H- and ^{13}C-nuclear magnetic resonance spectrometry (NMR), and mass spectrometry (MS).

III. CHROMATOGRAPHIC ANALYSIS OF DURABLE-PRESS AGENTS

A. Thin-Layer Chromatography

Player and Dunn [2] gave a brief introduction to TLC and offer information on several textile applications. More complete descriptions of this technique are available in books, such as those by Fried and Sharma [3] and Touchtone [4]. Because of their relatively high polarity, DP agents are not routinely analyzed by TLC. In an extensive analysis of a variety of DP precursors, Valk and co-workers [5] separated 15 different compounds, such as dihydroxyethyleneurea (DHEU), melamine, and urea, using a chloroform/methanol/water mobile phase and a cellulose stationary phase. Carbamate precursors were separated with carbon tetrachloride/methylene chloride/ethyl acetate/formic acid on silica gel. Moore and Babb [6] used alcohol/water mobile phases and cellulose as a TLC stationary phase to effect separations of 2-imidazolidinone (EU), 1,3-dihydroxymethyl-2-imidazolidinone (DMEU), 4,5-dihydroxy-2-imidazolidinone (DHEU), DMDHEU, tetramethylated DMDHEU, and 4,5-dimethoxyEU. Because the DP agents were reacted with the stationary phase (cellulose), this method was used to compare reactivity of these materials. Methyl carbamate (MC), N-hydroxymethyl methylcarbamate (MMMC), and N, N-dihydroxymethyl methylcarbamate (DMMC) (Fig. 5) were isolated by thick layer chromatography (chloroform, acetone, 2-propanol on Adsorbosil-1) [7]. In a preliminary portion of this work, Cashen visualized the eluted spots by first hydrolyzing with a sulfuric acid spray followed by acidic chromotropic acid spray.

Rennison [8] identified DMDHI in hydrolysates from fabrics finished with that reagent by eluting the products on silica with 1-propanol/water. Kantschev and Nesnakomova [9] monitored and optimized the synthesis of DHEU from glyoxal and urea with TLC. In an interesting application of TLC for studying DP chemistry, Chen [10] coated an aluminum plate with cellulose to study the interactions between tartaric acid and aluminum sulfate. The tartaric acid alone gave a different retention factor than the aluminum sulfate and mixtures of these two catalyst components. He concluded that there is an interaction between the aluminum ion and the hydroxyl groups of the acid.

B. Gas Chromatography

Basic information on gas chromatography and some qualitative and quantitative textile applications are given by Player and Dunn [2]. Instrumental and theoretical details as well as practical information on gas chromatography are presented by Baugh [11]. Because of the presence of hydroxyl and amido NH groups in their structure, DP agents are not sufficiently volatile to be analyzed by gas chromatography. Replacement of hydrogen in the HO and NH groups with trimethylsilyl renders the DP agents sufficiently volatile to be chromatographically separated. Bullock and Rowland [12] separated the N-SiMe$_3$, and N-CH$_2$OSiMe$_3$ derivatives of 1-methyl-2-imidazolidinone on a 2 ft $\times$ 0.25 in column packed with GE-XE-60. Divatia et al. [13] added acetone to solutions of DP agents to precipitate the active ingredients. These materials were then silylated in pyridine with N,O-bis(trimethylsilyl)acetamide (BSA) and analyzed on a 6 ft $\times$ 0.25 in column packed with SE-52. In preparation for mass spectrometric analysis of 4,5-dihydroxy-2-imidazolidinone (DHEU), 1-hydroxymethyl-4,5-dihydroxy-2-

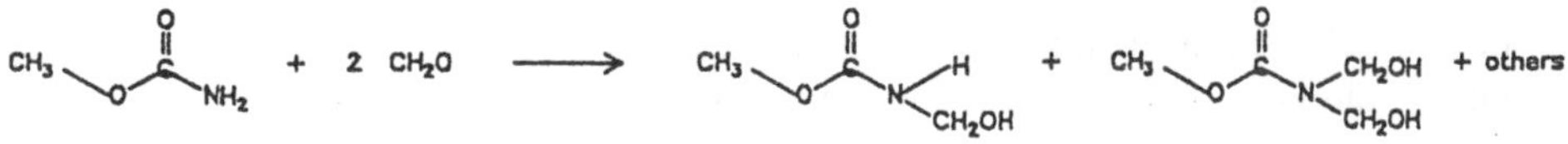

Figure 5   Methylolation of methyl carbamate.

imidazolidinone (MMDHEU), DMDHEU, several methylated derivatives of DMDHEU, and two commercial DP finishing agents, Beck and co-workers [14] dried the DP agents at room temperature and silylated them with N-methyl-N-trimethylsilyltrifluoroacetamide (MSTFA). MSTFA was selected because it is a stronger silylating agent than BSA and because the by-product, N-methyltrifluoroacetamide, is more volatile than acetamide from BSA. Chromatographic separation was effected on a 30-m DB-1701 fused silica column capillary column. A typical chromatogram of a DMDHEU-based commercial DP finish is shown in Figure 6. In this chromatogram, tetrasilylated DMDHEU eluted at 14.41 min. Figure 7 is a chromatogram of silylated glycolated DMDHEU. Compounds eluting from 26.89 to 30.88 are monoglycolated—that is, one DMDHEU -OH has been converted to -OCH₂CH₂OCH₂CH₂OH.

In a comparison of the cold sulfite method for determining formaldehyde released from DP-finished fabric with headspace gas chromatography (HGC), Kamath et al. [15] found the two methods to be comparable if the headspace fabric and titration temperature were the same. Low concentrations of formaldehyde in the headspace necessitated the use of a photoionization detector rather than the usual flame ionization detector. The use of HGC was described by these authors in an earlier publication [16].

Vail and Dupuy [17] extracted odor-causing materials from fabrics treated with trimethylolmelamine and analyzed them by GC. They found trimethylamine to be responsible for the fishy smells emanating from the finished fabric.

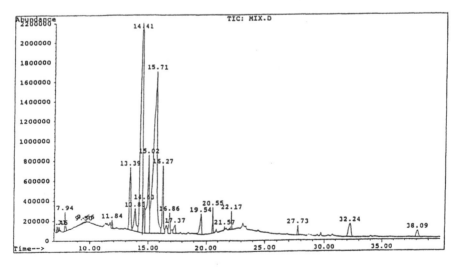

Figure 6   GC chromatogram of trimethylsilylated commercial DMDHEU finish. (Courtesy S. Yoon, Sequa Chemicals, Inc., Chester, S.C.)

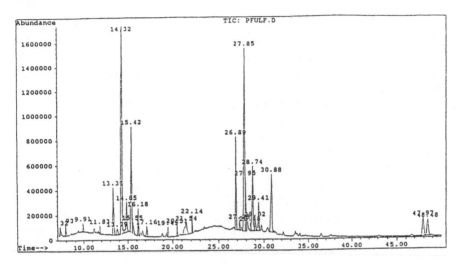

Figure 7   GC chromatogram of trimethysilylated commercial glycolated DMDHEU finish. (Courtesy S. Yoon, Sequa Chemicals, Inc., Chester, S.C.)

C.   High-Performance Liquid Chromatography

Compounds that are not sufficiently volatile to be separated by gas chromatography may be analyzed by HPLC. Player and Dunn [2] briefly discuss HPLC equipment and some textile applications. Instrumentation, separation mechanisms, and practical tips for use of HPLC may be found in McMaster's book [18] or other similar references. For analysis of DP agents, HPLC has the advantage that no derivatization is necessary as it is in GC. However, resolution in HPLC is not as good as that exhibited by capillary GC.

Early liquid chromatographic analysis of DP agents took advantage of the differences in interactions between mixture components and ion-exchange resins. Kumlin and Simonson [19] developed a liquid chromatographic method for analyzing the components in urea-formaldehyde resins using a mixed anion (DA-X8 as its sulfate salt)–cation (Aminex A-5 as its lithium salt) exchange resin stationary phase and an ethanol–water mobile phase with refractive index detection. With this system the authors were able to separate urea, N-hydroxymethylurea (MMU), N, N'-dihydroxymethylurea (DMU), some methylenediureas, and some oxymethylenediureas. In subsequent work [20] they used preparative liquid chromatography to separate N, N-DMU and N, N, N-trihydroxymethylurea (TMU). Structure of these materials was established by 270-MHz ^{1}H-NMR spectra. Symmetrical N, N'-DMU was formed in much smaller amounts than the unsymmetrical N, N-DMU. No evidence of tetramethylolurea was found.

Beck et al. [21] used a thermostated column packed with the lithium form of Aminex Q-15S (cation exchange resin) and water as the mobile phase to separate urea, formaldehyde, DHEU, MMDHEU, and DMDHEU. This technique was used to analyze several commercial DP agents. Beck and Pasad [22] used this method to quantitatively determine the amounts of DHEU, MMDHEU, and DMDHEU on fabrics that were partially cured. From these data at 70°C, 90°C, and 110°C, they determined both the rate constants and the energy of activation for the reaction between cellulose and DMDHEU. Beck and Pasad [23] also determined the effect of pad-bath pH and storage period on the hydrolysis of DMDHEU using the same chromatographic method. They concluded that the effect of pad-bath pH is much more important than the storage period on the stability of DMDHEU. DMDHEU was stable up to 55 days, providing the pad-bath pH did not exceed 6.

When reliable reversed-phase HPLC columns became available, they supplanted the ion exchange columns. Octadecylsilyl (C18) phases were most widely used for analysis of DP agents. In most cases, no organic modifier was required as water gave adequate separations. With these columns, plate counts as high as 100,000 plates/m were available and the chromatographer could purchase, rather than pack, a column. Figure 8 shows the structures of two potential cross-linking agents synthesized by Frick and Harper [24]. They followed both the synthesis from glyoxal and the corresponding diurea and the cis–trans isomerization of the ring hydroxyls by C18 reversed-phase HPLC. Frick and Harper [25] used C18 reversed-phase HPLC to determine the effect of pH on the synthesis of DMDHI from glyoxal and dimethylurea. They concluded that pH 8 was best for this reaction.

Beck and Pasad [26] determined the nature and amounts of reagent residues on fabric padded with 1,3-dihydroxymethyl-2-imidazolidinone (DMEU) using reversed-phase C18 HPLC. They found, as expected, that DMEU is much less stable than DMDHEU on fabric. Because DMEU reacts with itself, condensation products as well as 1-hydroxymethyl-2-imidazolidinone (MMEU) and EU were

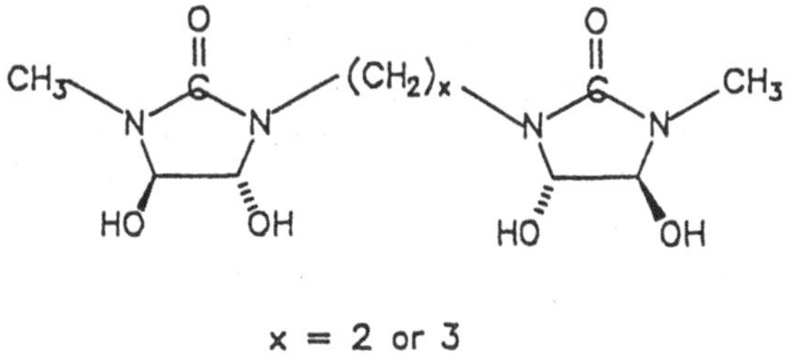

$x = 2$ or 3

Figure 8   Multifunctional dihydroxyimidazolidinone cross-linking agents.

observed in the chromatograms. In a related study of reagent residues on fabrics treated with N-hydroxymethylpyrrolidone (NMP), a monofunctional model compound for DP finishing, Beck and co-workers [27] used C18 HPLC to determine the amounts of those residues. Pasad et al. [28] used preparative HPLC to isolate a component from a commercial DMDHEU finish. ^{13}C-NMR supported a structure consistent with a dimer of DMDHEU.

In an effort to better understand the role of formaldehyde scavengers in DP finishing, Vail and Beck [29] used reversed-phase HPLC and ^{1}H-NMR to study the effect of EU and urea on formaldehyde released by DMMC-treated fabrics. The scavengers altered the equilibrium between DMMC and formaldehyde in the pad baths and diminished cross-linking in the fabric finished with those baths.

Two papers, each of which summarized the status of HPLC analysis of DP agents, were published in 1984. Beck and co-workers [30] compared results of cation-exchange columns and reversed-phase columns for analysis of DP agents. They also included reversed-phase chromatograms of partially and fully methylated DMDHEU finishes. In these chromatograms, the presence of both trans- and cis-isomers of 1,3-dimethoxymethyl-4,5-dihydroxy-2-imidazolidinone (DMMDHEU) was indicated. Retention times of the cis-isomers of DMDHI and DMMDHEU were about 3 min longer than those of the trans-isomers. Stronger interaction between the *cis*-hydroxyls and the stationary phase, possibly exposed silanol groups, is the most likely explanation for this elution behavior. Andrews [31] studied urea formaldehyde condensation products, DMDHEU, methylatd DMDHEU, and some other DP agents by reversed-phase HPLC to determine the effect of catalyst on hydrolysis of the finishes. It was also determined that the concentration of monomethylolated species decreases as formaldehyde concentration increases in these finish mixtures.

Ernes [32] developed an HPLC method for determination of formaldehyde in aqueous solutions generated in American Association of Textile Chemists and Colorants (AATCC) Test Method 112 [33]. Conversion to its 2,4-dinitrophenylhydrazone (2,4-DNP) allowed reversed-phase chromatographic analysis of formaldehyde from 2 to 10,000 μg/g of fabric. Results were comparable with those obtained using the Nash reagent specified in Method 112. The advantage of a chromatographic method such as this one is that the separation removes any interfering substances since their 2,4-DNP derivatives would have different retention times. Yoon [34] describes both normal and reversed-phase methods for analyzing the 2,4-DNP of formaldehyde. In that same reference, Yoon also describes the preparation and HPLC analysis of the dimedon derivative of formaldehyde.

Two recent developments in HPLC detectors are worthy of special note. The first, called LC Transform (by Lab Connections), is a simple interface that allows the eluent from an HPLC to be deposited on a small circular disk. The mobile-phase solvent is evaporated as the disk rotates, leaving the eluted analytes in a circular track. The disk is then placed in a device in the beam of an FTIR spectrometer. IR spectra are obtained as the disk rotates. Figure 9 shows the chromatogram of a commercial glycolated DMDHEU finish using this detector. Examples of spectral information from this technique are presented in the Section IV.C. This appears to be a powerful tool for monitoring analytes eluted from an HPLC column, regardless of the mechanism of separation.

A versatile, universal evaporative light scattering detector (ELSD) has been developed by Varex and is available through Alltech. Column eluent is nebulized with nitrogen gas to form a uniform dispersion of droplets. As the droplets pass through a heated tube, the solvent evaporates, leaving very fine particles of dried analyte in solvent vapor. As the particles pass through a flow cell, they scatter light from a laser diode. The scattered light is detected by a silicon diode, generating, after amplification, a chromatogram. Since solvent is evaporated, ELSD can be used with gradient elution and its response is not affected by changes in column or laboratory temperatures. It is more sensitive than refractive index (RI). A schematic diagram of the ELSD is shown in Figure 10. For comparison with the

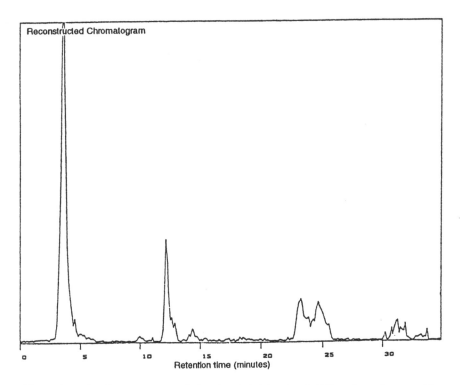

Figure 9 HPL Chromatogram of glycolated finish using LC Transform detector. (Courtesy S. Yoon, Sequa Chemicals, Inc., Chester, S.C.)

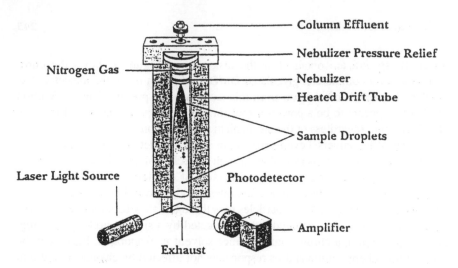

Figure 10 Schematic diagram of ELSD. (Diagram furnished by Alltech.)

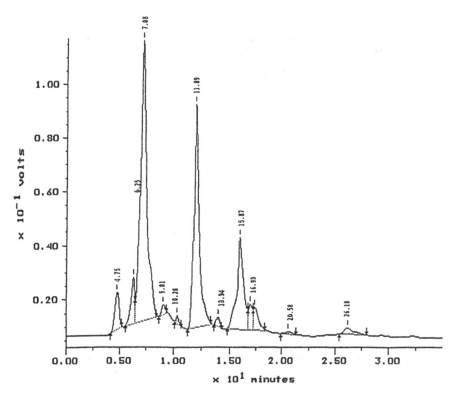

Figure 11 Refractive index chromatogram of glycolated DMDHEU. (Courtesy S. Yoon, Sequa Chemicals, Inc., Chester, S.C.)

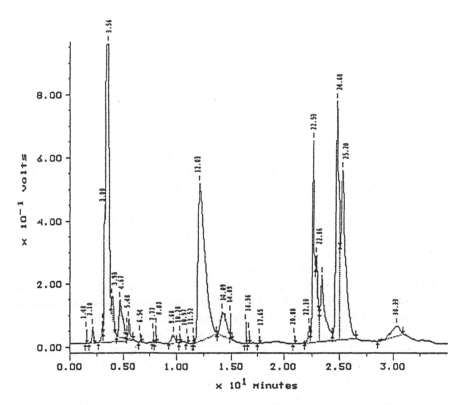

Figure 12   ELSD chromatogram of glycolated DMDHEU. (Courtesy S. Yoon, Sequa Chemicals, Inc., Chester, S.C.)

RI chromatogram (Fig. 11) and LC Transform responses (Fig. 9), Fig. 12 shows an ELSD chromatogram of the same sample.

IV.   SPECTROSCOPIC ANALYSIS OF DURABLE-PRESS AGENTS

A.   Ultraviolet-Visible Spectroscopy

General background information on UV-VIS spectroscopy can be found in any modern instrument analysis text book or in specific references, such as Perkampus [35].

Since the carbonyl group in most DP agents absorbs energy in the short wavelength ultraviolet region (~210 nm), UV-VIS spectroscopy is not particularly useful for their analysis. On the other hand, formaldehyde absorbs radiation at 397 nm, which should make its concentration measurable with visible radiation.

Unfortunately, the molar absorbtivity for this n $\rightarrow$ π^* is so small that this is not possible. To compensate for this weak absorption, formaldehyde is typically converted to a derivative which absorbs strongly in either the UV or VIS region. The use of several reagents, including acetylacetone (Nash reagent), chromotropic acid, 3-methyl-2-benzothiazolone hydrazone (MBTH), dimedon, and 2,4-dinitrophenylhydrazine, for this purpose is summarized by Yoon [34].

B. Near-Infrared Spectroscopy

When compared to UV-VIS and IR spectroscopy, NIR is a new and developing tool for both qualitative and quantitative applications. Background information and some practical aspects concerning the use of NIR spectroscopy may be found in Ref. 36. NIR is a secondary technique; that is, it requires calibration models to be developed from a reference analytical method. Ghosh et al. [37] developed NIR calibrations for amounts of DMDHEU and DMDHI on cotton fabrics. Kjeldahl nitrogen determinations were used as the reference data for regression. A three-wavelength model successfully predicted percent nitrogen with an $r^2 = .97$ and a standard error of prediction (SEP) of 0.198% for fabrics treated with either DMDHEU or DMDHI. Ghosh and Brodmann [38] developed a system for online monitoring of DMDHEU in a polyester/cotton fabric. Another three-wavelength regression model from Kjeldahl values and second derivative NIR spectra gave $r^2 = .98$ and SEP = 0.2%. Use of second-derivative spectra reduced baseline shifts by removing or reducing differences caused by surface effects.

Morris and co-workers [39] determined the amount of BTCA on finished fabric by NIR. The reference method involved measurement of the ratio of the carbonyl stretching and CH_2 bending absorbances by FTIR. This ratio varied linearly (r = .999) with percent wet pickup. The FTIR ratio was used to determine a NIR model. The model predicted percent BTCA with a SEP of 0.33 and r = .99. Use of second derivative NIR spectra only improved the SEP to 0.31.

C. Infrared Spectroscopy

General information on infrared spectroscopy can be found in any introductory organic chemistry text or in specific references, such as Colthup et al. [40]. Berni and Morris [41] provided information on experimental techniques useful in textile applications and briefly discuss IR analysis of DP agents.

The literature is replete with references to the use of IR spectroscopy as a tool for studying DP finishes. Most of the early IR investigations of DP finishes were by authors at the Southern Regional Research Center. The references discussed here are meant to be representative, rather than all-inclusive, of those publications. McCall et al. [42] described the following four techniques for obtaining infrared spectra of DP agents on finished cotton fabrics: potassium bromide disc, differential disc (KBr disc containing cotton placed in reference beam and disc containing

finished cotton placed in sample beam), acid hydrolysis, and multiple internal reflectance. The differential disc method was an early version of the subtractive
techniques that are now possible with Fourier-transform infrared (FTIR) instruments. In the acid hydrolysis method, the finished cotton fabric was hydrolyzed
with HCl and the hydrolysate mixed with KBr. This mixture was evaporated,
dried, and pressed into a pellet for spectral determination. In the reflectance
method, fabrics were pressed against a KRS-5 plate for spectral measurements. The
authors identified absorption bands that were characteristic of DMU, DMEU,
methylolated melamine, and some other finishes. In most cases, positive identification required spectra of known finishes for comparison with those obtained by
one or more of these methods. Vail et al. [43] obtained IR spectra of cellulose films
that had been reacted with either DMEU or DMMC. The authors concluded that
both of these reagents gave monomeric crosslinks when reacted with cellulose. In
a study of catalysis of the cross-linking reaction, Pierce and Vail [44] synthesized
a series of complexes between 2-imidazolidinone (EU) or tetrahydro-2-pyrimi-
done (propyleneurea, PU) and some metal perchlorates. They concluded that the
position of the carbonyl absorption band in these complexes could not be used to
predict the mode of bonding. Vail et al. [45] reported the formation of DMDHI
from DMU and glyoxal and characterized both the trans- and cis-isomers by IR
and proton NMR. Petersen [46] described the influence of structure on the position of the carbonyl absorption band in several precursors of DP finishes. In addition, he showed that there is a linear relationship between carbonyl wavenumber
and both the rate constant and energy of activation for the first methylolation of
these amides. Electron density on nitrogen atoms, steric effects, and planarity of
ring structures were reasons given for the differences in position of the carbonyl
absorbtion bands. Jung and co-workers [47] studied the fine structural changes in
cotton that accompany DP finishing and subsequent hydrolysis of the cross-linked
fabric. They found that all traces of finish absorbances were removed if fabrics finished with DMEU or MMEU were hydrolyzed with urea-phosphoric acid (UPac),
but DMDHEU signals remained even after several hydrolytic treatments. This information, coupled with data from strength measurements, led the authors to con-
clude that the DMDHEU cross-link is different from that of DMEU, either in its
nature or site of attachment to cellulose.

Morris et al. [48] determined the amount of DMDHEU on finished fabric
from FTIR spectra. By normalizing the calibration and sample spectra and applying a scaling factor, they developed a regression model (r = .9876) that predicted
percent DMDHEU within about 12% of the Kjeldahl value. In a subsequent publication [49] Morris compared the previously mentioned KBr disc and multiple internal reflectance techniques with diffuse reflectance infrared Fourier-transform
spectroscopy (DRIFTS). Samples investigated were cotton fabrics treated with
sodium hypophosphite (NaH_2PO_2, catalyst for cross-linking with BTCA) and
$BTCA/NaH_2PO_2$. Of the four DRIFTS sampling techniques investigated, the best

spectra were obtained using discs cut from fabric with no additional sample preparation, such as grinding or addition of KBr. The KBr disk technique gave more reproducible spectra than DRIFTS, which was better than the multiple internal reflectance method.

Investigations of both the mechanism of cross-linking with DMDHEU and BTCA or other polycarboxylic acids with cellulose, and the nature of those cross-links have utilized FTIR and other analytical techniques. Yang et al. [50] used photoacoustic spectroscopy (PAS) to determine the distribution of finishing agent (DMDHEU or methylated DMDHEU) in foam-finished and normally finished fabrics. They found that finish distribution was more uniform in foam-finished fabrics and that this led to higher wrinkle recovery angles. The PAS technique allows the acquisition of IR spectra from material near the surface (to a depth of a few micrometers) and it requires little sample preparation. A comparison of FTIR/PAS and DRIFTS for analysis of textile fibers and chemically modified fabrics is given by Yang [51]. DRIFTS showed an enhancement of band intensities for near-surface species compared to FTIR/PAS, but FTIR/PAS gave spectra from chromophores closer to the surface. Yang and Perenich [52] described information obtained by FTIR/PAS on DHDHEU and polycarboxylic acid treated fabrics. These studies compared the intensities of carbonyl bands in powdered samples and in the near-surface regions to determine finish distribution. Polycarboxylic acids studied were BTCA, all-*cis*-1,2,3,4-cyclopentanetetracarboxylic acid, and *trans*-aconitic acid. Finish distribution differences for the polycarboxylic acids were explained by the size and diffusion characteristics of the molecules. Using FTIR/PAS, Yang [53] has determined the degree of ester cross-linking in BTCA treated fabrics. This technique involves conversion of all carboxylate ions to carboxyl groups with dilute acid and then comparing the intensity of the acid and ester carbonyl peaks. This method has been used by Morris and co-workers [54] to determine the amount of BTCA or citric acid in polycarboxylic acid-finished cotton fabrics. These authors obtained spectral information from KBr discs.

Yoon (55) has employed FTIR as an HPLC detector to identify components of durable press mixtures. Figure 9 showed the liquid chromatogram reconstructed from infrared absorbance. The structure of one possible monoglycolated DMDHEU isomer is shown in Figure 13. Figures 14 and 15 are FTIR spectra of DMDHEU (peak at 3.85 min) and glycolated DHDHEU (peak at 23.23 min).

D.   Nuclear Magnetic Resonance Spectrometry

Nuclear magnetic resonance is an extremely powerful instrumental technique for determining structures of materials. General information on NMR principles may be found in any introductory organic chemistry text or in specific references, such as Sanders and Hunter [56]. Discussion in this chapter is divided into references

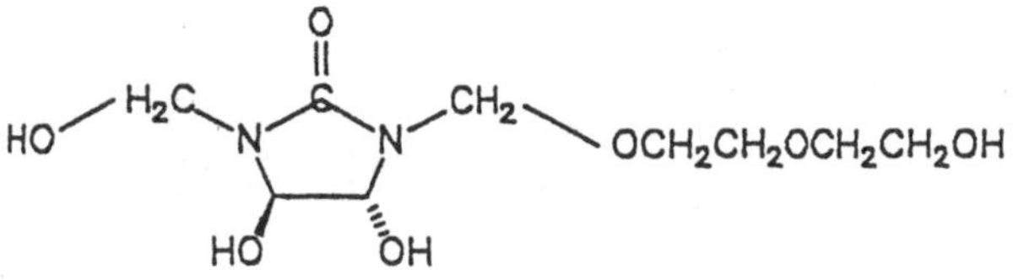

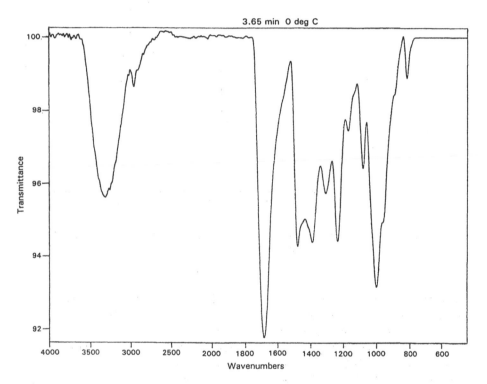

Figure 13   Glycolated DMDHEU.

Figure 14   FTIR spectrum of DMDHEU as eluted from a C18 HPLC column. (Courtesy S. Yoon, Sequa Chemicals, Inc., Chester, S.C.)

dealing with proton magnetic resonance (^{1}H-NMR) and carbon-13 magnetic resonance (^{13}C-NMR).

1. ^{1}H-NMR

As with infrared analysis of DP agents, many of the early ^{1}H-NMR studies were carried out at the USDA Southern Regional Research Center. References included

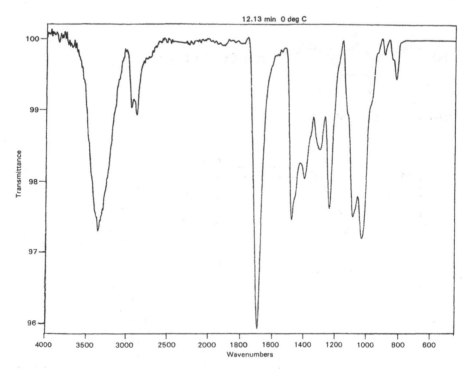

Figure 15 FTIR spectrum of glycolated DMDHEU as eluted from a C18 HPLC column. (Courtesy S. Yoon, Sequa Chemicals, Inc., Chester, S.C.)

here are representative, but not all-inclusive, of that work. Unless otherwise specified, all ^{1}H-NMR spectral data were measured in solution at 60 MHz.

Several groups investigated the structure of UF condensates with ^{1}H-NMR. Kumlin and Simonsen [19] identified and separated the products of urea formaldehyde reactions by HPLC and identified those compounds by NMR. Chemical shifts for MMU, N, N'-DMU, N, N-DMU, and trimethylolurea were given. These spectral characteristics allowed the symmetrical (N, N'-DMU) and unsymmetrical (N, N-DMU) disubstituted products to be differentiated. Chiavarini and co-workers [57] also characterized MMU, N, N'-DMU, methylenediurea, methoxymethylurea, dimethoxymethylurea, and dimethylolmethylenediurea by NMR in dimethyl sulfoxide-d_6 and dimethyl sulfoxide-d_6/CaCl$_2$ solutions. The latter solvent gave sharper peaks that led to better resolution, especially for NH and OH protons. Andrews [29] determined the distribution of MMU, N, N-DMU, N, N'-DMU, and TMU in mixtures of varying urea/formaldehyde ratios with ^{1}H-NMR.

In the course of studies with 1-(hydroxymethyl)-2-pyrrolidone (NMP) (Fig. 16) as a model monofunctional durable press agent, Beck et al. [58] isolated

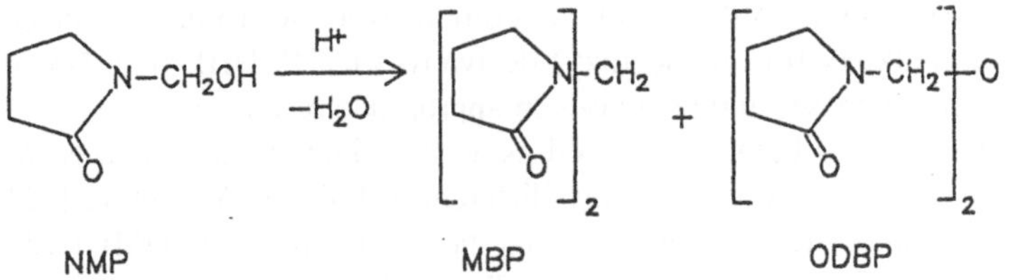

Figure 16 Dehydration products from N-(hydroxymethyl)-2-pyrrolidone.

both N, N'-methylenebis-2-pyrrolidone [MBP] and N, N'-(oxydimethylene)bis-2-pyrrolidone (ODBP) (Fig. 16). These species were characterized both by ^{1}H-NMR and ^{13}C-NMR. The oxydimethylene compound was converted to MBP and formaldehyde in the presence of acid at room temperature, on passage through a silica gel column, and on distillation, which indicates that this type of species may be relatively unstable in a fabric.

Vail and Beck [29] used both HPLC and ^{1}H-NMR to determine the effect that U and EU scavengers have on the composition of extracts from DMMC finished fabrics. HPLC was shown to be more reliable in the determination of low concentrations of MC.

Vail and co-workers [59] used ^{1}H-NMR to determine the composition, both qualitatively and quantitatively, of methyl carbamate/formaldehyde reaction mixtures. They also used ^{1}H-NMR to determine relative reactivity of the N-CH$_2$OH (N-methylol) and N-CH$_2$OCH$_3$ (methoxymethyl) moieties in acid-catalyzed reactions of the type found in cellulose cross-linking. From this work the authors concluded that the stem (atoms connected to the nitrogen) is more important in stabilizing the incipient carbocation (N-CH$_2$$^+$) than the nature of the leaving group (-OH or -OCH$_3$ in this example). In a related study, Xiang and co-workers [60] used ^{1}H-NMR to determine the rate constants for methylolation and demethylolation of methyl carbamate (MC) at different pH levels and ratios of MC/formaldehyde. To study the relationship between ease of hydrolysis of methoxy derivatives of DMDHEU and DMDHI, Vail [61] measured appropriate proton resonances at various temperatures up to 80°C. He concluded that bond cleavage occurs first at the carbon oxygen of the methylol group, regardless of the nature of the urea adduct. From NMR studies of methylated DMDHEU, Vail and Arney [62] concluded that the ring and pendant groups in DMDHEU are equal in reactivity toward small reagents, but steric hindrance reduces the reactivity of the ring groups toward larger reagents, such as cellulose. These data were confirmed on a 100-MHz instrument [63]. In further studies by Vail and Petersen [64] to determine the influence of leaving group effects on reactivity of substituted DMDHI molecules, ^{1}H-NMR showed that both the basicity of the leaving group and the inductive effects of remaining substituents and leaving group are determining factors. Rela-

tive rates of hydrolysis under mildly acidic conditions were determined by placing pairs of several alkoxy (or one acetoxy) derivatives of DMDHI in an NMR
tube and monitoring increases and decreases in appropriate resonances.

When DMDHEU was being developed as a cross-linking agent, it was in-
tensly studied by many people using many different techniques. Vail et al. [65]
concluded from proton NMR and other data that the ring hydroxyls in DMDHEU
are *trans*, and predominately *trans* in DMDHI. The *cis*-isomer of DMDHI can be
isolated, but it is easily converted to trans in solution by either acid or base catalysis. A summary of resonance assignments for ^{1}H-NMR spectra of a variety of substituted cyclic ureas, including EU, DHEU, DMEU, DMDHEU, DMDHI, and
DMPU, is given by Soignet and co-workers [66]. Spectral effects caused by substitution on the parent ring, splitting patterns, and hydrogen-bonding effects were
also discussed for these cyclic ureas.

Frick and Harper [67] used ^{1}H-NMR and HPLC to show that the glyoxal
adducts shown in Fig. 8 existed in aqueous solution as mixtures of the possible
cis–trans isomers. The substituted ethane (x = 2) was purified by recrystallization
and showed sharp NMR signals consistent with one pure compound in DMSO.
This compound in water gave multiple methyl resonances in the NMR spectrum
and showed five peaks in the liquid chromatogram. In a related study, these authors [67] reacted 1 mole of EU with 1 mole of glyoxal and obtained a water-
soluble adduct. Methylation of this adduct gave a solid that showed two sets of
multiple signals representing hydroxymethylene and methylene/methoxy protons.
Nine peaks were present in the liquid chromatogram of the water-soluble product.
These data plus those from ^{13}C-NMR, which are discussed later, led the authors to
assign the oligomeric structure shown in Figure 17.

With 300-MHz ^{1}H-NMR spectra, Chen [10] showed that tartaric acid does
complex with aluminum sulfate. These data supported the results he had obtained
on these durable-press catalysts with FTIR and TLC.

2.   ^{13}C-NMR

Development of Fourier-transform ^{13}C-NMR gave chemists an exceptionally versatile tool for determining molecular structures. Unless otherwise specified, all
spectra cited in this discussion were obtained in solution at 20 MHz. Frick and

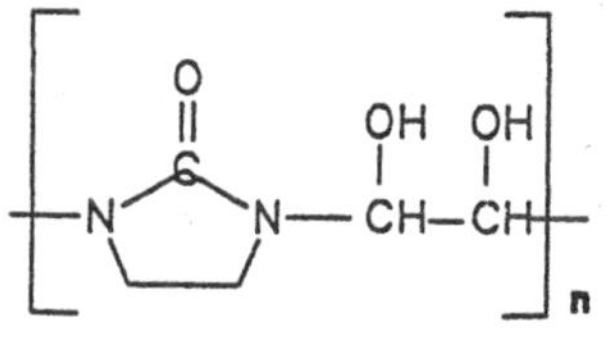

Figure 17   Adduct from glyoxal and EU.

Harper [67] observed two signals from methylene groups, four signals from hydroxymethylene groups, and three resonances for carbonyl groups in the ^{13}C-NMR spectrum of an aqueous solution of the adduct in Figure 17. Andrews [31] used this technique to determine the level of impurities in DMDHEU. Urea could be detected at 1 mol% by monitoring the carbonyl resonance. By comparison, HPLC was not able to detect urea in DMDHEU until the concentration reached 10 mol%. Beck et al. [58] reported ^{13}C-NMR chemical shifts for all three materials shown in Fig. 16.

Beck and Springer [68] reported ^{13}C-NMR chemical shift data for urea, MMU, monomethoxymethylurea, DMU, dimethoxymethylurea, methoxymethylhydroxymethylurea, DHEU, 4,5-dimethoxy-2-imidazolidinone, MMDHEU, DMDHEU, methoxymethylDHEU, DMDHEU methylated at one pendant group, DMDHEU dimethylated at both pendant groups and at both ring positions, both trimethylated DMDHEU isomers, tetramethylated DMDHEU, and DMDHI. Figure 18 shows the proton-decoupled ^{13}C-NMR spectrum of DMMDHEU. Two different tech-

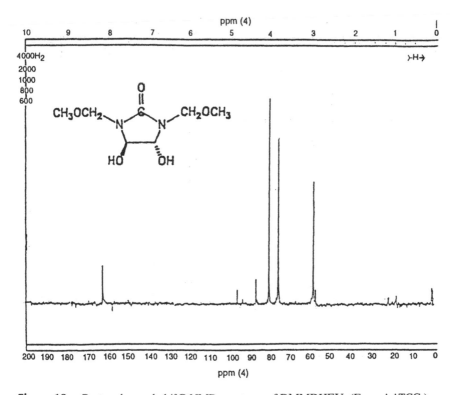

Figure 18.   Proton decoupled ^{13}C-NMR spectrum of DMMDHEU. (From AATCC.)

niques for structure elucidation [68], off-resonance and the attached proton test, are illustrated in Figures 19 and 20, respectively. In the off-resonance experiment, a small amount of coupling from protons to carbons is allowed. This splits the carbon resonance into n + 1 lines, where n is the number of attached protons. In the attached proton test, the experiment is run in a manner such that carbons bearing an even number (0 or 2) of protons exhibit an upward signal and carbons bearing an odd (1 or 3) number of protons exhibit a downward signal. Effects of structural changes, such as reaction with glyoxal, methylolation, and methylation, were also discussed. It was also concluded that the C_4 and C_5 carbons in *cis*-isomers of 4,5-dihydroxy-2-imidazolidinones resonate from 3 to 7 ppm upfield from the corresponding trans carbons. Steric crowding in the *cis*-isomer was given as the reason for this shielding effect, which was observed with both hydroxy and methoxy groups in these positions.

Hermanns et al. [69] used 50-MHz ^{13}C-NMR to determine the structure and composition of DMDHEU and DMDHI solutions. These authors [70] determined the materials extracted from DMDHEU- and DMDHI-finished fabrics by 50-MHz

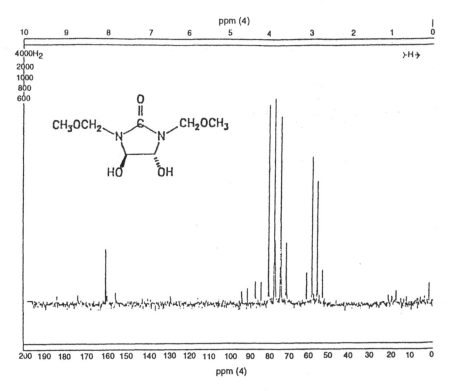

Figure 19 Off-resonance ^{13}C-NMR spectrum of DMMDHEU (From AATCC.)

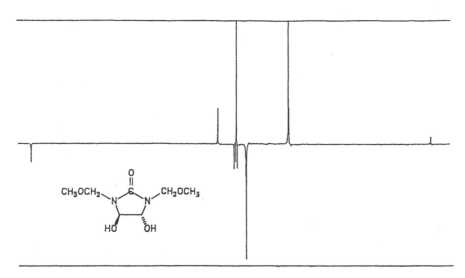

Figure 20 Attached proton test ^{13}C-NMR spectrum of DMMDHEU. (From AATCC.)

^{13}C-NMR. Swatches of finished fabric and water were sealed in ampules and heated at 40°C for 1 week or at 80°C for 4 days. No DMDHI was detected in the extract of the DMDHI-finished fabric. DMDHEU, MMDHEU, DHEU, and formaldehyde were all detected in the extract from DMDHEU-finished fabric. Generation of these materials was explained by hydrolysis of the cellulose cross-links and subsequent equilibration of the resulting products.

E.   Mass Spectrometry

The basic theory of mass spectrometry (MS) and interpretation of mass spectra can be found in any introductory organic chemistry text. More detailed information can be found in specific MS references, such as Chapman [71]. Representative references dealing with mass spectral analysis of the major classes of DP agents are discussed in this section.

Cashen [7] separated the methyl carbamate/formaldehyde reaction products by preparative TLC and analyzed each by MS. Based on MS fragmentation patterns and NMR data, structures for the products were suggested. Most were either oligomers or polymers of MC. Beck et al. [58] reported the major ions in the mass spectrum of ODBP (Fig. 16) as representative of this oxydimethylene structure. Similarities and differences in the electron impact (EI) mass spectra of 14 DP agents, including DMDHEU, DMDHI, and several alkylated derivatives of DMDHEU, were reported by Trask-Morrell et al. [72]. The presence of hydroxymethyl moieties was indicated by the presence of fragmentation of m/z 31 ions.

Electron impact and chemical ionization (CI) data for DMDHI, DHEU, DMDHEU, and several methylated derivatives of DMDHEU were described by Beck and co-workers [14]. Differences between fragmentation patterns caused by the two ionization techniques were discussed. These compounds were also silylated and analyzed by GC/MS in both EI and CI modes. Cleavage patterns that assisted in identification of some of the compounds were described. One such fragmentation allowed the identification of the *cis*-DMDHEU. Another fragmentation pattern led to the structure of the monoglycolated DMDHEU (Fig. 13) in a commercial DP finish. Mass spectra of trimethylsilylated DMDHEU and its glycolated derivative are shown in Figures 21 and 22.

Mass spectrometry was used by Trask-Morrell et al. [73] to indicate the presence of an anhydride intermediate in the polycarboxylic acid cross-linking of cellulose. Heating seven different materials, including two di-, two tri-, and three tetracarboxylic acids in the presence and absence of catalyst showed loss of water. This, and the presence of other ions in the mass spectra, suggested the generation of an intermediate acid anhydride in the cross-linking of cellulose with polycarboxylic acids.

V. SUMMARY

Modern analytical instrumentation has played a significant role in the development of DP finishing agents. These tools have been used for structure determina-

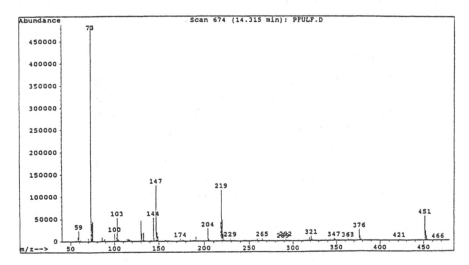

Figure 21 Mass spectrum of trimethylsilylated DMDHEU. (Courtesy S. Yoon, Sequa Chemicals, Inc., Chester, S.C.)

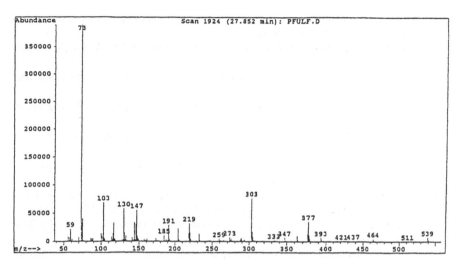

Figure 22 Mass spectrum of trimethylsilylated glycolated DMDHEU. (Courtesy S. Yoon, Sequa Chemicals, Inc., Chester, S.C.)

tion, mixture composition, properties, and mechanisms of cross-linking. The intent of this chapter has been to give illustrative, but not exhaustive, examples of the uses of TLC, GC, HPLC, UV-VIS spectroscopy, NIR spectroscopy, IR spectroscopy, NMR spectrometry, and mass spectrometry. Other analytical techniques, such as thermal analysis, have given valuable information about the means by which the cross-linking reaction occurs, but they were not the focus of this chapter. It is certain that any new DP agents that are discovered in the future will be studied by the techniques included here.

ACKNOWLEDGMENTS

The assistance of Paul Garwig in performing the computerized literature search that was a partial basis for this work is gratefully acknowledged. Special thanks go to Soon Yoon for helpful discussions about analytical techniques and for providing many of the chromatograms and spectra included in this chapter.

REFERENCES

1. H. B. Goldstein and J. M. May, Durably creased wash-wear cottons, *Textile Res. J.* *34*:325 (1964).
2. C. M. Player, Jr., and J. A. Dunn, Chromatographic methods, *Analytical Methods for a Textile Laboratory*, (J. W. Weaver, ed.), American Association of Textile Chemists and Colorists, Research Triangle Park, N.C., 1984.

3.   B. Fried and J. Sharma, *Thin Layer Chromatography: Techniques and Applications*, Marcel Dekker, New York, 1994.
4.   J. C. Touchtone, *Practice of Thin Layer Chromatography*, 3rd ed., Wiley, New York, 1992.
5.   G. Valk, K. Schliefer, and F. Klippel, Analysis of synthetic resin finishes, *Melliand Teltilber. 50*:449 (1969).
6.   D. R. Moore and R. M. Babb, Reactive thin-layer chromatography for the evaluation of cotton finishes, *Textile Res. J. 42*:500 (1974).
7.   N. Cashen, Chromatographic separation of the lyophilized reaction products of formaldehyde and methyl carbamate, *Textile Res. J. 43*:200 (1975).
8.   P. A. Rennison, Chromatographic detection of 4,5-dihydroxy-1,3-dimethyl imidazolidinone, *Textile Res. J. 51*:368 (1981).
9.   E. Kantschev and M. Nesnakomova, Investigation of the reaction between urea and glyoxal as intermediate stage to obtain products for resin finishing. III. Communication: Investigation of the reaction with quantitative thin-layer chromatography, *Textilvered. 16*:451 (1981).
10. C. Chen, Interaction of the components of mixed catalysts, *Textile Res. J. 60*:669 (1990).
11. P. Baugh, *Gas Chromatography: A Practical Approach*, Oxford University Press, New York, 1993.
12. A. L. Bullock and S. P. Rowland, Gas-liquid chromatographic study of selected derivatives of 2-imidazolidinone, *Anal. Chem. 42*:1783 (1970).
13. A. S. Divatia, J. J. Shroff, and H. C. Srivastava, Gas-liquid chromatography of trimethylsilyl derivatives of cellulose crosslinking agents, *Textile Res. J. 43*:701 (1973).
14. K. R. Beck, K. Springer, K. Wood, and M. Wusik, GC/MS analysis of durable press agents, *Textile Chem. Col. 20*:35 (1988).
15. Y. K. Kamath, S. B. Hornby, and H. D. Weigmann, Determination of free formaldehyde in durable press fabric: Comparison of cold sulfite method with headspace gas chromatography, *Textile Res. J. 56*:55 (1986).
16. R. U. Weber, Y. K. Kamath, and H. D. Weigmann, Headspace gas chromatography studies of formaldehyde release, *Book Paper Natl. Tech. Conf.—AATCC*, 1982, p. 154.
17. S. L. Vail and H. P. Dupuy, Determination of odors, *Textile Chem. Col. 11*:51 (1979).
18. M. C. McMaster, *HPLC: A Practical User's Guide*, VCH, New York, 1994.
19. K. Kumlin and R. Simonson, Urea formaldehyde resins I. Separation of low molecular weight components in urea-formaldehyde resins by means of liquid chromatography, *Angew. Macromol. Chem. 68*:175 (1978).
20. K. Kumlin and R. Simonson, Urea formaldehyde resins II. Formation of *N, N*-dimethylolurea and trimethylolurea in urea-formaldehyde mixtures, *Angew. Macromol. Chem. 72*:67 (1978).
21. K. R. Beck, B. J. Leibowitz, and M. R. Ladisch, Separation of methylol derivatives of imidazolidines, urea, and carbamates by liquid chromatography, *J. Chromatogr. 190*:226 (1980).
22. K. R. Beck and D. M. Pasad, Liquid chromatographic determination of rate constants for the cellulose-dimethyloldihydroxyethyleneurea reaction, *J. Appl. Pol. Sci. 27*:1131 (1982).

23. K. R. Beck and D. M. Pasad, The effect of pad-bath pH and storage period on the hydrolysis of DMDHEU, *Textile Res. J.* *52*:269 (1982).
24. J. G. Frick, Jr., and R. J. Harper, Jr., Multifunctional dihydroxyimidazolidinone crosslinking agents, *Ind. Eng. Chem. Prod. Res. Dev.* *21*:1 (1982).
25. J. G. Frick, Jr., and R. J. Harper, Jr., Reaction of dimethylurea and glyoxal, *Ind. Eng. Chem. Prod. Res. Dev.* *21*:599 (1982).
26. K. R. Beck and D. M. Pasad, Reagent residues of DMEU on cotton fabric as a function of pad-bath pH and storage period of the treated fabric, *Textile Res. J.* *53*:524 (1983).
27. K. R. Beck, D. M. Pasad, S. L. Vail, and Z. Xiang, Reagent residues on *N*-methylolpyrrolidone-treated cotton, *J. Appl. Polym. Sci.* *29*:3579 (1984).
28. D. M. Pasad, K. R. Beck, and S. L. Vail, Influence of reagent residues and catalysts on formaldehyde release from DMDHEU-treated cotton, *J. Appl. Polym. Sci.* *34*:549 (1987).
29. S. L. Vail and K. R. Beck, Evaluation of some side effects from the use of formaldehyde scavengers, *J. Appl. Polym. Sci.* *39*:1241 (1990).
30. K. R. Beck, D. M. Pasad, K. S. Springer, and C. M. Player, High performance liquid chromatographic analysis of durable press finishes, *Textile Chem. Col.* *16*:15 (1984).
31. B. A. K. Andrews, Use of reversed-phase high-performance liquid chromatography in characterization of reactants in durable press finishing of cotton fabrics, *J. Chromatogr.* *288*:101 (1984).
32. D. A. Ernes, An alternative method for formaldehyde determination in aqueous solutions, *Textile Chem. Col.* *17*:24 (1985).
33. Test Method 112, *AATCC Technical Manual*, American Association of Textile Chemists and Colorists, Research Triangle Park, N.C.
34. S. H. Yoon, Determination of formaldehyde, *Analytical Methods for a Textile Laboratory* (J. W. Weaver, ed.), American Association of Textile Chemists and Colorists, Research Triangle Park, N.C., 1984.
35. H. Perkampus, *UV-VIS Spectroscopy and Its Applications*, Springer-Verlag, New York, 1992.
36. B. G. Osborne, T. Fearne, and P. H. Hindle, *Practical NIR Spectroscopy with Applications in Food and Beverage Analysis*, Wiley, New York, 1993.
37. S. Ghosh, M. D. Cannon, and R. B. Roy, Quantitative analysis of durable press resin on cotton fabrics using near-infrared reflectance spectroscopy, *Textile Res. J.* *60*:167 (1990).
38. S. Ghosh and G. L. Brodmann, On-line measurement of durable press resin on fabrics using the NIR spectroscopy method, *Textile Chem. Col.* *25*:11 (1993).
39. N. M. Morris, S. Faught, E. A. Catalano, J. G. Montalve, Jr., and B. A. K. Andrews, Quantitative determination of polycarboxylic acids on cotton fabrics by NIR, *Textile Chem. Col.* *26*:33 (1994).
40. N. B. Colthup, L. H. Daly, and S. E. Wiberley, *Introduction to Infrared and Raman Spectroscopy*, Academic Press, Boston, 1990.
41. R. J. Berni and N. M. Morris, Infrared spectroscopy, *Analytical Methods for a Textile Laboratory* (J. W. Weaver, ed.), American Association of Textile Chemists and Colorists, Research Triangle Park, N.C., 1984.
42. E. R. McCall, S. H. Miles, and R. T. O'Connor, An analytical method for the identification of nitrogenous crosslinking reagents on cotton, *Am. Dyestuff Rep.* *56*:13 (1967).

43. S. L. Vail, J. G. Roberts, and R. Jeffries, Chemical structures of cross links from the reaction of *N*-methylolamides with cellulose, *Textile Res. J. 37*:708 (1967).

44. A. G. Pierce, Jr., and S. L. Vail, Amide-metal ion complex formation and its effect on the catalysis of cellulose cross-linking reactions, *Textile Res. J. 41*:1006 (1971).

45. S. L. Vail, R. H. Barker, and P. G. Mennitt, Formation and identification of *cis*- and *trans*-dihydroxyimidazolidinones from ureas and glyoxal, *J. Org. Chem. 30*:2179 (1965).

46. H. Petersen, In situ formation of polymers. A. Crosslinking chemical and the chemical principles of the resin finishing of cotton, *Chemical Aftertreatment of Textiles* (H. Mark, N. S. Wooding, and S. M. Atlas, eds.), Wiley Interscience, New York, 1971, p. 135.

47. H. Z. Jung, R. R. Benerito, E. J. Gonzales, and R. J. Berni, Urea-phosphoric acid hydrolysis of cotton modified with *N*-methylolated ethylene ureas, *Textiles Res. J. 44*:670 (1974).

48. N. M. Morris, R. A. Pittman, and R. J. Berni, Fourier transform infrared analysis of textiles, *Textile Chem. Col. 16*:43 (1984).

49. N. M. Morris, A comparison of sampling techniques for the characterization of cotton textiles by infrared spectroscopy, *Textile Chem. Col. 23*:19 (1991).

50. C. Q. Yang, T. A. Perenich, and W. G. Fately, Studies of foam finished cotton fabrics using FT-IR Photoacoustic spectroscopy, *Textile Res. J. 59*:562 (1989).

51. C. Q. Yang, Comparison of photoacoustic and diffuse reflectance infrared spectroscopy as near-surface analysis techniques, *Appl. Spectrosc. 45*:102 (1991).

52. C. Q. Yang and T. A. Perenich, Near-surface analysis of textile fabrics, yarns and fibers by FT-IR photoacoustic spectroscopy, Book Paper Natl. Tech. Conf. & Exhib., 1989, p. 235.

53. C. Q. Yang, Characterizing ester crosslinkages in cotton cellulose with FT-IR photoacoustic spectroscopy, *Textile Res. J. 61*:298 (1991).

54. N. M. Morris, B. A. K. Andrews, and E. A. Catalano, Determination of polycarboxylic acids on cotton fabric by FT-IR spectroscopy, *Textile Chem. Col. 26*:19 (1994).

55. S. Yoon, Private Communication, 1995.

56. J. K. M. Sanders and B. K. Hunter, *Modern NMR: A Guide for Chemists*, Oxford University Press, New York, 1993.

57. M. Chiavarini, N. Del Fanti, and R. Bigatto, Compositive characterization of urea-formaldehyde adhesives by NMR spectroscopy, *Angew. Makromol. Chem. 46*:151 (1975).

58. K. R. Beck, D. M. Pasad, and S. L. Vail, Synthesis, isolation, and characterization of *N, N'*-oxydimethylenebisamides, *J. Polym. Sci. Pt. A: Polym. Chem., 26*:725 (1989).

59. S. L. Vail, F. W. Snowden, and E. R. McCall, Use of nuclear magnetic resonance in studies of textile finishing agents. *N,N*-Bis(methoxymethyl) amides and *N, N*-dimethylolamides, *Am. Dyest. Rep. 56*:60 (1967).

60. Z. Xiang, K. Chung, J. H. Wall, and S. L. Vail, Investigating the reaction course of *N*-methylolation reaction of methyl carbamate by NMR, Book Paper Natl. Tech. Conf. & Exhib., 1983, p. 1.

61. S. L. Vail, The reactivity-hydrolysis relationship in chemical finishing of cotton, *Textile Res. J. 39*:774 (1969).

62. S. L. Vail and W. C. Arney, Jr., Reaction mechanisms of glyoxal-based durable press resins with cotton, *Textile Res. J. 41*:336 (1971).

63. S. L. Vail and W. C. Arney, Jr., Reactivity of Groups in Dimethyloldihydroxyethyleneurea, *Textile Res. J. 44*:400 (1974).

64. S. L. Vail and H. Petersen, Influence of leaving group effects on the reactivity of 4,5-dialkoxy-2-imidazolidinones, *I&EC Prod. Res. Dev. 14*:50 (1975).

65. S. L. Vail, G. B. Verburg, and A. H. P. Young, The 4,5-dihydroxy-2-imidazolidinone system for cross-linking cotton, *Textile Res. J. 39*:86 (1969).

66. D. M. Soignet, G. J. Boudreaux, R. J. Berni, and E. J. Gonzales, Nuclear magnetic resonance studies of substituted cyclic ureas, *Appl. Spectrosc. 24*:272 (1970).

67. J. G. Frick, Jr., and R. J. Harper, Jr., An imidazolidinone-glyoxal reactant for cellulose, *Textile Res. J. 53*:660 (1983).

68. K. R. Beck and K. S. Springer, ^{13}C-NMR analysis of durable press finishing agents, *Textile Chem. Col. 20*:29 (1988).

69. K. Hermanns, B. Meyer, and B. A. K. Andrews, ^{13}C-NMR Identification of cyclic ethyleneureas important in cellulosic textile finishing, *Ind. Eng. Chem. Prod. Res. Dev. 25*:469 (1986).

70. K. Hermanns, B. Meyer, and B. A. K. Andrews, Identifying extractible resin fragments in durable press cotton by ^{13}C-NMR spectroscopy, *Textile Res. J. 56*:343 (1986).

71. J. R. Chapman, *Practical Organic Mass Spectrometry: A Guide for Chemical and Biochemical Analysis*, John Wiley, New York, 1993.

72. B. J. Trask-Morrell, W. E. Franklin, and R. H. Liu, Thermoanalytical and mass spectrometric search for formaldehyde release markers in DP reagents, *Textile Chem. Col. 20*:21 (1988).

73. B. J. Trask-Morrell, B. A. K. Andrews, and E. E. Graves, Spectrometric analyses of polycarboxylic acids, *Textile Chem. Col. 22*:23 (1990).

7

Accessible Internal Volume Determination in Cotton

NOELIE R. BERTONIERE   Southern Regional Research Center, Agricultural Research Service, United States Department of Agriculture, New Orleans, Louisiana

I. INTRODUCTION

Techniques based on the principles of gel permeation chromatography have found wide application since the initial report by Porath and Flodin [1] on the use of cross-linked dextran gel media of graded permeability. Such gels swell in water that penetrates pores differing in size. These pores are selectively more permeable to solutes of decreasing molecular size. Gels with a low level of cross-linking are used to separate macromolecules such as proteins, while those having a high degree of cross-linking are used to separate smaller molecules such as sugars and oligosaccharides.

In early investigations Aggebrandt and Samuelson [2], in a study aimed at determining nonsolvent water in cut and beaten cotton fibers, used six polyethylene glycols (MW = 60–20,000) as solutes with a centrifuge technique. They noted that the value calculated for nonsolvent water (δ) increased as the molecular weight of the glycol increased. They rationalized this based on the presence of pores of various sizes in the cotton fibers and postulated that determination with polyethylene glycols of varying molecular weights could be used to elucidate the pore size distribution in cellulose fibers.

Later Stone and Scallan [3] conducted static measurements based on the principle of solute exclusion to study the structures of the cell walls of wood pulps and celluloses. Their criteria for suitable solutes were that they must (1) not be sorbed onto the cellulose, (2) be available in a wide range of molecular weights, (3) be available as narrow molecular weight fractions, (4) be uncharged, and (5) be of known size and shape, preferably spherical. Based on these criteria they selected the low-molecular-weight sugars (glucose, maltose, raffinose, and stachyose) and a series of 11 dextrans (MW = 2600–24,000) to characterize the distributions of

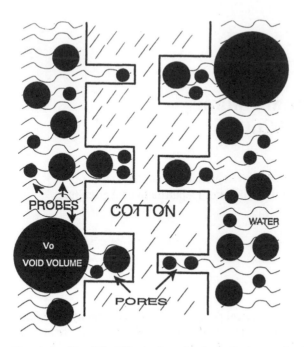

Figure 1   Simplified illustration of gel permeation mechanism.

pore sizes in wood pulps and celluloses. Based on evidence that predominantly linear dextrans behave as hydrodynamic spheres in solution, their molecular diameters, calculated from diffusion coefficients according to the Einstein-Stokes formula, were used. Calculations were based on changes in the concentration of the solute in the solution containing a known weight of the cellulose.

The basic principle upon which these, and the column chromatography method developed at the Southern Regional Research Center, are based is illustrated in an oversimplified manner in Figure 1. A very large molecule that cannot penetrate any of the pores in the cotton emerges from the cotton column first as it has a shorter path to traverse. In contrast, small molecules that can penetrate some of the accessible pores take a more circuitous route and thus emerge later. It is important to note that any of these methods can only assess the distribution of sizes in accessible pores. A pore that has no opening may exist but will not be detected by techniques based on gel filtration.

II.   DEVELOPMENT OF REVERSE GEL PERMEATION COLUMN CHROMATOGRAPHY METHOD

The column chromatography technique developed to assess pore size distribution in cotton cellulose is the reverse of most gel permeation chromatography methods.

Here the sample is the column packing and the materials of known size are the solutes. It was developed, and evolved slowly, over a period of years, at the Southern Regional Research Center.

A. Ball-Milled Cotton

Cotton cellulose is a highly crystalline material. Originally it was thought that a separation of solutes of different sizes could only be effected if the accessible, or amorphous, regions were increased by decrystallization. This was achieved by use of a vibratory ball mill by Martin and Rowland [4], who first reported that this decrystallized cotton cellulose had gel permeation properties comparable to highly cross-linked dextran. The sugars used as probes were erythrose, fructose, maltose monohydrate, raffinose pentahydrate, and stachyose tetrahydrate. Solutes emerging from their column were detected with a sensitive automatic polarimeter that distinguished between dextrorotatory and levorotary sugars. In a following report these authors [5] compared decrystallized cotton prepared from desized, scoured and bleached cotton printcloth with the decrystallized cotton after it had been cross-linked with formaldehyde in the swollen state. They found that although cross-linking reduced permeability to large molecules, the cross-linked material was more permeable than the untreated cellulose to compounds having molecular weights below 1000.

Work with ball-milled cotton continued [6] with a comparison of unmodified cotton, methylated cotton, cotton cross-linked with formaldehyde in the swollen and collapsed states, and microcrystalline wood cellulose. A differential refractometer replaced the polarimeter, and a siphon with a photoelectrically actuated mechanism for marking the recorder chart was added to the system. Fractions were collected and weighed. Relative elution volume was defined as the differences between the elution volume of the sugar and the void volume divided by the weight of cellulose in the column. Plots were made of the relative elution volumes against the molecular weights of characteristic crystalline hydrates of the sugars. The effective internal solvent volume (intercept where molecular weight equals zero) and the apparent limit of permeability (the molecular weight of a solute just large enough to be completely excluded from the gel) were extrapolated from this linear relationship. It was concluded that cellulose cross-linked in the swollen state exhibited increased permeability, whereas cross-linking under conditions that minimize swelling increased the internal volume while causing a decrease in the limit of permeability. Monofunctional substitution, with the methyl group here, increased the internal volume to the same extent as cross-linking in an unswollen state while increasing the limit of permeability. The microcrystalline wood cellulose was found to have as large an internal volume as decrystallized cotton cellulose, but a much higher limit of permeability. The large internal volume was surprising as the microcrystalline wood cellulose was a commercially modified material produced by controlled acid hydrolysis, which is assumed to have removed amorphous cellulose. Its "microcrystalline" nature was confirmed by x-ray scattering. The high permeability limit, almost twice as large as that calculated for decrystallized cotton cellulose, indicated that considerably less of the total internal volume is distributed in intermolecular spaces that are accessible only to molecules of smaller sizes.

This study was extended [7] to include the effects on the structure of decrystallized cotton produced by introduction of formaldehyde cross-linked under various reaction conditions. These included reaction in aqueous solution, in the vapor phase, in acetic acid, and in a bake-cure process. The experimental techniques and data handling for pore size distribution assessment remained the same. It was shown that accessibility was increased by reaction in aqueous solution, that reaction catalyzed by hydrochloric acid in the acetic acid medium formed products having larger internal volumes, but somewhat lower limits of permeability, and that both the internal volume and the permeability limit were decreased by the bake-cure process. A related study [8] reported changes in the permeation characteristics of cotton as a function of the levels of formaldehyde crosslinking achieved under bake-cure conditions and in the acetic–hydrochloric acid medium. Marked differences were found in the pore structures of the cotton cross-linked to progressively higher levels with both processes.

B.   Sephadex as a Model for Cellulose

One of the major experimental problems with columns made from cotton cellulose decrystallized by ball milling was column instability. The flow rate gradually decreased to the point where usable data could not be obtained. Because commercial, highly cross-linked dextrans behave like cellulose in the way they discriminate among low-molecular-weight sugars, two basic studies were conducted with Sephadex G-15 as a model for cellulose. In the first, Bertoniere et al. [9] studied the elution of sugars and sugar derivatives relative to glucose (R_g) to determine the effect of stereochemical and structural differences between molecules of approximately the same size. The following observations were made: (1) the gel could not distinguish between enantiomeric saccharides, (2) a decrease in R_g values was observed on going from a methylene to a hydroxyl to a methoxyl group in monosaccharides, (3) methylation or reduction of a particular hydroxyl group affects the R_g values selectively, (4) the substituted (methyl or glucosyl) α anomer is retained on the column longer than the corresponding β anomer, and (5) sugars having either one or no axially attached hydroxyl groups are eluted in the order: axially attached hydroxyl groups at C-4, at no carbon atom, at C-2, and at C-3. Thus the linear inverse relationship between the elution volumes and molecular weights of the characteristic hydrates of glucose, maltose, raffinose, and stachyose is fortuitous. It is nonetheless very useful as it permits the comparison of changes in the pore size distribution in cotton samples.

The original criterion of Stone and Scallan [3] was that solutes not be sorbed onto the cellulose if they are to be used as "feeler gauges." We therefore explored and reported [10] the interaction of several classes of solutes with the dextran gel Sephadex G-15, which was used as a model for cellulose. Solutes included a variety of sugars, methylated sugars, polyethylene glycols, polyethyleneimines, and derivatives of 2-imidazolidinone (ethyleneurea). The last class of compounds is of high practical interest because they form the basis for conventional cross-linking agents for cotton to impart easy care properties. It was found that low-molecular-weight polyethylene glycols were eluted in much smaller volumes than sugars having comparable molecular weights. The 15 2-imidazolidinone derivatives showed no simple relationship of elution volumes to molecular weights, but sorption via hydrogens on the ring nitrogen atoms appeared to be a factor. This had strong implications with respect to the interactions between these compounds and cotton cellulose in chemical finishing. Polyethyleneimines were sorbed so strongly on the column that they could not be eluted with water.

A third study [11] elucidated the interactions of several water-soluble solutes with both Sephadex G-15 and cotton. Water-soluble solutes included simple sugars, their completely methylated analogues, glucuronic and galacturonic acids, oligomers of ethylene glycols and their dimethyl ethers (glymes), and a series of 2-imidazolidinones. The following conclusions were drawn. Total pore water becomes available as solvent water to polysaccharides and polyethylene glycols as the molecular sizes of these solutes decrease toward and approach that of water; all water in a pore that is accessible to these polysaccharides or polyethylene glycols is available to the solutes as solvent water. Water-soluble solutes characterized by more limited hydrogen-bonding capabilities than saccharides and polyethylene glycols find only a fraction of the total water in accessible pores available as solvent water; the nonsolvent water is that which remains structured and bound on cellulosic or polysaccharidic surfaces. Water-soluble solutes that are characterized by hydrogen donor and acceptor strengths that are higher than those of saccharides and polyethylene glycols find all water in an accessible pore available as solvent water, and these solutes interact with the cellulosic or polysaccharidic surfaces in proportion to the strength of hydrogen bonding and the number of hydrogen-bonding sites in the solute. Permeation of water-soluble solutes into pores of cellulose or insoluble polysaccharides is influenced by electrostatic charge in the solutes, with cationic and anionic charges contributing to positive and negative sorption, respectively.

C. Chopped Cotton Fibers

Interactions between solutes in aqueous media and cellulose are the essence of chemical modification of cotton. Practical modifications of this fiber are usually conducted on the fabric where the desized, scoured and bleached fibers are intact.

Decrystallization by ball milling alters the molecular structure and reduces the degree of polymerization of cotton cellulose to approximately 500. Data on the pore size distribution in the cotton fiber would ideally be obtained on the whole, or at least minimally disturbed, fiber. In order to approach this ideal Blouin et al. [12] chopped the cotton fabric in a Wiley mill to pass successively though 20-, 40-, 60-, and 80-mesh screens. This shortened the fibers without causing significant decrystallization. Larger columns, holding three times as much cellulose, were used. They reported that crystalline cotton had an apparent internal volume approximately 53% of that of decrystallized cellulose. The molecular weight limit of permeability was 2900, compared to approximately 1900 for the decrystallized material. The work was extended with an investigation into the effect of mercerization, conducted on fibers that had been chopped to pass through the 20-mesh screen; reduction in size was complete on the mercerized fibers. Mercerization was reported to increase the apparent internal solvent volume by approximately 60%, but to decrease the limit of permeability to a molecular weight of 2200.

Following up on this initial report, Blouin et al. [13] used the new technique for evaluating fibrous cotton to study changes in structure from cross-linking with formaldehyde. In this study the cross-linking treatments were applied to fabric that was subsequently reduced by Wiley milling to the particle size required for uniform packing of the columns. The object of the study was to determine the pore structure of cotton cellulose following cross-linking with formaldehyde in typically wet-cure and bake-cure reactions. The state of distension of the accessible regions of the fibers at the time of cross-linking differs under the two reactions conditions and was expected to be reflected in the gel permeation properties of the cross-linked cottons. Cross-linked compositions were examined at progressively higher levels of formaldehyde contents, which were obtained under various reaction conditions. Cotton was cross-linked in a water-swollen state by both the Forms W and W' processes, which differ primarily in the higher concentration of reagent and lower concentration of water present in the latter. The fabric was cross-linked in the collapsed state by a bake-cure reaction, Form C. These cross-linked samples were prepared in a single curing step with $MgCl_2 \cdot 6H_2O$ as the catalyst. The wet-cure processes Form W and W' produced only limited alterations of the cellulose pore structure at the maximum levels of cross-linking. In contrast, large changes in pore structure resulted from cross-linking the cotton in a collapsed state by the bake-cure Form C process. Here, the permeability limit was reduced from a molecular weight of 2430 for untreated cotton to approximately 1250 at the lowest level of cross-linking achieved. No further decrease of this parameter was produced by cross-linking to the maximum level.

D.   Whole Fiber Cotton

The chemical modification of cotton fabric involves treatment of cotton cellulose in a whole fiber form. Wiley milling as described earlier, while producing a fibrous product, does introduce undesirable perturbations in this crystalline polymer. It was therefore desirable to develop a technique for preparing columns from whole-fiber cotton without chopping to pass an 80-mesh screen. The trial studies [14, 15] were conducted on sterile absorbent cotton, Soxhlet extracted sliver, roving, and yarn, and desized, scoured and bleached cotton printcloth. Among the various trial preparations of columns were (1) settling of loose fibers into a column, (2) random or ordered packing of chunks, balls, and sliver, (3) ordered wrapping of sliver, roving, yarn, and fabric prior to or during insertion into the column, and (4) ordered packing of die-cut disks. It is essential that packing be even. Methods 1–3 did not produce usable columns. Method 4 was first applied to cotton fabric. The die cut disks were exactly the same diameter as the interior of the column. The column was prepared with the fabric running perpendicular to the length of the column. Channeling was a problem, probably because the fabric construction restricted fiber swelling. The most satisfactory technique involved die-cut disks of batting. These dry disks of parallel fibers were pressed together into the column with a dowel rod, taking care to maintain fiber orientation perpendicular to the column's length. After wetting, additional compression allowed the addition of more disks. Such columns give usable data, but peaks are broader than those observed with either Sephadex G-15 or chopped cotton. The advantage is that one is able to assess the pore size distribution in cotton that has had minimal mechanical perturbation.

The technique was first used to assess changes in cotton effected by caustic mercerization and liquid ammonia treatment. The starting batting was sterile absorbent cotton batting. Caustic mercerization was with 23% NaOH at room temperature. Liquid ammonia treatment involved evaporation of almost all of the ammonia before quenching with and rinsing in water. Columns were characterized with the series of sugars, ethylene glycols, and glymes. Results showed that accessible solvent water in the pores was lowest for the untreated cotton, highest for the caustic mercerized cotton, and intermediate for the liquid ammonia-treated cotton. There were general similarities but small quantitative differences for results from chopped and whole-fiber cottons, as would be expected. This procedure represented a substantial advance in this endeavor and a real improvement in column stability.

E.   Fabric

Two different techniques have been described by other investigators for evaluating the pore size distribution in cotton fabric. The work of Bredereck et al. [16] is summarized in a recent review. These investigators cut discs of fabric that matched the column diameter exactly. The discs were introduced under gentle pressure into an HPLS steel column of 250 mm length and 4 mm internal diameter. The weight of the filling was approximately 2 g and the intermediate fiber volume 1.1–1.5 ml. This parameter was determined via the elution of Dextran T-2000. A double-piston pump with high pulsation and constant flow delivered the eluents. A Rheodyne

7125 sample application valve and LCD 202 RI detector completed the setup. The solutes were applied as 0.02 ml of 1% aqueous solutions (0.5% for dextrans of molecular weight greater than 40,000). The eluent was double-distilled water. Test samples were applied individually, and elution volumes were determined via peak maxima. Under normal circumstances columns were stable for 6 months. This technique was used to elucidate the effects of several finishing treatments on cotton fabric. These included cold and hot caustic mercerization, liquid ammonia treatment with ammonia removal both by rinsing in cold water and by dry evaporation on hot cylinders, and dry cross-linking with 1,3-dimethylol-4,5-dihydroxyethyleneurea.

Ladisch et al. [17] have reported a different liquid chromatography technique for studying the pore volume distribution in fabrics. They described a method for using a whole piece of fabric rolled into a cylinder. The fabric was rolled tightly along the warp direction and then pulled into a standard 7 mm i.d. × 300 mm LC column. The tightness of the packing was demonstrated via scanning electron microscopy. Experimentally the system includes a water reservoir, a pump adjusted to flow at a rate of 0.20 ml/min, a syringe loading sample injector (20 μl), the column immersed in a circulating water bath, a differential refractometer, and a chart recorder. These authors use different notations than those employed by Bertoniere et al. A brief summary follows.

Bertoniere et al.		Ladisch et al.	
V_0	Void volume; from Dextran T40	V_ϵ	External void volume; from Dextran T40
V_e	Elution volume	$V_{r,i}$	Elution volume
V_i	Internal water available as solvent to a specific solute	V_i	Void Volume; specific accessible internal void volume corresponding to probe of specific diameter

Results for cotton and ramie fabrics were reported in this initial publication. Nine polyethylene glycols (in a series) were used as molecular probes. The values of Nelson and Oliver [20] for their molecular diameters were used. The fabrics were evaluated at both 30 and 60°C. It was shown that cotton had 100–200% more void volume (V_i) than ramie. The total void volume (V_i) of both cotton and ramie was not sensitive to temperature changes in the range of 30–60°C. The results of Ladisch et al. are in agreement with published data [15]. Subsequent studies using this liquid chromatography method on whole fabric by Ladisch et al. involved dyeability. They obtained direct dye absorption isotherms using a rolled cotton fabric stationary phase in a liquid chromatographic column [18]. A frontal analysis technique gave results similar to those obtained with standard equilibrium adsorption measurements. In a later report [19] frontal analysis provided retention volumes of basic dyes on acrylic fabric, rolled and inserted into a liquid chromatographic column. The retention volumes indicated differences in adsorption and therefore in compatibilities of standard dye mixtures.

III. CURRENT METHODS USED AT SRRC TO CHARACTERIZE COTTON

A. Equipment

The system is assembled from separately purchased commercial units as shown in Figure 2. Component parts include columns, a sample injector, a pump, a differential refractometer detector, a recorder, a fraction collector, tared test tubes, and a precision balance. The columns used must be precision bore with dimensions

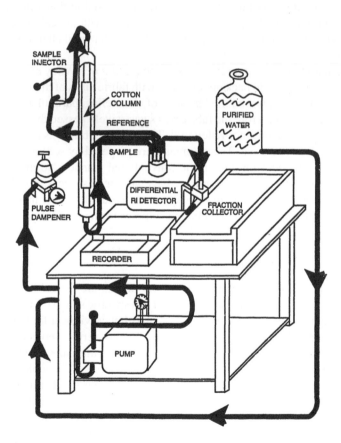

Figure 2   Diagram of assembled equipment used at SRRC.

2.54 (or 1.27) cm $\times$ 45–50 cm between top and bottom bed supports. The sample injector must be capable of delivering 0.5 ml to the column without interruption of flow. The pump should provide a flow rate of $\sim$26 ml/cm^2/hr; pulse dampening is usually necessary with the positive-displacement minipumps we have used.

B. Column Preparation

In order to prepare columns from whole cotton fibers a batting with parallelized fibers is prepared on a card. Disks are cut from the batt with a die having precisely the same diameter (2.54 or 1.27 cm) as the interior of the column. The column is packed with the die-cut disks in the dry state. These fiber disks are pressed together with a dowel rod, taking care to retain the configuration of fiber lengths running perpendicular to the length of the column. After the column is wetted down, the disks are further compacted with the dowel rod and additional fiber disks are added. It is essential that the maximum amount of cotton be used in the column and that both top and bottom bed supports make good contact with the cellulose. The column is placed in the system and water is pumped through it until all trapped air is removed. This can take several days.

In order to evaluate fabric we found it necessary to grind it in a Wiley mill as discussed earlier, as our attempts to pack our columns with fabric disks of the same diameter as the interior of the column failed because of channeling, particularly if the cotton was cross-linked. Therefore the fabric was successively passed through 20-, 40-, 60-, and 80-mesh screens in a Wiley mill. The ground fabric was placed in water and the slurry was degassed. The columns were prepared by settling the cotton slurries through an extension tube in the conventional manner. Normally the 2.54-cm-diameter columns were used, but we have also been successful with 1.27-cm-diameter columns, which require substantially less sample. Other investigators [16, 17] have reported success in preparing columns from intact fabric using other equipment as described earlier.

C. Evaluation

The water-soluble solutes used routinely as molecular probes are assembled in Table 1. Plots of the internal water available as solvent to a specific solute (V_i) versus its molecular weight are linear in the case of the sugars but curvilinear for both the ethylene glycols and the glymes. Within a homologous series, molecular weight is a good measure of relative molecular size. A molecular size basis is preferable when making comparison with different series of solutes. The molecular diameters of the sugars have been reported by Stone and Scallan [3]. Estimates of the molecular diameters of the lower molecular weight ethylene glycols were based on extrapolations from measurements of Nelson and Oliver [20]. Measurements of molecular diameters were not available for the glymes but were approximated by assuming that molecular sizes of the hydrated molecules are the same

Table 1 Molecular Probes Used in Reverse Gel Permeation Chromatography

Molecular probe	Molecular weight	Molecular diameter (Å)
Dextran T-40	40,000	
(void volume)		
Sugars		
Stachyose	666.58	14
Raffinose	504.44	12
Maltose	342.30	10
Glucose	180.16	8
Ethylene glycols, degree of polymerization		
6	282.33	15.6
5	238.28	14.1
4	194.22	12.7
3	150.17	10.8
2	106.12	8.4
1	62.07	5.5
Glymes, degree of polymerization		
1	222.28	13.8
2	178.22	12.1
3	134.17	9.9
4	90.12	7.4

as those of the parent glycols at the same molecular weight. Plots of V_i versus molecular diameter are linear for all three sets of molecular probes.

Dextran T-40, the sugars, and the ethylene glycols were applied individually as 2% solutions through a 0.5-ml sample loop. The flow rate was 26 ml/cm^2/hr. The eluate was monitored continuously with an LDC differential refractometer from Milton Roy. Elution volumes were determined gravimetrically by collecting the eluate in tared test tubes and summing the weights of fractions and proportional parts of fractions between the injection and the peak of the recorded elution curve for each solute. Gel permeation chromatographic results are obtained in terms of:

V_e, elution volume for each specific substrate
V_0, total void volume for the column
V_t, the total column volume
W, the weight (dry basis) of material in the column.

The total void volume V_0 is the elution volume of the high-molecular-weight solute Dextran T-40, molecular weight 40,000, which is totally excluded from the internal pore structure. Calculated results are expressed as accessible internal volume V_i (ml/g), specific gel volume V_g (ml/g), and internal water V_w (ml/g). These terms (per gram dry cellulose) were defined by the following equations [11]:

$$V_i = (V_e - V_0)/W$$
$$V_g = (V_t - V_0)/W$$
$$V_w = V_g - 0.629$$

The specific volume occupied by the solid cellulose in the water-wet fiber (required to calculate V_w) was taken as 0.629 ml/g, which corresponds to a density of 1.59 g/ml [21].

D. Plots

V_i values are typically the averages of six replicates. Standard deviations are generally in the order of 0.001 to 0.01. Data (V_i vs. the molecular diameter of the test solute) are fit to linear regression models. The three sets of molecular probes give similar but not identical results, as shown in Figure 3. Results for pores accessible to small or moderate-sized molecules consistently fall in the order sugars > ethylene glycols > glymes. The sugars, relatively stiff and bulky molecules, are very similar to cellulose itself in hydrophilicity and hydrogen bonding power [15]. These solutes are thus competitive with cellulose for bound water within the pores of cotton and find all of the internal water available as solvent. The ethylene glycols, more flexible, slender molecules, contain both a hydroxyl group and an ether oxygen. While their greater flexibility would result in better penetrating power, they would compete less successfully for internal bound water and find a smaller

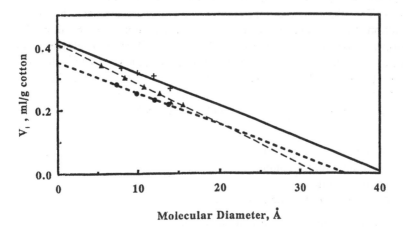

Figure 3 Differences in the internal volume of unmodified cotton available as solvent to the sugars (✦), ethylene glycols (▲), and glymes (●).

fraction of it available as solvent. The glymes contain only ether structures, and the internal water available to small molecules is substantially lower than that of the parent ethylene glycols.

Useful predicted values for V_2, V_{10}, V_{16}, C_{21}, and V_{27} have been calculated as the mean of V_i when molecular diameter was 2 (water), 10 (typical cross-linking agent), 16 (polyethylene glycol, molecular weight 300), 21 (polyethylene glycol, molecular weight of 600), and 27 (polyethylene glycol, molecular weight of 1000), respectively. Predicted values for molecular diameter when $V_i = 0$ gives M_x, the permeability limit. This is the smallest molecule indicated to be excluded from the fiber interior.

IV. STRUCTURES ELUCIDATED

A. Cotton Variety

Cotton fibers are seed hairs of plants belonging to the genus *Gossypium*. Each variety produces a characteristic type of fiber. *Gossypium barbadense* is a long staple type whereas *Gossypium hirsutum* is coarser. American upland cottons (*G. hirsutum*) account for most of the world fiber supply. Pima, a *G. barbadense* that is a complex cross of several cottons, contributes to the remainder. It is grown primarily in the southwestern United States. Different varieties of cotton are known to differ in many physical properties such as staple length, diameter, strength, elongation, toughness, and color. Potential differences in supramolecular structure have been less fully elucidated.

The reverse gel permeation chromatography technique was used to address differences in the pore structures of two varieties of cotton [22]. The varieties were DP-90 (a common upland variety) and NX-1 (a hybrid of upland and Pima cottons). Ginning was done on a small laboratory gin. Pore structure was assessed on the greige cottons in both the whole and ground states. The series of oligomeric sugars and ethylene glycols were used as molecular probes. Ground samples consistently had more accessible internal volume than the whole fibers across the entire range of pore sizes. This is attributed to damage effected by the grinding procedure. General differences between the two varieties were the same regardless of the physical state of the fiber in the column. The DP-90 had the more open structure across the whole range of pore sizes. This is illustrated in Figure 4 with data for the whole fibers. The chromatographic results were compatible with results from water of imbibition [23], which is related to internal volume in the water swollen state.

B. Bast Fibers

Another study [24] extended the use of this technique to jute, a lignocellulosic bast fiber. The pore structures of jute and purified cotton cellulose were compared and

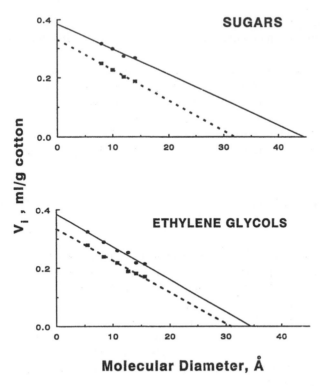

Figure 4 Differences in the pore size distribution between the cotton varieties NX-1 (■) and DP-90 (●).

the effect of scouring on jute was determined. Both native and scoured jute had greater pore volumes than purified cotton. Scouring effected an increase in the internal volume of the jute fiber over the measured range of pore sizes. Data on the fraction of the total internal water volume accessible to the water molecule itself indicated a similarity between cotton and scoured jute, but these and other data suggested a repelling interaction between the surfaces on the internal pores of native jute and the sugars used as test solutes. This repellant effect was attributed to the presence of lignin on these surfaces.

C. Pretreatments

Prior to chemical modification with various agents cotton fabric is routinely desized, scoured, and bleached. Scouring imparts the required wettability by removing the natural waxes and the sizing agent applied to facilitate fabric construction. Bleaching renders the fabric white so that it can be dyed true to color. In many instances swelling pretreatments are utilized to improve dimensional stability, dyeability, luster, softness, and textile performance. Such pretreatments include caustic mercerization and liquid ammonia treatment. To study changes in pore size distribution caused by these pretreatments, columns prepared from whole cotton fibers in the form of batting were used [25, 26]. The treatments included standard scouring/bleaching, caustic mercerization, and a liquid ammonia treatment with removal by volatilization at elevated temperature. Results were compared to data on the pore structure of the fiber in the greige state. The water holding capacities of these fibers as measured by water of imbibition, specific internal water V_w, and V_2 (sugars) as assembled in Table 2. Scouring/bleaching decreased the water holding capacity of the greige cotton as measured by water of imbibition and the column parameter V_w but increased it as indicated by gel permeation measurements. The internal volume accessible to water was substantially increased by caustic mercerization but was only slightly affected by liquid ammonia treatment. The gel permeation data are given in Figure 5. The relative accessibility of the cotton fibers to molecules of the size of durable press finishing agents (~ 10 Å) was slightly increased by scouring/bleaching, substantially increased by caustic mercerization, but moderately reduced by liquid ammonia treatment. Accessibility to molecules near the permeability limits of the fibers followed similar trends but differences were greater. Scouring/bleaching increased the permeability limit of the greige fibers, but subsequent mercerization or liquid ammonia treatment decreased it. There was a noteworthy difference in permeability limit and relative accessibility to large molecules. This is accounted for by a decrease in the rate of change in pore size on scouring/bleaching but substantial increases, to generally more than double, on subsequent caustic mercerization or liquid ammonia treatment of the scoured/bleached cotton.

Processing in liquid ammonia causes complex changes in the cotton fiber that are only partially understood. Liquid ammonia penetrates the cotton fiber, effect-

Table 2 Water-Holding Capacity of the Cotton Fibers as Indicated by Water of Imbibition and Gel Permeation Measurements (All Values in ml water/g Cotton)

Batting	Water of imbibition	V_w	V_2 sugars	V_2 ethylene glycols
Greige control	0.337 ± 0.004	0.340 ± 0.004	0.316 ± 0.003	0.291 ± 0.002
Scoured/ bleached	0.334 ± 0.006	0.297 ± 0.006	0.319 ± 0.004	0.307 ± 0.004
Caustic mercerized	0.540 ± 0.008	0.557 ± 0.002	0.598 ± 0.008	0.356 ± 0.001
Ammonia treated	0.348 ± 0.012	0.357 ± 0.004	0.357 ± 0.005	0.300 ± 0.002

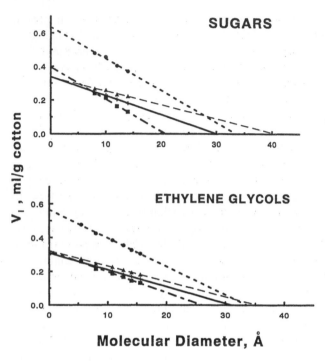

Figure 5 Changes effect in the pore size distribution of greige cotton (✚) effected by scouring/bleaching (▲), caustic mercerization (●), and liquid ammonia treatment with dry removal (■).

ing intra- and inter fibrillar swelling, which disrupts the hydrogen bond structures and leads to the formation of a cellulose–ammonia complex. Destruction of this complex by evaporative removal of the ammonia leads to the formation of cellulose III, but removal by water exchange results in regeneration of cellulose I. The removal technique alters the physical properties of the treated celluloses as well as its crystal structure. Its impact on the pore size distribution was assessed using the reverse gel permeation chromatography technique with whole fiber columns in an earlier investigation using sterilized medical cotton batting [27]. These purifed cotton battings were treated with liquid ammonia, which was removed by volatilization at ambient temperature, by volatilization at elevated temperature, and by water exchange. Results are given in Figure 6. All three liquid ammonia treatments increased the internal pore volumes accessible to small molecules in the purified cotton. The greatest increase was noted when the ammonia was removed by water exchange and the least when volatilization at elevated temperature was employed.

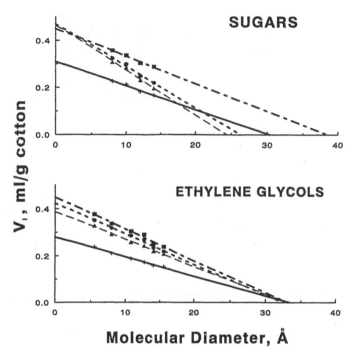

Figure 6   Internal water (V_i) in purified (+), ammonia–96°C (▲), ammonia–25°C (●), and ammonia–water (■) treated cotton battings accessible to sugars and ethylene glycols.

Ambient temperature volatilization had an intermediate effect. Decreases in the volumes of large pores were effected by ammonia treatment followed by volatilization at ambient and elevated temperature. Water exchange of the ammonia resulted in an increase in the volume of large as well as of the small pores.

D.   Cross-Linking

1.   DMDHEU

Cotton fabric must be cross-linked to impart the easy care properties required by the consumer. At present cotton fabric is cross-linked not with formaldehyde itself but with formaldehyde derivatives of amides. The most widely used of these is dimethyloldihydroxyethyleneurea (DMDHEU) and its various low formaldehyde-release modifications. Cross-linking with DMDHEU enhances resilience but with concomitant losses in strength. A study [28] was conducted to determine if these strength losses were associated with changes in the pore size distributions in the cotton fibers. Cotton printcloth was treated with DMDHEU to five add-on levels.

Magnesium chloride hexahydrate was the cross-linking catalyst. The reverse gel permeation chromatographic technique was used to follow changes in pore size distribution. Columns were prepared by settling water slurries of the Wiley-mill ground cotton as described earlier. The series of oligomeric sugars and ethylene glycols were used as molecular probes. These data are shown in Figure 7. Progressive losses in the accessible internal volume were observed with increasing degree of cross-linking across the entire distribution of pore sizes. Increases in re-silience were accompanied by the expected losses in strength, which in turn were associated with decreases in the accessible internal volume of the fibers.

Pores are voids between elementary fibrils or microfibrils. It is generally accepted that cross-linking tends to "fix" the cellulose structure in the state in which it exists during the cross-linking process. The cross-links hold the microfibrillar units in close association, which inhibits swelling. The closeness of the association is a function of the conditions and degree of cross-linking. The effect of reaction conditions has been demonstrated with this pore size distribution

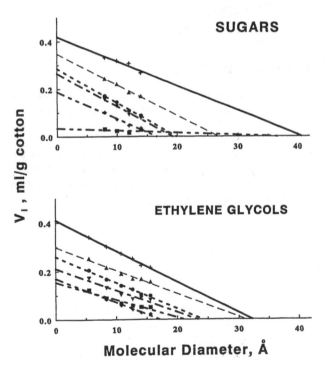

Figure 7   Internal water (V_i) in unmodified cotton (✚) and cotton cross-linked with 1% (▲), 2% (●), 4% (▼), 6% (◆), and 8% (■) DMDHEU accessible to sugars and ethylene glycols.

technique for cross-linking with formaldehyde [7]. It was shown that the accessibility of the cotton fiber was increased by reactions with formaldehyde in aqueous solution but decreased when a bake-cure technique was used. DMDHEU was applied in a pad-dry-cure process, which has "fixed" the fibers in the collapsed state that existed when covalent bond formation occurred during curing. The degree to which the cross-linking affected the pore size distribution must be related to the extent of reaction with DMDHEU as the processing conditions were constant. Small, medium, and large pores were comparably affected at the same level of DMDHEU application. Although strength losses were associated with collapse of the internal pore structure of the cotton, they were not associated with the loss of a specific pore size.

2. Dyeability

Cotton that has been cross-linked to impart durable-press properties is intrinsically dye-resistant because of the collapse of the internal structure as described earlier. For this reason cotton fabric is dyed before the cross-linking treatment is applied. Dyeing of textiles in garment form is now of economic interest to the American textile industry. As currently practiced, garments dyed in this manner are made of unmodified cotton and the resulting clothing thus has a rumpled appearance. In order to extend this process to include apparel with durable-press properties, technology had to be developed to overcome the dye-resistant properties of cross-linked cotton.

In the course of an investigation to develop durable-press reagents that do not contain formaldehyde, 4,5-dihydroxy-1,3-dimethyl-2-imidazolidinone (DHDMI) was evaluated as a cross-linking agent for cotton. The treated cotton fabrics, which contained only intralamellar cross-links, differ from conventionally cross-linked cottons, which contain both intra- and interlamellar crosslinks. They proved receptive to direct red 81 but exhibited the more usual dye resistance when larger dye molecules were used (Fig. 8). The internal structure of cotton cross-linked with this reagent was thus of interest.

The reverse gel permeation chromatography technique was employed to assess differences in the pore size distribution (Fig. 9) between cottons cross-linked with DHDMI and with DMDHEU to comparable levels of wrinkle recovery [29]. Results were compared to the receptivity of these cotton samples to the dye direct red 81 (Fig. 10). It was shown that although cross-linking of cotton with either DMDHEU or DHDMI reduced accessible internal volume, those treated with DHDMI retained substantially more accessible internal volume across the entire range of pore sizes. Increasing add-on of DMDHEU further reduced the accessible internal volume. In contrast, the accessible internal volume in DHDMI treated cotton was increased by additional add-on of this reagent. The trends with respect to relative receptivity to direct red 81 generally related better to the quantity of residual large pores (17 Å) than to remaining intermediate pores (10 Å).

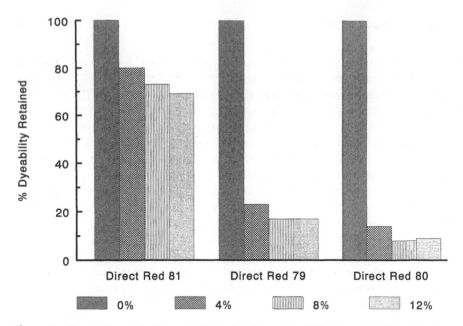

Figure 8 Dyeability with direct red 81 (molecular weight 676), direct red 79 (molecular weight 1049), and direct red 80 (molecular weight 1373) after cross-linking with DHDMI at levels indicated.

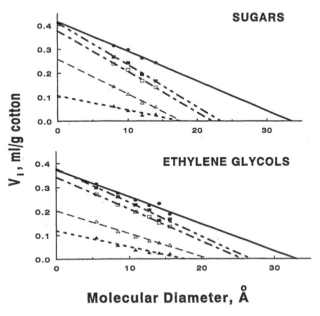

Figure 9 Internal water (V_i) in unmodified cotton (●) and cotton cross-linked with 3% DMDHEU (△), 8% DMDHEU (▲), 7.5% DHDMI (□), and 15% DHDMI (■).

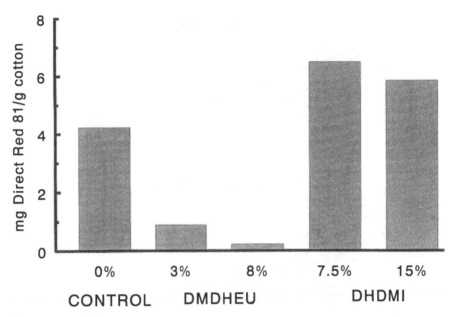

Figure 10 Effect of cross-linking with DMDHEU and DHDMI on uptake of direct red 81.

3. Formaldehyde-Free Reagents

OSHA regulations regarding allowable formaldehyde in air in the workplace have prompted research into the development of new cross-linking reagents that do not contain formaldehyde. Several formaldehyde-free cross-linking systems have been developed. The addition product of 1,3-dimethylurea and glyoxal (4,5-dihydroxy-1,3-dimethylimidazolidinone, DHDMI) is available commercially. Glyoxal with either ethylene glycols or 1,6-hexanediol as coreactive additives was another system explored [30, 31]. Systems based on polycarboxylic acids as cross-linking agents for cellulose [32, 33] with catalysis by alkali metal salts of phosphorus containing inorganic acids [34] are an ongoing area of research. The most effective of these organic acids to date is 1,2,3,4-butanetetracarboxylic acid (BTCA).

These formaldehyde-free reagents differ in the weight add-on required to impart easy care performance to cotton fabric. Generally it requires considerably more reagent to impart wrinkle resistance to cotton with the formaldehyde-free reagents than it does with the formaldehyde derivative DMDHEU. This suggests that cross-links imparted by the formaldehyde-free agents differ considerably from those from DMDHEU. A study was conducted to compare the selected formaldehyde-free cross-linking agents among themselves and with the conventional agent DMDHEU with respect to the degree to which they altered the pore size distribution in the crosslinked cotton.

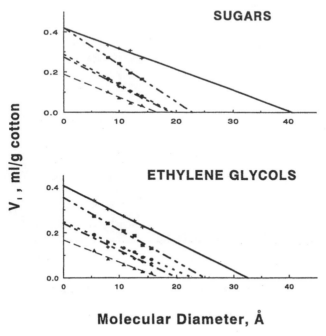

Figure 11 Internal water (V_i) in unmodified cotton (✦) and cotton cross-linked with DHDMI (■), BTCA (◆), glyoxal/glycol (●), and DMDHEU (▲).

The formaldehyde-free reagents were BTCA (butanetetracarboxylic acid), DHDMI (dihydroxydimethylimidazolidinone), and the glyoxal/glycol system. The fabric was an 80 × 80 cotton printcloth. Treatments were designed to impart the same degree of resilience to the fabric as measured by the conditioned wrinkle recovery angle. This was achieved with the exception of DHDMI, where a lower level of resilience was realized. The reverse gel permeation chromatographic technique was used to follow changes in pore size distribution. Columns were prepared by settling water slurries of the Wiley-milled cotton as described earlier. The series of oligomeric sugars and ethylene glycols were used as molecular probes. Results are given in Figure 11. It was concluded that formaldehyde-free cross-linking reagents effect a lower level of collapse of the internal pore structure of the cotton fiber than does DMDHEU at generally comparable levels of resilience.

4. BTCA Catalysts

The key to the success of cross-linking of cellulose with polycarboxylic acids was the development of new catalyst systems based on alkali metal salts of phosphorus containing inorganic acids [35, 36]. The most effective of these organic acids to date is 1,2,3,4-butanetetracarboxylic acid (BTCA). The level of textile performance that is realized with BTCA applied to cotton fabric is directly related to the catalyst used. An investigation [37] was therefore conducted in which we evaluated cotton cross-linked with BTCA via catalysis by six alkali metal salts of phosphorus acids and by sodium carbonate. The catalysts included in this study were $NaH_2PO_2 \cdot H_2O$, $NaH_2PO_3 \cdot 2.5H_2O$, $Na_2HPO_3 \cdot 5H_2O$, $NaH_2PO_4 \cdot H_2O$, Na_2HPO_4, $Na_4P_2O_7$, and Na_2CO_3. The treatments were applied to all-cotton printcloth via a pad-dry-cure process. Pore size distribution was assessed on Wiley milled fabric via the reverse gel permeation chromatographic technique. The water-soluble molecular probes employed were sugars and ethylene glycols. Plots of V_i against molecular diameter are given in Figure 12, and a comparison of the effects on small and medium sized pores is given in Figure 13. Definite patterns were observed in textile performance realized with the different catalysts. It was shown that the total volume in residual small pores was inversely related to the resilience level achieved and that retained breaking strength was directly related to the volume in

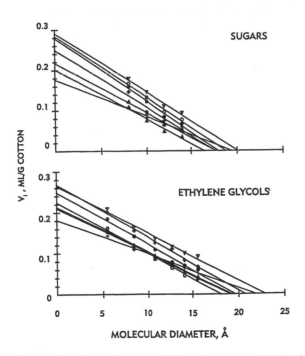

Figure 12   Internal water (V_i) in cotton cross-linked with BTCA via Na_2HPO_2 (+), NaH_2PO_3 (▲), Na_2HPO_3 (+), NaH_2PO_4 (△), Na_2HPO_4 (○), $Na_4P_2O_7$ (●), and Na_2CO_3 (▽) catalysis as functions of the molecular diameters of sugars and ethylene glycols.

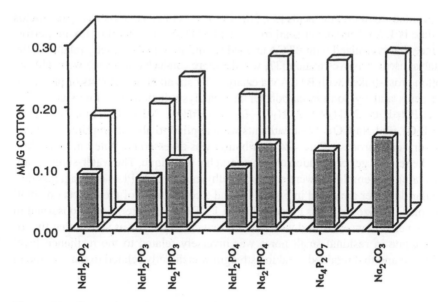

Figure 13   Comparison of residual small (V_2, open bars) and medium (V_{10}, hatched bars) pores in cotton cross-linked with BTCA; sugars were used as molecular probes.

residual small pores. Patterns with respect to abrasion resistance were more complex. As BTCA add-ons were comparable, the data suggest that the more effective catalysts, NaH_2PO_2 and NaH_2PO_3, are either effecting a greater number of cross-links in the cotton or producing cross-links that differ in actual structure.

V.   FUTURE WORK

Research in this area continues. It is being used to study differences in accessibility of different cross-linking agents for cotton to the interior of the cotton fibers. Work is also being initiated into changes in the pore size distribution effected by treatment with cellulase enzymes, which is now a commercial practice.

REFERENCES

1.   J. Porath and P. Flodin, Gel filtration: A method for desalting and group separation, *Nature 183*:1657 (1959).
2.   L. G. Aggebrandt and O. Samuelson, Penetration of water-soluble polymers into cellulose fibers, *J. Appl. Polym. Sci. 8*:2801 (1964).
3.   J. E. Stone and A. M. Scallan, A structural model for the cell wall of water swollen wood pulp fibres based on their accessibility to macromolecules, *Cellulose Chem. Technol. 2*:343 (1968).

4. L. F. Martin and S. P. Rowland, Gel permeation properties of decrystallized cotton cellulose, *J. Chromatogr. 28*:139 (1967).

5. L. F. Martin and S. P. Rowland, Gel permeation properties of cellulose. I. Preliminary comparison of unmodified and crosslinked, decrystallized cotton cellulose, *J. Polym. Sci. Part A-1 5*:2563 (1967).

6. L. F. Martin, F. A. Blouin, N. R. Bertoniere, and S. P. Rowland, Gel permeation technique for characterizing chemically modified celluloses, *Tappi 52*:708 (1969).

7. L. F. Martin, N. R. Bertoniere, F. A. Blouin, M. A. Brannan, and S. P. Rowland, Gel permeation properties of cellulose II: Comparison of structures of decrystallized cotton crosslinked with formaldehyde by various processes, *Textile Res. J. 40*:8 (1970).

8. L. F. Martin, F. A. Blouin, and S. P. Rowland, Characterization of the internal pore structures of cotton and chemically modified cottons by gel permeation, *Separation Sci. 6*:287 (1971).

9. N. R. Bertoniere, L. F. Martin, and S. P. Rowland, Stereoselectivity in the elution of sugars from columns of Sephadex G15, *Carbohydrate Res. 19*:189 (1971).

10. L. F. Martin, N. R. Bertoniere, and S. P. Rowland, The effects of sorption and molecular size of solutes upon elution from polyhydroxylic gels, *J. Chromatogr. 64*:263 (1972).

11. S. P. Rowland and N. R. Bertoniere, Some interactions of water-soluble solutes with cellulose and Sephadex, *Textile Res. J. 46*:770 (1976).

12. F. A. Blouin, L. F. Martin, and S. P. Rowland, Gel-permeation properties of cellulose. Part III: Measurement of pore structure of unmodified and of mercerized cottons in fibrous form, *Textile Res. J. 40*:809 (1970).

13. F. A. Blouin, L. F. Martin, and S. P. Rowland, Gel-permeation properties of cellulose. Part IV: Changes in pore structure of fibrous cotton produced by crosslinking with formaldehyde, *Textile Res. J. 40*:959 (1970).

14. C. P. Wade, Preparation of whole-fiber cotton gel-filtration chromatography columns, *J. Chromatogr. 268*:187 (1983).

15. S. P. Rowland, C. P. Wade, and N. R. Bertoniere, Pore structure analysis of purified, sodium hydroxide-treated and liquid ammonia-treated cotton celluloses, *J. Appl. Polym. Sci. 29*:3349 (1984).

16. K. Bredereck and A. Blüher, Determination of the pore structure of cellulose fibers by exclusion chromatography. Principles and use examples for swelling treatments and resin finishing of cotton fabrics, *Melliand Textilberichte 73*:297(English), 652 (German) (1992).

17. C. M. Ladisch, Y. Yang, A. Velayudhan, and M. R. Ladisch, A new approach to the study of textile properties with liquid chromatography—Comparison of void volume and surface area of cotton and ramie using a rolled fabric stationary phase, *Textile Res. J. 62*:361 (1992).

18. C. M. Ladisch and Y. Yang, A new approach to the study of textile dyeing properties with liquid chromatography—Part I: Direct dye absorption on cotton using a rolled fabric stationary phase, *Textile Res. J. 62*:481 (1992).

19. Y. Yang and C. M. Ladisch, A new approach to the study of textile dyeing properties with liquid chromatography—Part II: Compatibility of basic dyes for acrylic fabric, *Textile Res. J. 62*:531 (1992).

20. R. Nelson and D. W. Oliver, Study of cellulose and its relation to reactivity, *J. Polym. Sci. Part C 36*:305 (1968).

21. P. H. Hermans, *Physics and Chemistry of Cellulose Fibers with Particular Reference to Rayon*, Elsevier, New York, 1949, p 20.

22. N. R. Bertoniere, W. D. King, and S. E. Hughs, Effect of variety on the pore structure of the cotton fiber, *Lignocellulosics—Science, Technology, Development and Use* (J. F. Kennedy, G. O. Phillips, and P. A. Williams, eds.), Ellis Horwood Limited, Chichester, UK, 1992, p. 457.

23. H. M. Welo, H. M. Ziffle, and A. W. McDonald, Swelling capacities of fibers. Part II: Centrifuge studies, *Textile Res. J. 22*:261 (1952).

24. N. R. Bertoniere, S. P. Rowland, M. Kabir, and A. Rahman, Gel permeation characteristics of jute and cotton, *Textile Res. J. 54*:434 (1984).

25. N. R. Bertoniere and W. D. King, Effect of scouring/bleaching, caustic mercerization and liquid ammonia treatment on the pore structure of cotton textile fibers, *Textile Res. J. 59*:114 (1989).

26. N. R. Bertoniere, Pore structure analysis of cotton cellulose via gel permeation chromatography, *Cellulose—Structure and Functional Aspects* (J. F. Kennedy, G. O. Phillips and P. A. Williams, eds.), Ellis Horwood Limited, Chichester, United Kingdom, 1989, p. 99.

27. N. R. Bertoniere, W. D. King, and S. P. Rowland, Effect of mode of agent removal on the pore structure of liquid ammonia treated cotton cellulose, *J. Appl. Polym. Sci. 31*:2769 (1986).

28. N. R. Bertoniere and W. D. King, Residual pore volume, resilience and strength of crosslinked cotton cellulose, *Textile Res. J. 60*:606 (1990).

29. N. R. Bertoniere and W. D. King, Pore structure and dyeability of cotton crosslinked with DMDHEU and with DHDMI, *Textile Res. J. 59*:608 (1989).

30. C. M. Welch, Formaldehyde-free durable press finishing of cotton, *Textile Chem. Color. 16*:265 (1984).

31. C. M. Welch and J. G. Peters, Low, medium, and high temperature catalysts for formaldehyde-free durable press finishing by the glyoxal-glycol process, *Textile Res. J. 57*:351 (1987).

32. S. P. Rowland, C. M. Welch, M. A. F. Brannan, and D. M. Gallagher, Introduction of ester cross links into cotton cellulose by a rapid curing process, *Textile Res. J. 37*:933 (1967).

33. S. P. Rowland, C. M. Welch, and M. A. F. Brannan, Cellulose fiber crosslinked and esterified with polycarboxylic acids, U. S. Patent 3,526,048, September 1, 1970.

34. C. M. Welch and B. K. Andrews, Catalysis for processes for formaldehyde-free durable press finishing of cotton textiles with polycarboxylic acids, U. S. Patent 4,820,307, April 11, 1989.

35. C. M. Welch, Tetracarboxylic acids as formaldehyde-free finishing agents. Part I: Catalyst, additive, and durability studies, *Textile Res. J. 58*:480 (1988).

36. C. M. Welch, Durable press finishing without formaldehyde, *Textile Chem. Color 22*:13 (1990).

37. N. R. Bertoniere, W. D. King, and C. M. Welch, Effect of catalyst on the pore structure of cotton cellulose cross-linked with butanetetracarboxylic acid, *Textile Res. J. 64*:247–255 (1994)

8

Pore Structure in Fibrous Networks as Related to Absorption

LUDWIG REBENFELD, BERNARD MILLER, and ILYA TYOMKIN
TRI/Princeton, Princeton, New Jersey

I. INTRODUCTION

The textile production process is remarkably flexible, allowing the manufacture of fibrous materials with widely diverse physical properties. All textiles are discontinuous materials in that they are produced from macroscopic subelements (finite-length fibers or continuous filaments). In woven and knitted textiles, the fibers or filaments are first formed into spun or multifilament yarns prior to either weaving or knitting. In nonwoven materials, the fibers or filaments are processed directly into the final planar structure, and then either chemically or physically bonded or mechanically interlocked. The chemical, physical, and mechanical properties of textile materials depend on the inherent properties of the component fibers and on the geometric arrangement of the fibers in the structure.

The discontinuous nature of textile materials means that they have void spaces or pores that contribute directly to some of the key properties of textiles, for example, thermal insulating characteristics, liquid absorption properties, and softness and other tactile characteristics. In physical terms, textile materials have solidities less than unity and therefore finite porosities. This chapter considers the evaluation of the pore structure of textile materials, particularly as that structure relates to liquid absorption and to the porous barrier characteristics of these and related planar materials.

II. POROSITY

A. Basic Concepts

Porosity is one of the important physical quantities that is used to describe textile materials. Porosity, ϵ, represents the fraction of the nominal bulk volume of a material that is occupied by void space. In terms of textile properties, it can be expressed as

$$\epsilon = 1 - \frac{W}{dh} \tag{1}$$

where W is the areal density (mass per unit area) of the fabric, h is its nominal thickness, and d is the bulk density of the fibers from which the fabric is produced. Porosity can range widely, depending on product design and processing techniques.

The cross-sectional shapes of natural and man-made textile fibers vary widely, and the pore structure of a fiber assembly is strongly influenced by this geometrical characteristic. The manner in which fibers can pack in an assembly is largely determined by their cross-sectional shape. If we model fibers as being circular in cross section, then the packing of uniform cylinders provides a good example of how this factor limits the lowest porosity that can be achieved. The closest possible packing of parallel cylinders with uniform radii is in rhombohedral (hexagonal) packing, shown schematically in two dimensions in Figure 1. The porosity of such a system is 0.093, which is the minimum possible value of porosity for cylinders of equal radii. This value is independent of the cylinder radius. To the extent that this model represents an idealized two-dimensional textile material composed of identical fibers with circular cross-sectional shapes, the lowest possible porosity of a textile material is 0.093. If the fibers were not all equal in radius or if they deviated from being perfectly circular in cross-sectional shape, the minimum porosity would increase significantly. On the other hand, if the fibers were square or rectangular in cross-sectional shape, the porosity of the closest packed structure

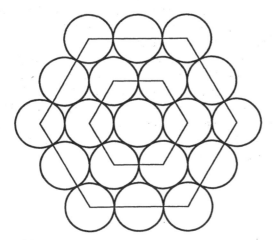

Figure 1 Two-dimensional representation of close-packed cylinders of equal and uniform radii.

could approach zero. In contrast to cylinders, the closest rhombohedral packing of spheres with uniform radii would produce a porosity of 0.259 [1].

The porosities of real textile materials are surprisingly high, reflecting the fact that the component fibers are not densely packed and that the fibers are not uniform in shape and diameter. In fact, textile materials are normally designed to have relatively high porosities. Open and bulky woven and knitted fabrics and certain air-laid nonwovens have porosities in the range of 0.95 and higher. Even those fabrics that appear dense and solid will have porosities in the range of 0.6 to 0.7. Porosity is an especially important property in connection with those textile materials that are used as liquid absorption media. Absorption can be defined as the process wherein liquid displaces the air in the void spaces or the pores in the material, either spontaneously as in wicking or under an external driving pressure. The amount of liquid that can be absorbed is a direct function of the fabric porosity.

The porosity of a material can be experimentally determined by a number of methods, including those based on direct gravimetric and volumetric measurements, optical techniques, liquid imbibition, and gas expansion [2,3]. For textile materials, direct gravimetric and volumetric measurements are normally used to obtain the quantities required in Eq. (1). In quantifying porosity, it is important to distinguish between porosity values that are based on pores that are effective and those that are isolated. Effective pores are defined as those that form a continuous and interconnected phase that reaches to the nominal surface of the network. Isolated pores, on the other hand, are completely enclosed by the solid material and are not a part of the continuous phase. Obviously, only the effective interconnected pores contribute to the sorptive capacity of a material. Direct measurements and optical techniques provide estimates of the total porosity of a material, while imbibition measurements provide estimates of porosities based only on the effective interconnected pores.

B. Porosity and Compression

The porosity of a textile material is strongly affected by lateral compressive forces to which the material is subjected. This is due to the fact that textiles are highly porous and therefore compliant, as shown in Figure 2, where the thickness of three glass-fiber nonwovens differing in areal density is plotted as a function of compressive pressure [4]. The thickness of each of these materials decreases rapidly at first and then levels off with increasing pressure. These results also indicate that the compressed thickness depends on the areal density of the material; that is, at a given compressive pressure the thickness increases with the areal density. Calculating the porosities of the three nonwovens at each compressive pressure, the relationship shown in Figure 3 is obtained. The data for the three materials fall on the same line, indicating that fabrics with different areal densities may not compress to the same thickness, but they do compress to the same porosity. It is also

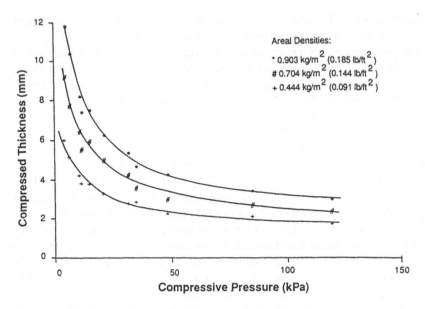

Figure 2   Thickness of three glass fiber mats as a function of compressive pressure. (From Ref. 4.)

interesting to note the strong dependence of porosity on low levels of compressive pressure, and the apparent leveling off of porosity with increasing compressive pressure. Beyond a certain level of compressive pressure, the fibers in the nonwoven become so tightly packed that further pressure cannot cause a further decrease in fabric thickness since the fibers themselves must be considered incompressible. The exact relationships between fabric areal density, thickness, porosity, and compressive pressure, and the shape of curves such as those shown in Figures 2 and 3, will depend on the type of fiber used, the fiber cross-sectional shape and dimensions, and the structure of the material, that is, whether it is a woven or knitted fabric or a specific type of nonwoven material. Nevertheless, the porosities of all textile materials are strongly dependent on compressive pressure. This is particularly important in understanding the pressure dependence of the liquid absorption characteristics of textiles.

It must be emphasized that porosity is an average property of a fibrous material that describes the structure of the material in a very limited way. It provides no description or information about the nature of the void space or about the structure of the fiber network. Entirely different materials can have the same porosity values, and we must look for other means of describing and quantifying the pore structure of textile materials.

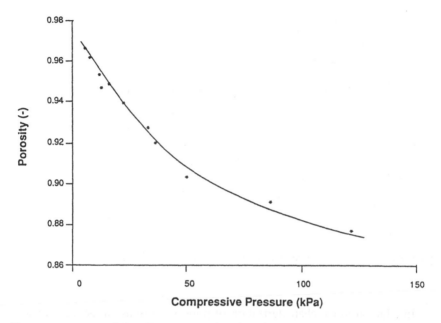

Figure 3   Porosity of glass fiber mats as a function of compressive pressure. (From Ref. 4.)

III.   PORE STRUCTURE

A.   Pore Dimensions

While porosity is an important physical quantity, a more descriptive way of characterizing the porous nature of a network is by quantifying the dimensions of the pores. Considering the pore shape shown in Figure 4, pore dimensions can be described in many terms, for example, their volumes, surface areas, average diameters, and minimum diameters, frequently referred to as pore throats. Each of these dimensions could be critical in controlling a specific kind of behavior. Pore volume is the dominant factor that determines the capacity for absorption of liquid. The total surface area would be the critical property if adsorption phenomena were of primary interest. Pore throat dimensions would be most important if one were concerned with porous barriers and flow-through processes (e.g., filtration), where particle capture and resistance to liquid flow would be directly relatable to the size of these throats. Entrance or exit pore dimensions (end diameters) would be crucial in predicting the probability of particulates penetrating and being retained by the interior of a structure. An average pore diameter could be used as a general characterization of a porous network.

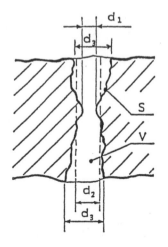

d_1 = throat diameter

d_2 = average diameter

d_3 = entry or exit diameter

S = surface of pore

V = volume of pore

Figure 4   Schematic of a pore in a fibrous network.

In addition to the characterization of pore dimensions on the basis of geo-metrical considerations, pore dimensions can be quantified on the basis of various permeability models. For example, the Carman–Kozeny flow model yields a hy-draulic radius that is related to the volume-to-surface ratio of a pore or capillary. Discussion of permeability models is outside the scope of this chapter, and the reader is referred to standard texts on this subject [2,3].

B.   Distribution of Pore Sizes

While average values of geometric quantities describing pore dimensions provide valuable information about network structure, they are only a little more informa-tive than average porosity values. Fibrous networks are invariably heteroporous (i.e., the dimensions of the pores are not equal), and it is therefore important to consider the distribution of these quantities. Depending on product design and the processing technologies, pore size distributions can be broad or narrow, unimodal, bimodal, or even trimodal.

Pore size distributions in nonwovens are typically unimodal and relatively broad, with a range of values that may cover several orders of magnitude. Non-wovens with bi- and trimodal distributions can also be produced. Woven fabrics manufactured from continuous monofilaments, sometimes referred to as screen-ing fabrics, have pore dimensions that are essentialy monodisperse. The pores in such fabrics are those formed between the monofilaments and at monofilament crossover points, and their dimensions are determined by the filament diameters and the weave pattern.

Woven fabrics manufactured from spun or multifilament yarns have bimodal pore size distributions. A system of large pores is formed by the interlacing yarns as determined by the weave pattern, just as in the case of monofilament woven fabrics. These interyarn pores may be relatively uniform in size and shape. However, another system of pores is formed within the yarn structure between the component fibers or filaments. These intrayarn or interfiber pores are much smaller than the interyarn pores and are in most cases somewhat more polydisperse. Their size and shape are determined by the fiber or filament diameters and cross-sectional shapes, and by the degree of twist that is imposed to impart yarn cohesion and mechanical integrity.

C. Liquid Porosimetry

Liquid porosimetry, also referred to as liquid porometry, is a general term to describe procedures for the evaluation of the distribution of pore dimensions in a porous material based on the use of liquids. There are many forms of liquid porosimetry, but we restrict our discussion to two major methods that are designed to characterize the pore structure in terms of pore volumes and in terms of pore throat dimensions. Both pore volumes and pore throat dimensions are important quantities in connection with the use of fiber networks as absorption and barrier media. Pore volumes determine the capacity of a network to absorb liquid, that is, the total liquid uptake. Pore throat dimensions, on the other hand, are related to the rate of liquid uptake and to the barrier characteristics of a network.

IV. PORE VOLUME DISTRIBUTION ANALYSIS

A. Basic Concepts

Liquid porosimetry evaluates pore volume distributions (PVD) by measuring the volume of liquid located in different size pores of a porous structure. Each pore is sized according to its effective radius, and the contribution of each pore size to the total free volume of the porous network is determined. The effective radius R of any pore is defined by the Laplace equation:

$$R = \frac{2\gamma \cos \theta}{\Delta P} \qquad (2)$$

where

γ = liquid surface tension
θ = advancing or receding contact angle of the liquid
ΔP = pressure difference across the liquid/air meniscus

For liquid to enter or drain from a pore, an external gas pressure must be applied that is just enough to overcome the Laplace pressure ΔP.

In the case of a dry heteroporous network, as the external gas pressure is decreased, either continuously or in steps, pores that have capillary pressures lower than the given gas pressure ΔP will fill with liquid. This is referred to as liquid intrusion porosimetry and requires knowledge of the advancing liquid contact angle. In the case of a liquid-saturated heteroporous network, as the external gas pressure is increased, liquid will drain from those pores whose capillary pressure corresponds to the given gas pressure ΔP. This is referred to as liquid extrusion porosimetry and requires knowledge of the receding liquid contact angle. In both cases, the distribution of pore volumes is based on measuring the incremental volume of liquid that either enters a dry network or drains from a saturated network at each increment of pressure.

B. Instrumentation

Until recently, the only version of this type of analysis to evaluate PVDs in general use was mercury porosimetry [5]. Mercury was chosen as the liquid because of its very high surface tension so that it would not be able to penetrate any pore without the imposition of considerable external pressure. For example, to force mercury into a pore 5 μm in radius requires a pressure increase of about 2 atm. While this might not be a problem with hard and rigid networks, such as stone, sand structures, and ceramics, it makes the procedure unsuitable for use with fiber materials that would be distorted by such compressive loading. Furthermore, mercury intrusion porosimetry is best suited for pore dimensions less than 5 μm, while important pores in typical textile structures may be as large as 1000 μm. Some of the other limitations of mercury porosimetry have been discussed by Winslow [6] and by Good [7].

A more general version of liquid porosimetry for PVD analysis, particularly well suited for textiles and other compressible planar materials, has been developed by Miller and Tyomkin [8]. The underlying concept was earlier demonstrated for low-density webs and pads by Burgeni and Kapur [9]. Any stable liquid of relatively low viscosity that has a known $\cos \theta > 0$ can be used. In the extrusion mode, the receding contact angle is the appropriate term in the Laplace equation, while in the intrusion mode the advancing contact angle must be used. There are many advantages to using different liquids with a given material, not the least of which is the fact that liquids can be chosen that relate to a particular end use of a material.

The basic arrangement for liquid extrusion porosimetry is shown in Figure 5. In the case of liquid extrusion, a presaturated specimen is placed on a microporous membrane, which is itself supported by a rigid porous plate. The gas pressure within the closed chamber is increased in steps, causing liquid to flow out of some of the pores, largest ones first. The amount of liquid removed at each pressure level

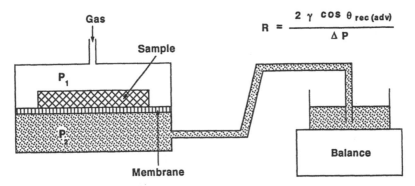

Figure 5 Basic arrangement for liquid porosimetry to quantify pore volume distributions. (From Ref. 10.)

is monitored by the top-loading recording balance. In this way, each incremental change in pressure (corresponding to a pore size according to the Laplace equation) is related to an increment of liquid mass. To induce stepwise drainage from large pores requires very small increases in pressure over a narrow range that are only slightly above atmospheric pressure, whereas to analyze for small pores the pressure changes must be quite large. These requirements are illustrated in Figure 6. In early versions of instrumentation for liquid extrusion porosimetry, pressurization of the specimen chamber was accomplished either by hydrostatic head changes or by means of a single-stroke pump that injected discrete drops of liquid into a free volume space that included the chamber [8]. In the most recent instrumentation developed by Miller and Tyomkin [10], the chamber is pressurized by means of a computer-controlled, reversible, motor-driven piston/cylinder arrangement that can produce the required changes in pressure to cover a pore radius range from 1 to 1000 μm. The pressure is monitored by one of two transducers (a pair is used to maintain sufficient accuracy at both low and high pressures), and the signal is fed to the computer, which, through feedback logic, adjusts the piston position to set the target pressure almost instantly. A schematic of the complete assembly is shown in Figure 7.

The computer also monitors the output of the balance according to a program that identifies when the weight-change rate at a given pressure, corresponding to a given radius, has dropped to an insignificant level. It then activates the instrumentation to reach the next step in the pressurizing sequence, as the number and magnitude of the pressure changes have been programmed as desired beforehand. After drainage at the final pressure level is complete, the instrument can act in reverse to perform a set of liquid intrusions. Multiple drainage/uptake cycles can be programmed to run automatically on the same specimen.

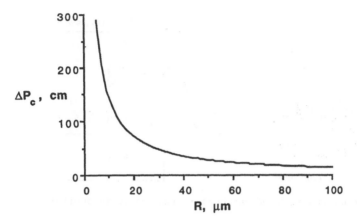

Figure 6   General form of relationship for water between pore size and capillary pressure necessary to either fill or drain that size pore. (From Ref. 10.)

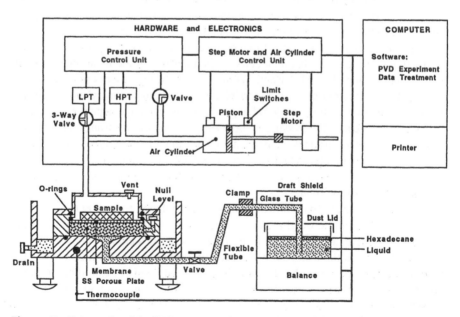

Figure 7   Schematic of the TRI Autoporosimeter for determining pore volume distributions. (From Ref. 10.)

C.   Data Analysis and Applications

Prototype data output for a single cycle incremental liquid extrusion run is shown in Figure 8. The experiment starts from the right as the pressure is increased, draining liquid first from the largest pores. The cumulative curve represents the amount of liquid remaining in the pores of the material at any given level of pressure. The first derivative of this cumulative curve as a function of pore size becomes the pore volume distribution, showing the fraction of the free volume of the material made up of pores of each indicated size.

PVD curves for two typical fabrics woven from spun yarns are shown in Figure 9. The bimodal nature of these curves, discussed previously, is evident. PVD curves for several nonwoven materials are shown in Figure 10. These materials normally have unimodal PVD curves, but generally the pores are larger than those associated with typical woven fabrics. PVD curves for one of the glass fiber nonwovens described in Figure 2 are shown in Figure 11 at three different levels of compression corresponding to the mat thicknesses indicated. Several interesting points can be noted. First, the pore volumes become smaller with increasing compression (decreasing mat thickness), and at the same time the structures appear to have become less heteroporous; that is, the breadth of the PVD curves decreases with increasing compression. Also, the total pore volume (area under each curve), and therefore the sorptive capacity, decreases with compression. Since in many end-use applications fibrous materials are used under some level of compression,

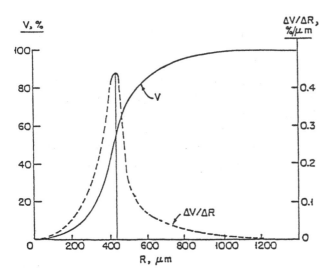

Figure 8   Prototype data output for a liquid extrusion experiment: measured cumulative volume and the corresponding first derivative, which is the PVD curve. (From Ref. 10.)

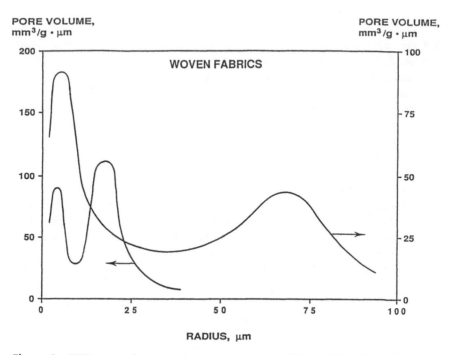

PORE VOLUME,
mm³/g · µm

PORE VOLUME,
mm³/g · µm

WOVEN FABRICS

RADIUS, µm

Figure 9 PVD curves for two typical spun yarn woven fabrics. (From Ref. 10.)

it is particularly important to evaluate pore structure under appropriate compression conditions.

The PVD instrumentation described here, referred to as the TRI Autoporosimeter, is extremely versatile and can be used with just about any porous material, including textiles, paper products, membrane filters, particulates, and rigid foams. It also allows quantification of interlayer pores, surface pores, absorption/desorption hysteresis, uptake and retention capillary pressures, and effective contact angles in porous networks [10]. The technique has also been used to quantify pore volume dimensions and sorptive capacity of artificial skin in relation to processing conditions [11].

V. PORE THROAT ANALYSIS

A. Basic Concepts

Knowledge of the minimum diameters of continuous or connected pores, commonly referred to as pore throats, is important for many types of applications of porous media. These dimensions of pores are critical in various porous barriers and

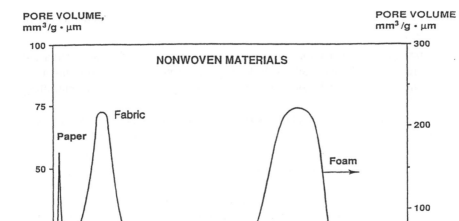

Figure 10 PVD curves for some nonwoven fabrics. (From Ref. 10.)

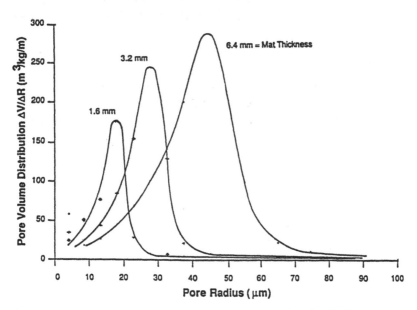

Figure 11 PVD curves for one of the glass fiber mats described in Fig. 2 at three levels of thickness. (From Ref. 4.)

flow-through phenomena, for example, in the use of fabrics as filtration media and in several geotextile applications. In terms of absorption media, pore throat dimensions are not related to sorption capacity, but they do play a role in controlling rates of liquid uptake.

The method used to quantify pore throat dimensions is based on the well-known "minimum bubble pressure" principle, which is operative when gas pressure is applied to one side of a wetted fabric while the other side is in contact with a liquid [12]. As the applied pressure is increased, a critical pressure is reached when the first gas (typically air) bubble emerges through the largest pore available within the sample. This is illustrated schematically in Figure 12. The effective radius of this largest pore is obtained based on Eq. (2), using a liquid with a receding contact angle close to 0°, so that

$$R_{max} = \frac{2\gamma}{\Delta P} \tag{3}$$

where ΔP is the pressure gradient across the liquid/gas interface.

A single bubble pressure experiment identifies only the largest pore within a scanned area of the material. However, if the specimen were cut into smaller parts, as shown schematically in Figure 13, and each part were characterized separately, then the collected data would present the largest pore in every subpart and would provide the necessary information to construct a pore throat distribution. Small pores have little chance of being detected until the scanned areas become small enough.

B. Instrumentation

Miller et al. [12] designed a multiport chamber shown schematically in Figure 14 to carry out bubble pressure measurements over different wetted areas. Six sets of six holes with different cross-sectional areas were drilled vertically through the

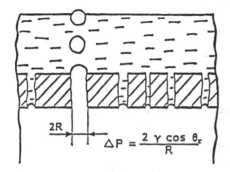

Figure 12   Principle of the minimum bubble pressure experiment. (From Ref. 12.)

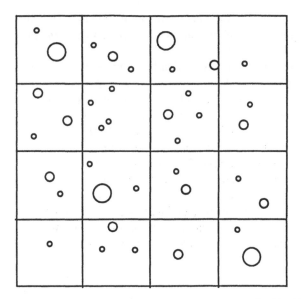

Figure 13   Hypothetical distribution of large pores in a fibrous network. (From Ref. 12.)

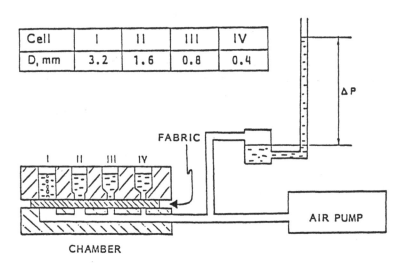

Figure 14   Schematic of apparatus for multiple-scan minimum bubble pressure measurements. (From Ref. 12.)

lower section of the apparatus so that they connect to the air inlet below. The same size holes were drilled in the upper plate above the specimen, coincident with their counterparts below.

The entire cell is placed in enough liquid so that the fabric specimen is wetted and liquid is present in each hole above it. Air pressure is increased by pumping until the first bubbles appear in one of the holes. This pressure is recorded to give the effective pore radius through Eq. (3), and that hole is closed with a plug. Additional pressure is then applied until bubbles appear at another hole. The pressure is recorded, the hole is plugged, and the process is continued with the remaining holes. The fabric specimen is then moved so that the holes are located over another portion of the fabric, and the bubble pressures are determined in the same manner. The process is repeated until a sufficiently large area of the material has been scanned. The data are then analyzed statistically to provide a distribution starting with the largest pore (throat) in the sample, and going down to about the mean pore size. Smaller pores cannot be analyzed, but they contribute little to any transport or absorption process.

C. Applications

The technique can be used to quantify the distribution of the larger pores in a wide range of planar porous materials. In Figure 15 is shown the distribution of large pores in a Millipore membrane filter rated as an 8-μm filter. As can be seen, the material did not reveal the presence of any pores with diameters larger than about 4.8 μm, indicating the margin of safety of this product. It is also noteworthy that the distribution of these large pores is extremely narrow. Similarly narrow pore throat distributions are observed in Figure 16 for a glass fiber mat at three levels of compression.

In Figure 17 are shown the distributions of large pores in a typical woven cotton fabric and in its durable-press treated counterpart. The original purpose of this comparison was to determine whether DP finishing with a cross-linking resin would reduce the pore dimensions. However, the results show that the pores in the DP cotton are actually somewhat larger. This is a direct consequence of the fact that both materials were laundered using conventional home laundry equipment before being analyzed. The untreated cotton fabric shrank more than the DP-treated one, and the relative change in fabric dimensions caused the pore dimensions to be greater in the treated material.

VI. CONCLUSIONS

Pore size distributions are the principal factors controlling the extent and rate of absorption of liquids by fibrous networks. Analytical techniques and instrumentation are now available that can determine the effective volumetric capacities and

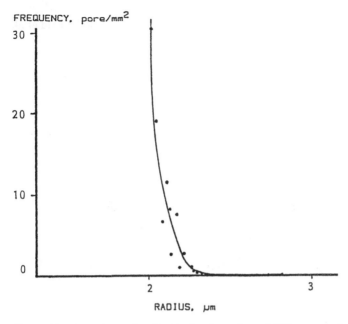

Figure 15 Pore throat size distribution for an 8-μm Millipore membrane. (From Ref. 12.)

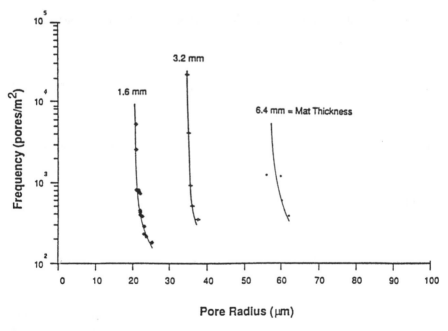

Figure 16 Pore throat distributions for one of the glass-fiber mats described in Fig. 2 at three levels of thickness. (From Ref. 4.)

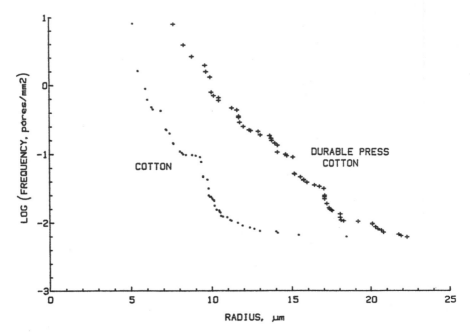

Figure 17 Pore throat distributions for a cotton fabric and for the same fabric treated with a durable-press resin. (From Ref. 12.)

throat dimensions of available pores. These techniques are especially useful since they describe the structure of the porous network as it is after exposure to a specific liquid. Porosity values are of little or no use as predictors for absorption performance, since they provide no information about pore structure, and they may overestimate total useful absorption capacities by including inaccessible free volume elements.

REFERENCES

1. M. Muskat, *The Flow of Homogeneous Fluids Through Porous Media*, J. W. Edwards, Ann Arbor, Mich., 1946.
2. F. A. L. Dullien, *Porous Media—Fluid Transport and Pore Structure*, Academic Press, New York, 1979.
3. A. E. Scheidegger, *The Physics of Flow Through Porous Media*, Macmillan, New York, 1974.
4. D. E. Hirt, K. L. Adams, R. K. Prud'homme, and L. Rebenfeld, In-plane radial fluid flow characterization of fibrous materials, *J. Thermal Insulation 10*:153–172 (1987).
5. M. A. Ioannidis, I. Chatzis, and A. C. Payatakes, A mercury porosimeter for investigating capillary phenomena and microdisplacement mechanisms in capillary networks, *J. Colloid Interface Sci. 143*:22–36 (1991).

6. D. N. Winslow, Advances in experimental techniques for mercury intrusion porosimetry, *Surface and Colloid Science*, Vol. 13 (E. Matijevic and R. J. Good, eds.), Plenum Press, New York, 1984.
7. R. J. Good, The contact angle of mercury on the internal surfaces of porous bodies, *Surface and Colloid Science*, Vol. 13 (E. Matijevic and R. J. Good, eds.), Plenum Press, New York, 1984.
8. B. Miller and I. Tyomkin, An extended range liquid extrusion method for determining pore size distributions, *Textile Res. J. 56*:35–40 (1986).
9. A. A. Burgeni and C. Kapur, Capillary sorption equilibria in fiber masses, *Textile Res. J. 37*:356–366 (1967).
10. B. Miller and I. Tyomkin, Liquid porosimetry: New methodology and applications, *J. Colloid Interface Sci. 162*:163–170 (1994).
11. D. M. Klein, I. Tyomkin, B. Miller, and L. Rebenfeld, Pore volume distribution of artificial skin by liquid extrusion analysis, *J. Appl. Biomater. 1*:137–141 (1990).
12. B. Miller, I. Tyomkin, and J. A. Wehner, Quantifying the porous structure of fabrics for filtration applications, *Fluid Filtration: Gas*, Vol. I, ASTM STP 975 (R. R. Raber, ed.), American Society for Testing and Materials, Philadelphia, 1986.

9
Micromeasurement of the Mechanical Properties of Single Fibers

SUEO KAWABATA   The University of Shiga Prefecture, Hikone City, and Kyoto University, Kyoto, Japan

I. INTRODUCTION

There is a strong anisotropy in the mechanical properties and strength of fibers. This is caused by the strong orientation of molecular chains along the direction of the fiber's axis. Clarification of the details of these fiber properties is necessary for both the science of oriented polymers and the engineering needed to apply these fibers to various fibrous structures and fiber-composite materials. Because of the difficulty in directly measuring single fibers due to the very small size of the fibers, fiber mechanical properties are usually estimated by an indirect method such as the measurement of a fiber bundle or fiber/resin composites. Even though this difficulty exists, direct measurement is desirable and necessary for more precise research on fibers and fiber assembly bodies.

   The single-fiber measurement eliminates the uncertainty of measurement caused by the indirect method. One difficulty in direct measurement is, however, the measurement of the very small force and deformation caused by the small size of the fiber. Recently, a system of directly measuring the mechanical properties of single fibers was developed [1], and the anisotropy in the mechanical properties and the strength of various fibers have been clarified using this system [1,3,4,7–16]. This new measurement was named *micromeasurement* by the author and is introduced in this chapter.

II. ANISOTROPY IN MECHANICAL PROPERTIES

Consider an elastic body, being referred to an orthogonal set of cartesian axes X_1, X_2, X_3 as shown in Figure 1, whose mechanical properties are represented by the following linear equations [2]:

311

$$\sigma_i = \sum_{j=1}^{6} C_{ij} e_j \tag{1}$$

or

$$e_i = \sum_{j=1}^{6} S_{ij} \sigma_j \tag{2}$$

where $\sigma_i (i = 1–6)$ is the engineering component of stress and e_i ($i = 1–6$) is the engineering component of strain. The terms σ_1, σ_2, and σ_3 correspond to the normal components of the stress acting on the X_1, X_2, and X_3 planes, respectively, and σ_4, σ_5, and σ_6 are the shear stress components of the stress acting on these planes, respectively. The terms e_1, e_2, and e_3 are the normal strains along the X_1, X_2, and X_3 axes, respectively and e_4, e_5, and e_6 are the shear strain on the X_1, X_2, and X_3 planes, respectively. The terms C_{ij} and S_{ij} ($i,j = 1–6$) are elastic constants representing material properties and are called the stiffness constants and compliance constants, respectively. This relationship, Eq. (1) or (2), is called the generalized Hooke's law. When the body is isotropic, S_{ij} is represented by the following matrix, where S_{ij} is the value of the compliance of the ith row and jth column:

$$S_{ij} = \begin{bmatrix} S_{11} & S_{12} & S_{12} & 0 & 0 & 0 \\ S_{12} & S_{11} & S_{12} & 0 & 0 & 0 \\ S_{12} & S_{12} & S_{11} & 0 & 0 & 0 \\ 0 & 0 & 0 & 2(S_{11} - S_{12}) & 0 & 0 \\ 0 & 0 & 0 & 0 & 2(S_{11} - S_{12}) & 0 \\ 0 & 0 & 0 & 0 & 0 & 2(S_{11} - S_{12}) \end{bmatrix} \tag{3}$$

There are only two independent parameters among the S_{ij}, S_{11} and S_{12}. In the case of uniaxial extension along the X_1 axis ($\sigma_2 = \sigma_3 = 0$) in Fig. 1, we have

$$e_1 = S_{11} \sigma_1 \tag{4}$$

Hooke's law is expressed as

$$e_1 = (1/E) \sigma_1 \tag{5}$$

From Eqs. (4) and (5),

$$S_{11} = 1/E \tag{6}$$

where E is Young's modulus.

Poisson's ratio ν is defined by the strain ratio under the uniaxial extension along the X_1 axis ($\sigma_2 = \sigma_3 = 0$) as follows:

$$\nu = -e_2/e_1 \tag{7}$$

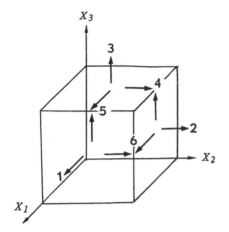

Figure 1 Coordinate system and the suffix numbers indicating the engineering components of stress and strain.

From Eqs. (2), (3), and (5),

$$S_{12} = -\nu/E \tag{8}$$

The term $2(S_{11} - S_{12})$ is equal to $1/G$ where G is the shear modulus and, from Eqs. (6) and (8), there is a relationship such that

$$G = E/[2(1 + \nu)] \tag{9}$$

In the case of the isotropic body, there are two independent mechanical constants, any two of E, G, and ν.

Because of the molecular orientation along the fiber axis (the X_3 axis in the coordinates in Figure 2), fibers have strongly anisotropic mechanical properties. Young's modulus along the fiber axis is different from that in the direction transversing the fiber axis. In the fiber cross-sectional plane the X_1–X_2 plane, the property is isotropic because of the symmetric structure of the fiber about its axis. Such an anisotropy is called *fiber symmetry*; the modulus along the fiber axis is normally higher than the modulus along the transverse direction.

The compliance constants S_{ij} of the fiber symmetric body are shown by Eq. (10). This matrix form may be derived by the symmetric condition

$$S_{ij} = \begin{bmatrix} S_{11} & S_{12} & S_{13} & 0 & 0 & 0 \\ S_{12} & S_{11} & S_{13} & 0 & 0 & 0 \\ S_{13} & S_{13} & S_{33} & 0 & 0 & 0 \\ 0 & 0 & 0 & S_{44} & 0 & 0 \\ 0 & 0 & 0 & 0 & S_{44} & 0 \\ 0 & 0 & 0 & 0 & 0 & 2(S_{11} - S_{12}) \end{bmatrix} \tag{10}$$

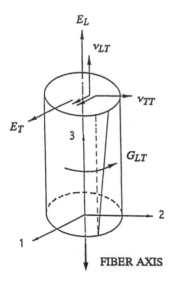

Figure 2   Fiber symmetric anisotropy of fiber.

The S_{ij} of Eq. (10) may be expressed with moduli as follows.

$$S_{ij} = \begin{bmatrix} 1/E_T & -\nu_{TT}/E_T & -\nu_{LT}/E_L & 0 & 0 & 0 \\ 0 & 1/E_T & -\nu_{LT}/E_L & 0 & 0 & 0 \\ & & 1/E_L & 0 & 0 & 0 \\ & & & 1/G_{LT} & 0 & 0 \\ & \text{(Symmetric)} & & & 1/G_{LT} & 0 \\ & & & & & 2(1 + \nu_{TT})/E_T \end{bmatrix} \quad (11)$$

There are five independent elastic constants that represent the material property of the fiber symmetry body. They are

E_L, longitudinal modulus ($= 1/S_{33}$)
E_T, transverse modulus ($= 1/S_{22} = 1/S_{11}$)
G_{LT}, longitudinal shear modulus ($= 1/S_{44}$)
ν_{LT}, longitudinal Poisson's ratio ($= -S_{13}E_L$)
ν_{TT}, transverse Poisson's ratio ($= -S_{12}E_T$)

The terms E_L and E_T are the modulus along the fiber axis and its transverse direction, respectively, and G_{LT} is the shear modulus related to the torsion of the fiber about the fiber axis. The longitudinal Poisson's ratio ν_{LT} is defined by $-e_1/e_3$ (or $-e_2/e_3$) in the uniaxial extension of the fiber in the longitudinal direction—the X_3 axis direction in Fig. 2. The transverse Poisson's ratio ν_{TT} is the Poisson's ratio in the fiber cross-sectional plane and defined by the strain ratio $-e_2/e_1$ in the uniaxial extension in the X_1 direction, or $-e_1/e_2$ in the uniaxial extension in the X_2 direction. It is necessary to measure these five constants for the characterization of fiber mechanical property even if the linear elasticity of the fiber is assumed. When a fiber has nonlinearity in its mechanical behavior, we have to measure these nonlinearities for at least the three deformation modes: the longitudinal, transverse, and torsional deformation modes. The in-plane shear deformation in the cross-sectional plane corresponds to the shear modulus G_{TT} $(= 1/S_{66})$, and this modulus may be replaced by $2(S_{11} - S_{12})$ in the linear case as shown in Eq. (10). However, this relation is not valid in the nonlinear case. When the fiber property is viscoelastic, we have to characterize it for each of these deformation modes. These characterizations of the fiber properties are important. The mechanical anisotropy of the fiber strictly reflects the microstructure of the fiber. In the research on the micromechanics of fiber/resin composites, all of the compliance constants in Eq. (10) are necessary for the stress analysis of the composite, especially the matrix/fiber in the interface region. In this chapter, this micromeasurement of single fibers is introduced.

III.   MICROMEASUREMENT

A.   The Longitudinal Modulus E_L

The mechanical noise of the tensile tester must be kept to a minimum [1]. A single fiber approximately 5–10 cm in length is reinforced at both ends by gluing pieces of paper with adhesive so that it can be clamped by a chuck in a tensile tester as shown in Figure 3.

The E_L is measured from the slope of the stress–strain curve of, for example, a constant rate of extension. The E_L of a fiber in the longitudinal compression mode is not necessarily the same as that in the extension mode. The tensile E_L of most organic fibers is normally larger than the compressional E_L. The measurement of the longitudinal compression property was carried out for Kevlar 29 (aramid fiber) using a microcomposite method [3,4,17]. A uniaxially oriented fiber composite of Kevlar 29 and epoxy resin, 5 mm in length, 1 mm in diameter, was prepared and compressed in its fiber direction as shown in Figure 4. The compression modulus of the composite was converted into a fiber modulus by applying the simple mixture law as follows:

$$E_c = V_f E_f + V_m E_m \tag{12}$$

where E_c, E_f, and E_m are the compression modulus of the composite, fiber, and matrix resin respectively, and V_f and V_m are the volume fraction of the fiber and matrix, respectively, where $V_f + V_m = 1$.

This simple equation is reliable with a high-volume fraction of fiber. The V_f of the specimen used in this experiment was around 0.85–0.9. This small size of

316   Kawabata

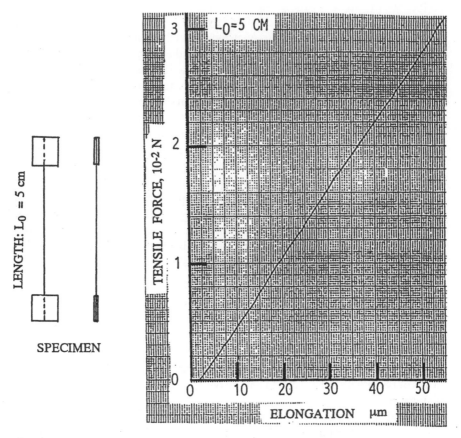

Figure 3 Specimen of a single fiber for the tensile testing and the initial region of a load-elongation curve of Kevlar 29 single fiber measured by a low-mechanical-noise tensile tester. (From Ref. 1.)

the composite enables such a high fraction. The complete longitudinal property of the Kevlar 29 fiber is shown in Figure 5. The tensile region was measured by the single-fiber extension measurement, and the compression region was measured by the microcomposite method. Note that both curves are smoothly connected to each other even though they were measured separately. As seen in this curve, the compression strength is much weaker than the tensile strength.

B. Transverse Modulus E_T

In order to investigate the fiber transverse property, the transverse compression method of single fiber was applied, and a new tester was built as shown in Fig-

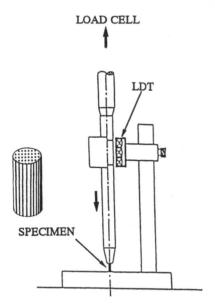

Figure 4 Longitudinal compression testing of fiber using the microcomposite method. (From Refs. 3, 4 and 17)

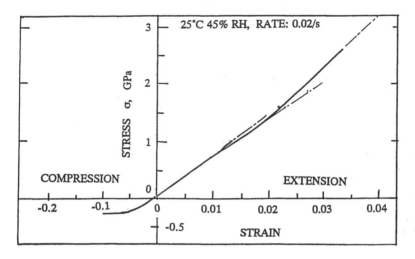

Figure 5 The longitudinal property of Kevlar 29 fiber. (From Refs. 3, 4 and 17)

ures 6 and 7 [1,7]. A single fiber is placed on a flat steel bed that has a mirror-finish surface. The fiber is compressed by a hard steel compression rod. The tip of the rod has a contact plane of 0.2×0.2 mm^2. Its surface is also given the same type of mirror finish as the bottom plane. The compression rod is connected to an electromagnetic power driver with a load capacity of 50 N. A force transducer connected to the compression rod detects the compressional force. The linear differential transformer (LDT), which is also connected to the compression rod, directly detects very small changes in the diameter of the fiber without error arising from the compliance of the force transducer, as the transducer is mounted outside the deformation-detecting system. The resolution of the LDT is 0.05 μm.

An equation that describes the diametral change U in a fiber with a circular cross section as a function of the transverse compressional force per unit length of fiber F has been derived for an anisotropic body by Ward et al. [2,5,6] and is based on the equation derived by McEwan in 1949 for an isotropic body as a solution to the contact problem. The equation derived by Ward is as follows:

$$U = (4F/\pi)[(1/E_T) - (\nu^2_{LT}/E_L)][0.19 + \sinh^{-1}(R/b)] \tag{13}$$

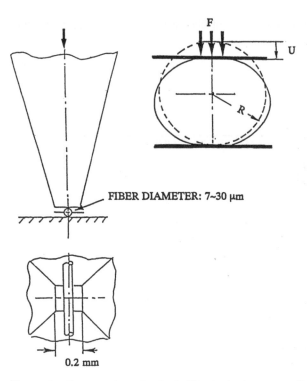

FIBER DIAMETER: 7~30 μm

0.2 mm

Figure 6   Compression of a single fiber in the transverse direction.

Figure 7   Transverse compression tester.

where b is given by

$$b^2 = (4FR/\pi)[(1/E_T) - (\nu_{LT}^2/E_L)] \tag{14}$$

and R is radius of the fiber. When $\nu_{LT}^2/E_L \ll 1/E_T$, the term ν_{LT}^2/E_L in Eqs. (13) and (14) may be eliminated and these equations become simpler. From our investigation [6], it was found that the degree of error caused by this equation simplification is about 1% of the exact value of E_T. From these equations, E_T is obtained by measuring the relation between F and U by transverse compression and solving Eq. (13).

The transverse compression curves of Kevlar 149, Kevlar 49, and Kevlar 29 are shown in Figure 8 [6].

As seen in this figure, the Kevlar fibers have a clearly ductile property in their transverse direction, while the yielding does not appear in the tensile property of these Kevlar fibers in the longitudinal direction. Also, the yielding stress is much

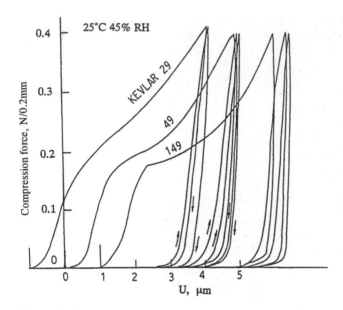

Figure 8   Transverse compression properties of the Kevlar fibers. (From Ref. 6.)

lower than the longitudinal tensile strength, as shown later for Kevlar 29 in
Table 1. Carbon fibers and ceramic fibers, however, do not exhibit a behavior of
yielding in the transverse compression property, and show a tendency for brittle
fracture as well in their longitudinal properties. Also, their transverse strength is
much higher than that of organic fibers, as shown in Figure 9. The relationships
between E_L and E_T for various fibers and between E_T and breaking stress or yielding stress of the same fibers are shown in Figures 10 and 11, respectively [6].

C.   Shear Modulus G_{LT}

The shear modulus G_{LT} is obtained from the torsion of the fiber (Fig. 12) about the
fiber axis. For a cylindrical rod, the G_{LT} is obtained as follows:

$$G_{LT} = TL/(\theta I_p) \tag{15}$$

where T is torque, L is length of specimen, θ is torsional angle (rad), and I_p is torsional moment of inertia of area, given for a cylindrical rod of diameter D by

$$I_p = \pi D^4/32 \tag{16}$$

The shear strain at the fiber skin is $\gamma = \theta D/(2L)$ and the skin stress is
$\sigma = \theta D G_{LT}/(2L)$.

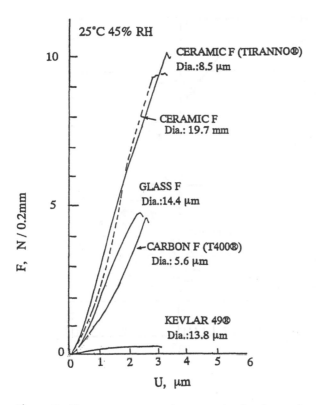

Figure 9 Transverse compression property of carbon and ceramic fibers. (From Ref. 6.)

When the torsion of a single fiber of $D = 14\,\mu m$, $L = 5\,mm$, and $G_{LT} = 2\,GPa$ is measured, the torque T is approximately 70 nN m (0.7 mg cm) at torsion angle $\theta/L = 6\pi \times 10^3$ rad/m, which is the torsion angle range normally measured. This small torque is caused by the small diameter of a single fiber. A highly sensitive torque transducer for this torque range has been developed [1]. The mechanism of the torsion tester is shown in Figure 13 and the tester is shown in Figure 14 [1].

A typical torque–torsional angle relation is shown in Figure 15 for a Kevlar 49 single fiber. A constant rate of torsion was applied at a rate of 0.53π rad/s. The shear modulus was obtained from the initial slope of the curve.

As seen from this figure, yielding is observed also in the torsional property. In the region larger than the yielding torque, optical microscopy observation of the fiber surface reveals diagonal lines. After repeated torsion, many such lines are observed and fiber splitting along the fiber axis is initiated from these lines that leads the fiber to a state of reduced shear stiffness, and then fiber failure [8].

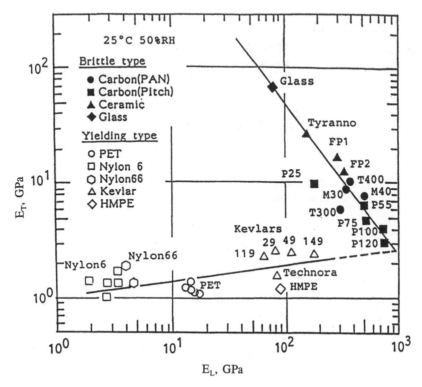

Figure 10   Relation between E_L and E_T. The brittle type group and ductile group meet where E_L is approximately 1000 GPa, which is near the modulus of diamond, the ultimate material. (From Ref. 6.)

D.   Measurement of Poisson's Ratios, ν_{LT} and ν_{TT}

The ν_{LT} may be measured by measuring the change in fiber diameter that occurs with fiber longitudinal extension. A special tester with a resolution of 0.01 μm was designed for detecting fiber diameter change as shown in Figure 16 [15]. The ν_{LT} is measured from the slope of the e_L–e_T relation curve as shown in Figure 17 [15]. A single-fiber measurement for ν_{TT} is not possible at present. We have estimated this parameter from ν_{TT} by measuring the Poisson's ratio of a unidirectional fiber composite plate by the ordinary strain gauge method.

E.   Anisotropy in the Mechanical Properties of Fibers

Table 1 lists a full set of elastic constants for Kevlar 29 fiber [3]. These constants were measured in an atmospheric condition of 25°C, 45% RH. The G_{LT} values of some fibers are shown with their E_L and E_T in Table 2 [3,4,7,10–13,16].

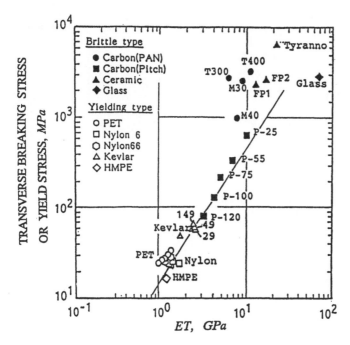

Figure 11 Correlation of breaking stress or yield stress and E_T. (From Ref. 6.)

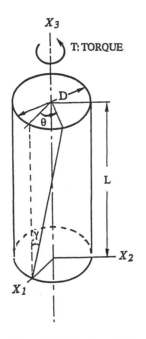

Figure 12 Torsion of fiber.

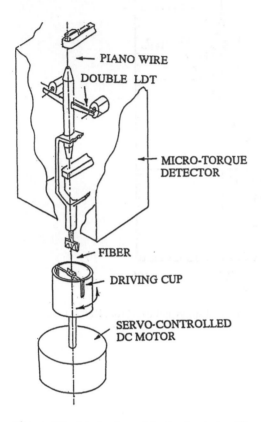

Figure 13   Mechanism of the torsion tester. (From Ref. 1.)

III.   CONCLUDING REMARKS

As seen from Figures 10 and 11 and Table 2, apparel fibers [16] as well as high-strength fibers exhibit strong anisotropy in their mechanical properties. These mechanical properties are closely related to the mechanical properties of fiber assembly bodies. It is important for textile scientists and engineers to have a good understanding of these fiber properties for future advanced research on fibers and textiles.

Figure 14 Torsion tester.

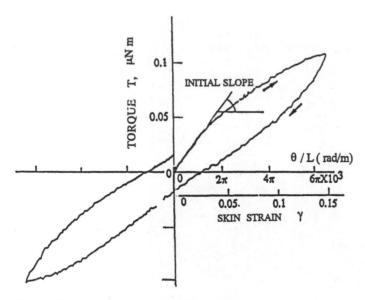

Figure 15 Torsion property of a Kevlar 29 fiber.

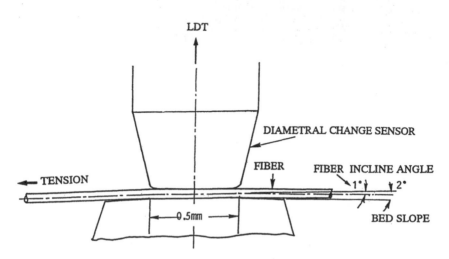

Figure 16 Tester measuring Poisson's ratio ν_{LT}. (From Ref. 15.)

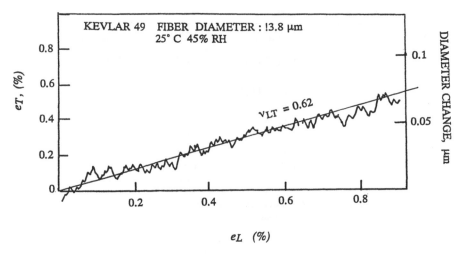

Figure 17 The e_L–e_T curve of a Kevlar 49 single fiber.

Table 1 Mechanical Properties of Kevlar 29 in 25°C, 45% RH

Elastic constants		
E_L(GPa)		
Tensile	79.8	Strain at 0.005
	69.5	Strain at 0.02
	98.4	Breaking strain region
Compression	60.0	Strain at −0.001
	45.0	Breaking strain region
E_T (GPa)	2.59	
G_{LT} (GPa)	2.17	
ν_{LT}	0.63	
ν_{TT}	0.43	

Strength	Stress (GPa)	Strain
Longitudinal		
Tensile	2.55	0.037
compression	0.31 (yielding)	0.091
Transverse		
compression	0.056 (yielding)	0.007
Torsion	0.101 (yielding)	0.047

Table 2 E_L, E_T, and G_{LT} of Fibers

Fiber	E_L (GPa)	E_T (GPa)	G_{LT} (GPa)
Carbon (T-300, PAN)	234.6 (308.1)*	6.03	18.2
Ceramic (Tiranno)	159.7	26.5	45.80
Kevlar 49	113.4 (129.6)*	2.49	2.01
HMPE	89.3	1.21	1.90
Kevlar 29	79.8 (98.4)*	2.59	2.17
Glass	77.4	67.9	42.5
PET	15.6	1.09	0.63
Nylon	2.76	1.37	0.55
Wool	3.33	1.09	1.47

Source: Data accumulated from Kawabata's experiments [3].
*The value in the breaking region is shown in parentheses.

REFERENCES

1. S. Kawabata, Proc. 4th US-Japan Conf. on Composite Materials at Washington DC, June 27–29, 1988, Technomic, Lancaster, Pa., 1989, pp. 253–262.
2. I. M. Ward, *Mechanical Properties of Solid Polymers*, 2nd ed., John Wiley & Sons, Chichester, New York, 1985.
3. S. Kawabata, Annual Report of the Research Institute, for Chemical Fibers, Department of Polymer Chemistry, Kyoto University, Kyoto, Japan (Japanese ed.), 1992, pp. 17–24.
4. T. Kotani and S. Kawabata, Proc. 15th Composite Materials Symposium, Society for Composite Materials, Japan, 1990, pp. 113–116.
5. D. W. Hadley, I. M. Ward, and J. Ward, *Proc. R. Soc. A285*:275 (1965).
6. P. R. Pinnock, I. M. Ward, and J. M. Wolfe, *Proc. R. Soc. A291*:267, 1966.
7. S. Kawabata, *J. Textile Inst. 81*:432 (1990).
8. S. Kawabata and M. Niwa, Proc. 9th ICCM, Madrid, Vol. 6, 1993, pp. 671–677.
9. S. Kawabata and M. Sera, Proc. Advanced Composites, Woolongong, Australia, 1993, pp. 797–802.
10. S. Kawabata and K. Katsuma, Proc. 21st Textile Res. Symp. at Mt. Fuji, 1992, pp. 1–4.
11. C. Muraki, M. Niwa, and S. Kawabata, Proc. 21st Textile Res. Symp. at Mt. Fuji, August 1992, pp. 10–13.
12. C. Muraki, M. Niwa, and S. Kawabata, *J. Textile Inst. 85*:12 (1994).
13. S. Kawabata, Abstract 2nd Int. Conf. on Advanced Materials & Technology, Hyogo Pref., Kobe, 1991, pp. 51–58.
14. S. Kawabata, N. Amino, K. Katsuma, M. Sera, T. Kotani, and M. Kakiuti, Proc. 22nd Textile Res. Symp. at Mt. Fuji, August 1993, pp. 54–60.
15. S. Kawabata, Proc. 18th Textile Res. Symp. at Mt. Fuji, August 1989, pp. 1–6.
16. S. Kawabata, C. Muraki, and M. Niwa, 18th Textile Res. Symp. at Mt. Fuji, August 1989, pp. 7–12.
17. S. Kawabata, T. Kotani, and Y. Yamashita, *J. Textile Inst. 86*:347 (1995).

10

Objective Measurement of Fabric Hand

SUEO KAWABATA   The University of Shiga Prefecture, Hikone City, and
Kyoto University, Kyoto, Japan
MASAKO NIWA   Nara Women's University, Nara, Japan

I.   INTRODUCTION

There are two types of clothing fabric performance. One type is utility performance, such as strength, color durability, shrinkage resistance, etc. While this type of performance is, of course, very important for clothing materials, consumers are generally satisfied with fabrics that meet these criteria to a certain extent. Beyond this the consumers' attention turns to higher level performance factors, such as improved quality from the standpoint of garment appearance and comfort. This second factor of fabric quality-type performance factors is related to the idea of "better fit" to the human body, and is also an essential requirement in clothing material. The evaluation of fabric quality performance is, however, more difficult than the evaluation of utility performance [1].

The quality of clothing fabric with regard to the second type of performance has been evaluated by consumers and textile producers subjectively by means of the hand touch of fabric from the mechanical-comfort viewpoint. This evaluation is called *hand evaluation* and the fabric property relating to this evaluation is *fabric hand* ("handle" in England). The subjective judgment of fabric hand is based on human sensitivity and experience. It is true that this subjective method is the most direct method for evaluating fabric mechanical comfort, as the human body and sensitivity feel the comfort of clothing.

This subjective evaluation is essential and becomes highly refined with the accumulation of experience. However, a problem exists in that it is a subjective method, which restricts the scientific understanding of fabric hand for those who wish to design high-quality fabrics by engineering means. Because of the importance of the scientific understanding of fabric hand, many trials for replacing the subjective method with an objective method have been carried out by many researchers in the textile field, beginning with the trials by Peirce in 1930 [2]. Peirce proposed a correlation between fabric hand and fabric mechanical properties. Many textile scientists conducted research in this field after Peirce. Because of the difficulty in linking human sensitivity to fabric properties, progress in this field has been slow. However, the importance of the understanding of fabric hand has been considered throughout.

Kawabata and his co-workers began researching fabric hand around 1969 based on their concepts of fabric hand and on the work of many of their predecessors in this field. The research focused on the analysis of the judgment of fabric hand as carried out by experts in textile mills, especially finishing mills for wool textiles. The first step of the research was to standardize the fabric hand expressions that were traditionally used by the experts in wool textile mills. Based on this standard, a numerical expression of fabric hand became possible; then subjective hand judgment was transferred to an objective evaluation system based on fabric mechanical properties [3]. In this chapter, we take a look at the objective evaluation system of fabric hand.

II. SUBJECTIVE HAND JUDGMENT

When we touch a clothing fabric and inspect its hand by finger sensitivity, we not only enjoy the fabric touch itself, but we also think about the performance of the fabric as a clothing fabric on the basis of our experience. In the end, we must determine if the fabric is good for clothing from the viewpoint of the second performance type described in the preceding section. The criterion in this judgment is therefore not simply a like or dislike of the feeling, but rather judgment based on the comfort and beautiful appearance of the clothing.

Professionals working in textile mills must daily produce good fabrics for consumers, and by doing so have accumulated many years of knowledge about consumer fabric preference; this information is passed and transferred from professionals to professionals by means of their professional hand judgment. Although the basis for the criteria is consumer preference, the individual consumer is not necessarily a good judge, and his or her criteria for good fabric are not necessarily reliable or consistent due to lack of experience. Experts in textile mills have come to understand many consumer preferences and use semi-objective criteria although making a subjective judgment. This is a major advantage in asking the advice of experts. This research on fabric hand focused on the analysis of the experts' subjective hand judgments, men's suiting materials in particular (Fig. 1).

In order to develop an objective hand evaluation system, in 1972 Kawabata organized the Hand Evaluation and Standardization Committee in Japan, and 12 experts were invited to join the committee. Progress toward an objective evaluation system has been made possible by the cooperation of this group of experts.

Figure 1 Hand touch in suiting.

A. Analysis of Expert Hand Judgment

When experts touch a fabric, they primarily inspect the mechanical properties including surface properties. Then they summarize these properties with hand expressions such as "smoothness," "stiffness," etc. Each of these expressions summarizes a fabric property that is closely related to the fabric performance with respect to comfort and beautiful appearance, we call this the essential performance as garment material. Next, the experts again summarize these fabric properties to evaluate the overall hand in terms of an expression, as good or poor, or a grade with quality rank. Thus, there are two steps in performing the hand evaluation.

1. Evaluation of the fabric hand, which summarizes the specific fabric properties that express fabric characteristics in relation to fabric quality.
2. Evaluation of the overall hand expressing the fabric quality with regard to the essential fabric performance of the garment or clothing that will be made from the fabric.

There are not many hand expressions for type 1. Among these, three for winter/autumn-use suiting and four for summer-use suiting have been selected as

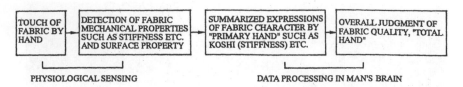

| TOUCH OF FABRIC BY HAND | DETECTION OF FABRIC MECHANICAL PROPERTIES SUCH AS STIFFNESS ETC. AND SURFACE PROPERTY | SUMMARIZED EXPRESSIONS OF FABRIC CHARACTER BY "PRIMARY HAND" SUCH AS KOSHI (STIFFNESS) ETC. | OVERALL JUDGMENT OF FABRIC QUALITY, "TOTAL HAND" |

PHYSIOLOGICAL SENSING   DATA PROCESSING IN MAN'S BRAIN

Figure 2   Subjective judgment of fabric hand by experts.

important hand expressions, called "primary hand". The hand of type 2 is an over-all hand called "total hand". Figure 2 shows the experts' hand-evaulation process.

B.   Primary Hand and Its Grading

The three primary hands for winter/autumn suiting and the four for summer suiting have been defined by the experts as follows.

For winter/autumn suiting:

Stiffness (*koshi*): A feeling related mainly to bending stiffness. A springy property promotes this feeling. A fabric having a compact weave density and made from springy and elastic yarn gives a high value.

Smoothness (*numeri*): A mixed feeling coming from a combination of smooth, sup-ple, and soft feelings. A fabric woven from a cashmere fiber gives a high value.

Fullness (*fukurami*): A feeling coming from a combination of bulky, rich, and well-formed impressions. A springy property in compression and thickness, ac-companied by a warm feeling, is closely related with this property. (The Japanese word literally means "swelling.")

For summer (meaning midsummer) suiting:

Stiffness (*koshi*): The same as *koshi* in winter/autumn suiting.

Crispness (*shari*): A feeling coming from a crisp and ridged fabric surface. This is found in a woven fabric made from a hard and strongly twisted yarn. This gives a cool feeling. (The Japanese word means crisp, dry, and a sharp sound caused by rubbing the fabric surface with itself.)

Fullness (*fukurami*): The same as fullness in winter/autumn suiting.

Antidrape (*hari*): The opposite of limp conformability, whether the fabric is springy or not. (The Japanese word means "spread.")

One primary hand, stiffness, is related to a moderate space between the hu-man body and the outer garment to allow freer body movements. A moderate stiff-ness also brings a beautiful drape of the garment. Smoothness is the hand most important to fabric quality. In general, consumers like the strong feeling of smoothness and get a sense of quality from this feeling. This is due to the smooth contact of fabric with human skin. This smooth contact is necessary to prevent skin injury; by instinct people defend themselves from injury. This instinct is related to a feeling of comfort. On the other hand, the smooth fabrics stick to sweaty skin in tropical climates. Consumers prefer a rough rather than smooth surface in hot climates because of the comfortable feel in this situation. This is the crispness hand. Although the crispness hand is essential in tropical climates, most western countries have temperate climates, and as such do not traditionally prefer this hand. Only the consumers in Japan and a few other Asian countries such as China, Korea, etc. prefer this hand for midsummer suiting. This is one of their traditional hands. Accordingly, only winter/autumn primary hands may be applicable to western country consumers.

The grading of feeling intensity was applied to each of these Primary hands. Standard samples were selected for each grade and expressed numerically on a scale of 1–10. This number was named "primary hand value," or hand value (HV) for short, as shown in Table 1.

C.   Total Hand and Its Grading

The total hand was also standardized by the group of experts for the fabric categories winter/autumn and midsummer. Standard samples that express the grade of total hand were selected and graded as shown in Table 2 [4–7].

Fabric characteristics are expressed by the hand values of the three primary hands, and its quality by a total hand value (THV). Fabric hand is clearly expressed by these hand values. For example, the hand of a worsted suiting is expressed by values such as stiffness = 3.6, smoothness = 6.7, fullness = 5.3, and THV = 3.7.

D.   Extension to Other Types of Fabrics

The first trial for the standardization of fabric hand was conducted primarily with suiting materials, regardless of fiber kind. Worsted fabrics are traditionally a major material used in suits due to a preference for wool from the standpoint of suit

Table 1   Hand Value of the Primary Hand

Hand value	Feeling grade
10	The strongest
.	
.	
.	
5	Medium
.	
.	
.	
1	The weakest
0	No feeling

Table 2   Total Hand Value (THV)

Grade	THV
Excellent	5
Good	4
Average	3
Fair	2
Poor	1
Not useful	0

quality. The hand of a suit is traditionally based on the hand of worsted fabrics, and precise criteria have been derived from the worsted-base fabrics. We must therefore consider whether the traditional hand criteria may be applicable to other fabrics of suiting regardless of fiber kind. With this in mind, we included fabric specimens woven from various kinds of fibers in the hand assessment. The only condition for the assessment was that the specimen be a material used in suits.

A similar procedure for hand assessment was carried out for women's garment fabrics, and it was discovered that the criteria for the hand of women's suiting had much in common with that of men's suiting. We could apply the standard of men's suiting hand to the women's suitings as shown in Section III.B. Women's thin dress fabrics, however, have a little different hand from that of suitings. We had to add two more primary hands, as follows [3,4,5,6,7]:

Stiffness (*koshi*): The same definition as stiffness of men's suiting.
Antidrape (*hari*): The same definition as antidrape of men's suiting.
Crispness (*shari*): The same definition as crispness of men's suiting
Fullness (*fukurami*): The same definition as fullness of men's suiting.
Scrooping feeling (*kishimi*): Silk fabric possesses this feeling strongly.
Flexibility with soft feeling (*shinayakasa*): Soft, flexible, and smooth feeling.

Standard samples for each primary hand were selected, and the feeling intensity of each primary hand was graded using a scale of 1–10 in the same manner as men's suiting. It was unfortunately difficult to find experts to conduct the THV evaluation for this fabric category. We are still continuing the assessment at present.

III.   OBJECTIVE EVALUATION OF HAND VALUE AND TOTAL HAND VALUE [1, 3]

As seen in Fig. 1, human fingers can detect the fabric bending stiffness, surface properties, compression property, and shearing property of a fabric. In addition, another important action of the inspection is streching of fabric at low strain levels. The mechanical response of the fabric is transferred to the person's brain, where the fabric hand properties are evaluated as previously mentioned, based on the person's experience. The concept of the objective evaluation system is as follows. Instead of using the touch of a fabric by a hand or finger, we measure fabric

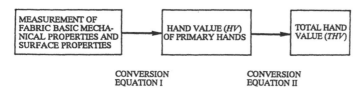

Figure 3   Objective system for hand evaluation.

mechanical properties and express them with mechanical parameters. Then these parameters are converted into the hand values with a conversion equation (equation type I). Next, these hand values are converted into a total hand value with the second conversion equation (equation type II) as shown in Fig. 3.

A. Mechanical Parameters

Based on our preliminary investigation, fabric mechanical properties and surface properties related to fabric hand and applicable to this objective system were selected. Deformation modes selected here were the basic deformations of fabric and the complex modes were avoided, considering the future application of a system to the design and control of fabric hand. As is well known, fabric mechanical properties in a low-load region possess a peculiar nonlinearity in their properties. These properties must be measured exactly and expressed by parameters. One example of the nonlinearity is hysteresis behavior in the load-deformation relation. This hysteresis plays an important part in objective evaluation of fabric hand. The selected properties and parameters are introduced in this section. Fabric samples of 20 cm × 20 cm were used for all measurements. The standard measuring condition is shown here. There are some other conditions, such as nonwoven conditions [11], high sensitivity condition for thin fabrics, etc. [15].

1. Tensile Property

As shown in the top of Fig. 4, a 20 cm × 5 cm sample is cramped and extension is applied along the 5 cm direction up to a maximum load 500 N/m. Rate of tensile strain is 4.00×10^{-3}/s. This is a type of biaxial extension called *strip biaxial extension*. This deformation mode is much easier to use than simple uniaxial extension for theoretical property prediction. This simplicity is important for further fabric design for controlling the fabric hand. There are three parameters expressing this nonlinear property in the warp direction, and another set of three is necessary for the weft direction. For hand value derivation, these two directional values are averaged.

2. Bending Property

Pure bending (Fig. 5) is applied to a fabric 1 cm in length with a constant rate of curvature, 5.0×10^{-3} m^{-1}/s. The stiffness (slope) and hysteresis are measured.

3. Shearing Property

A rate of shear strain of 8.34×10^{-3}/s (shear deformation 1.46×10^{-4} degrees/s), is applied under a constant extension load 10N/m up to a maximum shear angle of 8 degrees (Fig. 6). The stiffness (slope) and hysteresis are measured.

4. Compression Property

A fabric specimen is compressed in the direction of thickness to a maximum pressure of 5 KN/m^2 (50 gf/cm^2), at a constant velocity, 20 μm/s (Fig. 7). The shape

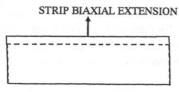

STRIP BIAXIAL EXTENSION

TENSILE DEFORMATION

LT (Linearity of load-extension curve)
      = $WT/$ [area of triangle Q_1 Q_2 Q_3]
WT (Tensile energy) = [area Q_1 A Q_2 Q_3]
RT (Tensile resilience) = [area Q_1 B Q_2 Q_3] /WT

EM Extensibility, strain at 500 N/m

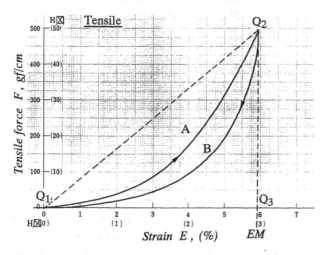

Figure 4   Tensile property. The standard measuring condition is shown. Scale in parenthesis is the high-sensitivity condition for thin fabrics. Curve A is extension process and B is recovery process. Linearity of curve A is defined by the ratio of WT to the area of a triangle shown by additional dotted lines.

of the load thickness is similar to the shape of the tensile property curve, and the same parameters are used with the identification C(LC, etc.).

5.   Surface Property

Surface geometrical smoothness and frictional smoothness are measured. The sensors for these measurements are shown in Fig. 8(a) and (b), respectively. The contact surface of the frictional sensor is 10 parallel piano wires 0.5 mm in diameter, and the surface shape is similar to that of a human fingerprint, as shown in Fig. 8(c). A weight is used to apply 0.5N (50 gf) contact force during measurement. The rough surface of the fingerprint shape is sensitive to fabric surface roughness.

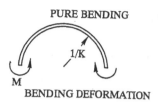

PURE BENDING

1/K

M

BENDING DEFORMATION

B   (Bending rigidity) Measured from mean slope in the range
       K=0.5~1.5 cm^{-1}

2HB (Hysteresis of bending moment) Measure at K=0.5 cm^{-1}

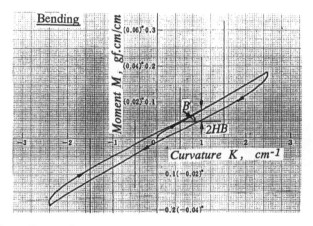

Figure 5   Bending property. The standard measuring condition is shown. Scale in parenthesis is applied to the high-sensitivity condition for thin fabrics.

For the geometrical smoothness sensor, a single wire of the same diameter is used to measure geometry more accurately. The signals from these sensors pass a frequency filter with a second-order high-pass response. The frequency response is shown in Fig. 9. The sweep velocity is 1 mm/s. When we touch a fabric and sweep our finger across the fabric surface, the sweep velocity is normally 5 cm/s; that is, the 1 Hz in the measurement corresponds to about 50 Hz in an actual sweep. A frequency component higher than about 250 Hz in an actual sweep is naturally eliminated by the fingerprint surface and the transducer mechanism. The most sensitive frequency range of human sensation is 50–200 Hz [8], and a filter is used to detect only this range, eliminating the noise component from surface sensing.

The parameters representing surface properties are MIU, MMD, and SMD, which are measured for a 2-cm return sweep. They are defined as

MIU, mean frictional coefficient (for a 2-cm return sweep)
MMD, mean deviation of frictional coefficient
SMD, mean deviation of surface contour

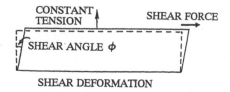

G (Shear stiffness)
2HG   (Hysteresis of shear force at 0.5° of shear angle)
2HG5 (Hysteresis of shear force at 5° of shear angle)

Shearing

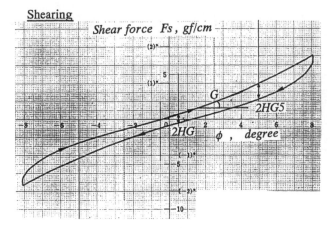

Figure 6   Shearing property. The Standard measuring-condition is shown Scale in parenthesis is applied to the High Sensitivity condition for thin fabrics

The mechanical and surface parameters are shown in Table 3.

In the beginning of the development of the objective system, a system of four
machines was used to measure these parameters (Fig. 10). A fabric specimen of 20
cm × 20 cm was used consistently throughout the system. This system was later
named KESF 1, 2, 3, and 4, or simply, the KESF system. Recently, an automatic
system was developed; however, the principle of the measurement is the same as
with the original machines, with all measurement operations fully automated.

B.   Equations for the Calculation of Hand Values

The equation used to derive primary hand values was assumed to be linear. The
equation for THV was also assumed to be nonlinear, considering the existence of
the optimum value of HV contributing to the highest value of THV. The HV equation is as follows:

COMPRESSION DEFORMATION

LC (Linearity of compression curve)
   $=WC/$ [area of triangle Q_1 Q_2 Q_3]
WC (Compressional energy) = [area Q_1 A Q_2 Q_3]
RC (Compressional resilience) = [area Q_1 B Q_2 Q_3] $/WC$

T_0 (Initial thickness used as "Thickness", defined as the
   fabric thickness at pressure $P=50$ N/m^2 [0.5 gf/cm^2])

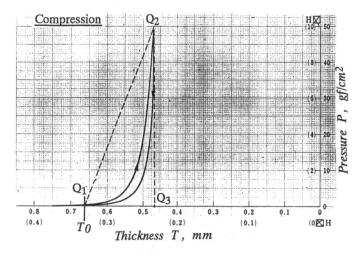

Figure 7 Compression property. The Standard measuring-condition is shown. Scale in parenthesis is applied to the High Sensitivity condition for thin fabrics.

$$Y_k = C_0 + \Sigma\, C_{ki}x_i \tag{1}$$

where Y_k is the kth hand value such that, $k = 1$ is stiffness, $k = 2$ is smoothness, and $k = 3$ is fullness for winter/autumn suiting, and $k = 1$ is stiffness, $k = 2$ is crispness, $k = 3$ is fullness, and $k = 4$ is antidrape stiffness for summer suiting. The term x_i is the normalized ith ($i = 1$–16) mechanical parameter, normalized as

$$x_i = (X_i - M_i)\,/\sigma_i \tag{2}$$

where X_i is the mechanical parameter shown in Table 4. Note that a logarithm is used for some parameters. M_i and σ_i are the mean and standard deviation of X_i for

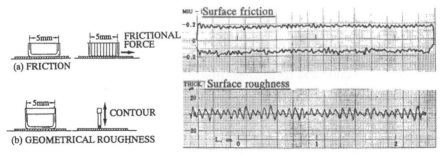

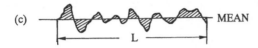

(a) FRICTION

(b) GEOMETRICAL ROUGHNESS

MIU (Coefficient of friction)
MMD (Mean deviation of MIU, frictional roughness)
SMD (Mean deviation of thickness, geometrical roughness)

(c) —————— MEAN

L

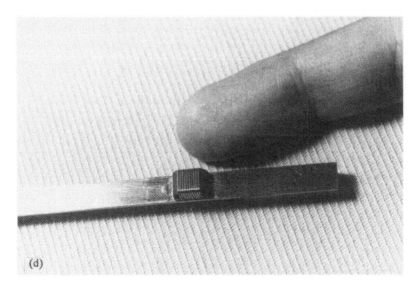

(d)

Figure 8 Surface properties: (a) measurement of surface friction, (b) Surface geometry measurement, (c) mean deviation (MD) = [hatched area/L], and (d) the appearance of the contact surface of friction detector.

the men's suiting population. C_0 and C_{ki} are constant coefficients, shown in Table 4 with M_i and σ_i.

The value of THV value is derived by substituting Y_k that are derived from Eq. (1) into Eq. (3) as follows:

$$\text{THV} = C_0 + \Sigma\, Z_k \tag{3}$$

where

$$Z_k = C_{k1}(Y_k - M_{k1})\,/\,\sigma_{k1} + C_{k2}(Y_k^2 - M_{k2})\,/\,\sigma_{k2} \tag{4}$$

Thus Z_k is the contribution of the kth primary hand to THV. The constants M_{k1} and σ_{k1} are population means and standard deviations of Y_k, and M_{k2} and σ_{k2} are population means and standard deviations of Y_k^2, respectively, shown in Table 4 with the constant coefficients C_{k1} and C_{k2}, and the constant C_0.

The primary hand equations have been derived on the basis of experts' judgment for the men's suiting population and were later named KN-101-Winter and KN-101-Summer for the primary hand of winter/autumn and summer suiting respectively, and the THV equations were KN-301-Winter and KN-301-Summer, respectively.

C. Extension to a New Object Population

The equations introduced in the preceding sections are applicable to the men's suiting population, and the coefficients in the equations were derived by correlat-

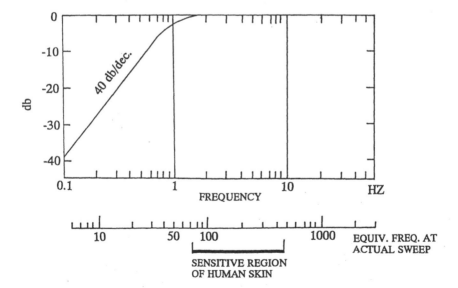

Figure 9 Frequency response of the filter.

Table 3 Mechanical and Surface Properties

Parameters	Description	Unit
Tensile[a]		
LT	Linearity of Load/extension curve	None
WT	Tensile energy	N/m (gf cm/cm^2)
RT	Tensile resilience	%
EM[b]	Extensibility, strain at 500 N/m	
	(gf/cm of tensile load)	None
Bending[a]		
B	Bending rigidity	10^{-4} Nm (gf cm^2/cm)
2HB	Hysteresis of bending moment	10^{-2} N (gf cm/cm)
Shearing[a]		
G	Shear stiffness	N/m deg. (gf/cm degree)
2HG	Hysteresis of shear force at 0.5 degrees	N/m (gf/cm)
	of shear angle	
2HG5	Hysteresis of shear force at 5 degrees	N/m (gf/cm)
	of shear angle	
Compression		
LC	Linearity of compression/thickness	None
	curve	
WC	Compressional energy	N/m (gf cm/cm^2)
RC	Compressional resilience	%
Surface[a]		
MIU	Coefficient of friction	None
MMD	Mean deviation of coefficient of	None
	friction (frictional roughness)	
SMD	Geometrical roughness	μm
Construction		
T	Fabric thickness	mm
W	Fabric weight/unit area	10 g/m^2 (mg/cm^2)

[a]Average of the values in warp and weft directions is applied. The warp and weft directional values are identified by 1 and 2, respectively, such as MMD-1, B-2, etc.
[b]EM is not used for the conversion equation to HV.
Source: Refs. 1 and 3.

ing subjective judgment by the experts with the mechanical parameters of fabric. If a new population is the goal of the objective measurement, a similar procedure to that described earlier is necessary, to derive the coefficients for the new population. However, the construction of new equations is not necessary in some cases. It may be possible to apply the men's suiting equation to other categories of fabrics. This is based on the fact that primary hand may be applied commonly to many

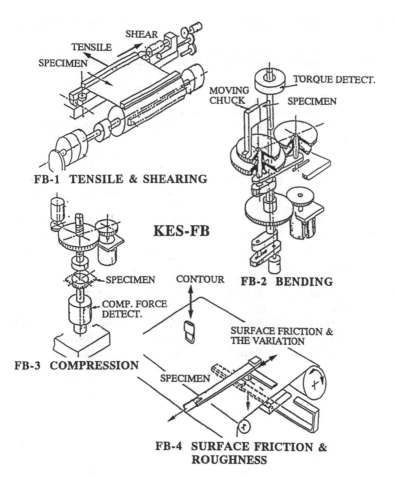

FB-1 TENSILE & SHEARING

KES-FB

FB-2 BENDING

FB-3 COMPRESSION

FB-4 SURFACE FRICTION & ROUGHNESS

Figure 10 The KESF system.

different categories of fabrics. Even in the case of total hand, it may be possible to apply the men's suiting equation for THV. This is because there is a possibility of a common criterion in human interactive materials.

An example is the application of men's suiting equations to women's suiting. We use the same coefficients as the coefficients of men's suiting equations for both HV and THV equations, and apply a minor modification as follows. The mechanical parameters X_i are normalized by Eq. (5) with the M_i' and σ_i', the population mean and standard deviation of women's suiting, where

$$x_i = (X_i - M_i') / \sigma_i' \tag{5}$$

Table 4 Equations Converting Mechanical Parameters into HV of Primary Hand and THV of Total Hand

A. HV equation for evaluating the primary hand value of suiting (Equation KN101-W-Series) for men's winter suiting

	C_i			Population parameters men's winter suitings	
Mechanical parameters	Smoothness (*numeri*) $C_0 = 4.7533$	Stiffness (*koshi*) $C_0 = 5.7093$	Fullness (*fukurami*) $C_0 = 4.9799$	($n = 214$) M_i	σ_i
Tensile	(5)	(4)	(3)		
LT	−0.0686	−0.0317	−0.1558	0.6082	0.0611
log WT	0.0735	−0.1345	0.2241	0.9621	0.1270
RT	−0.1619	0.0676	−0.0897	62.1894	4.4380
Bending	(4)	(1)	(6)		
log B	−0.1658	0.8459	−0.0337	−0.8673	0.1267
log 2HB	0.1083	−0.2104	0.0848	−1.2065	0.1801
Shear	(3)	(2)	(4)		
log G	−0.0263	0.4268	0.0960	−0.0143	0.1287
log 2HG	0.0667	−0.0793	−0.0538	0.0807	0.1642
log 2HG5	−0.3702	0.0625	−0.0657	0.4094	0.1441
Compression	(2)	(5)	(1)		
LC	−0.1703	0.0073	−0.2042	0.3703	0.0745
log WC	0.5278	−0.0646	0.8845	−0.7080	0.1427
RC	0.0972	−0.0041	0.1879	56.2709	8.7927
Surface	(1)	(6)	(2)		
MIU	−0.1539	−0.0254	−0.0569	0.2085	0.0215
log MMD	−0.9270	0.0307	−0.5964	−1.8105	0.1233
log SMD	−0.3031	0.0009	−0.1702	0.6037	0.2063
Construction	(6)	(3)	(5)		
log T	−0.1358	−0.1714	0.0837	−0.1272	0.0797
log W	−0.0122	0.2232	−0.1810	1.4208	0.0591

[a]Order of importance.

B. Suiting THV equation parameters (Equation KN301-W) for men's winter suiting, where $C_{00} = 3.1466$

k	Y_k	C_{k1}	C_{k2}	M_{k1}	M_{k2}	σ_{k1}	σ_{k2}
1	Smoothness	−0.1887	0.8041	4.7537	25.0295	1.5594	15.5621
2	Stiffness	0.6750	−0.5341	5.7093	33.9032	1.1434	12.1127
3	Fullness	0.9312	−0.7703	4.9798	26.9720	1.4741	15.2341

Table 4 (Continued)
C. HV equation for evaluating the primary hand values of suiting (Equation KN101-S-Series) for men's summer suiting

	Crispness (shari)	Stiffness (koshi)	Fullness (fukurami)	Antidrape stiffness (hari)	Population parameters men's summer suiting (n = 156)	
			C_i			
Mechanical parameters	$C_0 = 4.7480$	$C_0 = 4.6089$	$C_0 = 4.9217$	$C_0 = 5.3929$	M_i	σ_i
Tensile	(3)[a]	(5)	(1)	(6)		
LT	0.2012	−0.0031	−0.4652	0.0156	0.6286	0.0496
log WT	0.1632	0.1154	−0.1793	−0.1115	0.8713	0.0977
RT	0.1385	0.0955	0.0852	0.0194	66.4557	5.4242
Bending	(2)	(1)	(6)	(1)		
log B	0.4260	0.7727	−0.0209	0.8702	−0.9641	0.1081
log 2HB	−0.1917	0.0610	0.0201	0.1494	−1.4150	0.1635
Shear	(5)	(2)	(3)	(3)		
log G	0.0400	0.2802	0.0567	0.0643	−0.0662	0.1079
log 2HG	−0.0573	−0.1172	0.0361	−0.0938	−0.0533	0.1769
log 2HG5	0.1237	0.1110	−0.0944	0.2345	0.3536	0.1678
Compression	(4)	(4)	(5)	(4)		
LC	0.0828	−0.0193	−0.0388	−0.1153	0.3271	0.0660
log WC	−0.0486	−0.1139	0.1411	−0.0846	−0.9552	0.1163
RC	−0.2252	−0.1164	0.0440	−0.0506	51.5427	8.8275
Surface	(1)	(3)	(4)	(2)		
MIU	−0.2712	−0.2272	−0.1157	−0.3662	0.2033	0.0181
log MMD	0.1304	0.0472	−0.0635	0.1592	−1.3923	0.1707
log SMD	0.9162	0.1208	−0.0560	0.1347	0.9155	0.1208
Construction	(6)	(6)	(2)	(5)		
log T	0.0001	0.0245	−0.0591	0.0067	−0.3042	0.0791
log W	0.0824	0.0549	0.2770	0.0918	1.2757	0.0615

[a]Order of importance.

D. Suiting THV equation parameters (Equation KN301-S) for men's mid-summer suiting, where $C_{00} = 3.2146$

k	Y_k	C_{k1}	C_{k2}	M_{k1}	M_{k2}	σ_{k1}	σ_{k2}
1	Crispness	1.1368	−0.5395	4.7480	24.8412	1.5156	14.9493
2	Stiffness	−0.0004	0.0066	4.6089	22.4220	1.0860	11.1468
3	Fullness	0.5309	−0.3741	4.9217	25.2704	1.0230	10.1442
4	Antidrape stiffness	0.3316	−0.4977	5.3929	30.7671	1.2975	14.1273

The M_i' and σ_i' for the women's suiting population are shown in Table 5. The equation for the THV derivation of women's suiting is exactly the same as that of the men's THV equation. In the case of women's suiting, one additional hand, "soft feeling," is also important. This is not a primary hand, but rather one segment of the total hand value, and is used frequently. This hand is derived from mechanical parameters in the same manner as primary hand. The coefficient for the conversion equation is shown in Table 6.

We may apply the basic men's equation to a very wide range of fabric categories. The inspection of the validity of this extension method is continuing as of this writing.

Table 5 M_i' and σ_i' of Women's Suiting Population (Equation KN201-MDY-Series)

Mechanical parameters	Population parameters of women's suitings ($n = 220$)	
	M_i'	σ_i'
Tensile		
LT	0.6177	0.0823
log WT	1.1511	0.2166
RT	42.0564	6.9586
Bending		
log B	−0.8722	0.2565
log 2HB	−1.1444	0.3473
Shear		
log G	−0.0745	0.2099
log 2HG	0.1312	0.2966
log 2HG5	0.4217	0.2596
Compression		
LC	0.4070	0.1061
log WC	−0.6211	0.2380
RC	52.2626	9.1288
Surface		
MIU	0.2416	0.0431
log MMD	−1.7248	0.1926
log SMD	0.5696	0.3521
Construction		
log T	−0.0446	0.1693
log W	1.3550	0.1270

Source: Ref. 9.

Table 6 Coefficients for the Converting Equation for Soft Feeling of Women's Suiting (Equation KN201-MDY-Series)

Mechanical parameters	Soft feeling (*sofutosa*), C_i for $C_0 = 3.2881$
Tensile	(4)[a]
LT	−0.1783
log WT	0.0102
RT	−0.3573
Bending	(5)
log B	−0.3073
log 2HB	0.0159
Shear	(3)
log G	−0.4214
log 2HG	0.0146
log 2HG5	−0.0326
Compression	(2)
LC	−0.0472
log WC	0.5641
RC	0.4741
Surface	(1)
MIU	−0.2159
log MMD	−0.9211
log SMD	0.3479
Construction	(6)
log T	−0.0657
log W	0.0340

[a]The order of importance.

Table 7 Mechanical Parameters of Samples 1, 2, and 3

		Sample		
i	X_i	1	2	3
Tensile				
1	LT	0.526	0.565	0.653
2	log WT	1.058	1.017	0.826
3	RT	66.9	74.4	59.9
Bending				
4	log B	−1.096	−1.177	−0.733
5	log 2HB	−1.529	−1.699	−1.076
Shearing				
6	log G	−0.158	−0.213	0.224
7	log 2HG	−0.105	−0.398	0.151
8	log 2HG5	0.227	−0.034	0.614
Compression				
9	LC	0.312	0.293	0.242
10	log WC	−0.682	−0.821	−0.780
11	RC	59.1	54.5	51.2
Surface				
12	MIU	0.182	0.174	0.220
13	log MMD	−2.027	−1.879	−1.747
14	log SMD	0.389	0.344	0.632
Construction				
15	log T	−0.122	−0.233	−0.284
16	log W	1.406	1.307	1.400

D. HV and THV Derivation Example

The mechanical parameters of a fabric specimen are measured as shown in Table 7. We then substitute these parameters into Eqs. (1) and (2) to obtain the HV of the three (or four) primary hands, then substitute these HV values into Eqs. (3) and (4) to obtain THV. Samples 1, 2, and 3 are suiting for winter/autumn use. Mechanical parameters are shown in Table 7.

This fabric is for winter/autumn suiting. We then obtain the HV and THV of these specimen as shown in Table 8.

E. Analysis of Fabric Hand and Quality

It is useful to plot the HV of the primary hand and the THV on a hand chart, as shown in Fig. 11. The shaded area is the high-quality zone, derived from statistical survey of commercial suitings. When the hand values of a sample fall into this zone, the sample is evaluated as a high-quality fabric. Figure 12 is the same chart

Table 8 The Primary Hand and THV

k	Primary hand	$Y_k(=$ HV) and THV for sample		
		1	2	3
1	Stiffness (*koshi*)	4.00	3.61	7.67
2	Smoothness (*numeri*)	7.65	6.57	3.70
3	Fullness (*fukurami*)	6.94	5.35	4.09
	Total hand value (THV)	4.47	3.70	2.70

HAND CHART

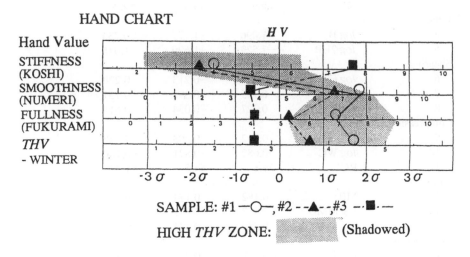

SAMPLE: #1 —○—, #2 - -▲- -, #3 — · ■ · —

HIGH *THV* ZONE: (Shadowed)

Figure 11   Hand chart for winter/autumn suiting.

HAND CHART

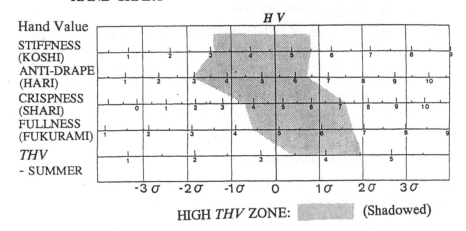

HIGH *THV* ZONE: (Shadowed)

Figure 12   Hand chart for summer suiting.

for summer suiting. Plotting a hand chart is a convenient and simple method of fabric hand analysis. In order to relate the analysis to fabric design, it is important to cover all mechanical parameters on the mechanical parameter chart shown in Fig. 13. The parameters of samples 1, 2, and 3 are plotted on this chart. The shaded area is a good fabric zone that is derived statistically. The scales on the horizontal axes are normalized x_i axes. For convenience, the raw value is also scaled on each axis. This chart can be used to find extremely abnormal mechanical properties.

F. How to Construct the Equations

The coefficient C in the conversion equations was derived on the basis of the experts' subjective judgments. When one wants to create a new equation for any object population, one must locate reliable judges to perform the subjective judgments. The physical parameters that are the basis of an objective evaluation must be those that express the related fabric properties as accurately as possible. For example, if the number of parameters used for creating an equation is 10, the number of samples correlating to the subjective values must be at least 10 times the number of parameters—more than 100 in this example.

When the subjective data matrix $[Y_i]$ ($i = 1–n$, n = number of samples) and the mechanical parameter matrix $[X_{ij}]$ (i = 1–n, j = 1–m, m = number of parameters) are correlated with the linear equation, the constant coefficients C_j are obtained by solving Eq. (6) for C_j in the condition so that the regression error is kept to a minimum.

$$[Y_i] = [C_j] [X_{ij}] \tag{6}$$

$$\sum_{i=1}^{n} (Y_i - Y_i')^2 = \text{minimum} \tag{7}$$

where Y_i' is the regressed value calculated from X_{ij} and a known C_j. This procedure is a regular multivariable regression, which is common in statistics [10].

In the application of multivariable regression, however, it is necessary to use the regression method most suitable to the circumstances. In the case of fabric hand, there are strong correlations between some of the parameters and hand values. When two variables have a strong correlation, we usually eliminate one of them. However, even though B and 2HB may have a strong mutual correlation, for example, we can not eliminate either one because both parameters are important to fabric design in characterizing fabric bending property. We considered them both necessary unless a perfect correlation exists between them. In the case of primary hand, for example, smoothness and fullness have a strong correlation. From a statistical standpoint, we may eliminate one of them; however, both have been used by many experts for a long time. Even though a strong correlation exists between them, they each make a different, important contribution to fabric quality.

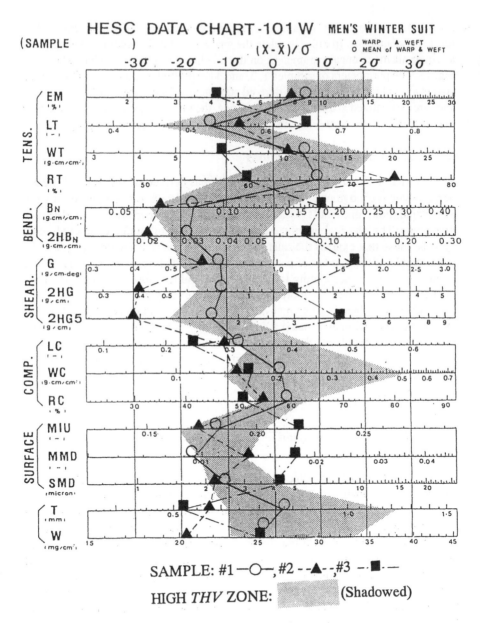

Figure 13   Mechanical parameter chart for winter suiting.

Another situation is when two or three mechanical parameters represent the same property of a fabric—for example, when B and 2HB both represent the bending property. From a fabric design point of view, these two are not separate parameters but are considered to be one group. From these considerations for correlation between parameters and the group concept, the following "block stepwise regression" was used for the regression of primary hand, the KN-101 series [3, 11]. This block stepwise regression is a modification [12] of the step-wise regression.

Variables are grouped into six blocks, with each block corresponding to a fabric property, such as tensile, bending, etc. In the first step, each variable group is regressed separately with Y and the block with the highest regression accuracy is chosen. The resulting regression error is then regressed with each of the remaining blocks in the same manner. The first and second regression equations are added to form a new regression equation in which the two blocks are regressed. The same procedure is repeated until the last block is completed. The rank of the step also gives us information on the ranking of the importance of the blocks to the Y value. After the regression equation is complete, we again apply stepwise regression to variables in the first block to reconstruct the regression equation for the first block; then the variables of the second block are regressed stepwise, and the new regression equations of the first and second blocks are added. This procedure continues, following the order of the block already determined by the first stepwise block regression. In this stepwise method, a significance inspection was done at every step to determine if the new step was necessary. In this case, we did not eliminate any blocks and used all parameters.

For the derivation of the THV equation, we have no blocks, and the ordinary multivariable regression method was applied to constructing the KN-301 series equation, where the square-term variables are included as shown in Eqs. (3) and (4).

IV. DIRECT APPLICATIONS OF THE MECHANICAL PARAMETERS

The mechanical parameters used for the objective evaluation of fabric hand are useful not only for the fabric hand evaluation but also for evaluating the fabric performance from different viewpoints. One example is the application to tailoring process control in suit manufacturing [1, 13]. The control chart is shown in Fig. 14. The mechanical parameters used in this control are those of the tensile and shearing properties of fabric. When the parameters of a fabric fall into the central zone indicated as the "noncontrol" zone, the tailoring of this fabric does not require any control on the suit manufacturing line. When only some of the parameters fall into this zone, tailoring control is necessary, for example, careful handling of the fabric during sewing, the use of reinforcement tape, etc., as indicated.

While using this chart, it was discovered empirically that the parameters of high-quality suiting from the mechanical comfort viewpoint in wearing fall into

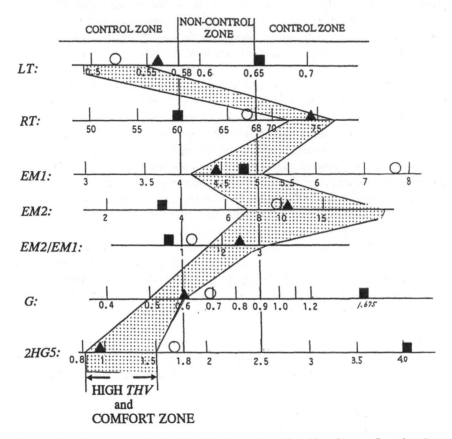

Figure 14 Process control chart for tailoring and comfortable suit zone. Sample # 2 (▲) satisfies the comfort condition. Sample # 1 (○) satisfies almost the comfort zone except for warp directional extensibility, *EM1*, which is too high.

the snake-shaped zone shaded on this chart. We call this zone the "comfortable suit zone."

V.   CONCLUDING REMARKS

In this chapter, we introduced the objective hand evaluation method. This method was developed for the objective measurement of fabric hand; however, the method may be applied to other materials that interact with the human senses, such as leather, artificial leather [14], and even to study the effects of cosmetics on human hair softness, food technology, etc. The authors have named these materials "human interactive materials." Some actual applications have been reported.

The mechanical parameters used in this analysis have been applied not only to the objective hand evaluation system but also to many other fields such as tailoring process control, the prediction of the making up of a suit, the prediction of comfort wearing properties based on these parameters, etc. The delicate nonlinear mechanical properties in the low-load region are important for the characterization of human interactive materials.

REFERENCES

1. S. Kawabata and M. Niwa, Fabric performance in clothing and clothing manufacture, *J. Textile Inst. 80*:19–50 (1989).
2. F. T. Peirce, The "handle" of cloth as a measurable quantity, *J. Textile Inst. 21*:T377–416 (1930).
3. S. Kawabata, *The Standardization and Analysis of Hand Evaluation*, 2nd ed., Hand Evaluation and Standardization Committee, Textile Machinery Society of Japan, Osaka, 1980.
4. S. Kawabata, ed., *HESC Standard of Hand Evaluation (HV Standard for Men's Suiting)*, HESC, Textile Machinery Society of Japan, Osaka, 1975.
5. S. Kawabata, ed., *HESC Standard of Hand Evaluation*, Vol. 1, *HV Standard for Men's Suiting*, 2nd ed., HESC, Textile Machinery Society of Japan, Osaka, 1980.
6. S. Kawabata, ed., *HESC Standard of Hand Evaluation*, Vol. 2, *HV Standard for Women's Thin Dress Fabric*, HESC, Textile Machinery Society of Japan, Osaka, 1980.
7. S. Kawabata, ed., *HESC Standard of Hand Evaluation*, Vol. 3, *HV Standard for Men's Winter Suiting*, HESC, Textile Machinery Society of Japan, Osaka, 1982.
8. V. B. Mountcastle, R. H. LaMotte, and G. Carli, Detection thresholds for vibratory stimuli in humans and monkeys: Comparison with threshold events in mechanoreceptive afferent nerve fibers innervating the monkey hand, *J. Neurophysiol. 35*:122–136 (1972).
9. M. Niwa, Analysis of Fabric Hand of High-Quality Apparel Fabrics on the Basis of Objective Evaluation Technique and the Design and Development of the High-Performance Fabrics, Report of Research Project, Grant-in-Aid for Co-operative Research in Japan, 1988, pp. 125–155.
10. P. G. Hoel, *Introduction to Mathematical Statistics*, 4th ed., John Wiley and Sons, New York, 1971.
11. S. Kawabata, M. Niwa, and W. Fumei, Objective hand measurement of nonwoven fabrics, Part I: Development of the equations, *Textile Res. J. 64*: 597–610 (1994).
12. N. R. Draper and H. Smith, *Applied Regression Analysis*, John Wiley and Sons, New York, 1966.
13. S. Kawabata, K. Ito, and M. Niwa, Tailoring process control, *J.Textile Inst. 83*: 361–373 (1992).
14. M. Niwa, C. Liu, and S. Kawabata, Application of the objective fabric-hand evaluation technology to the other materials such as artificial leather, foam and tissue paper,

Proc. 18th Textile Technology Symposium at Mount Fuji, Textile Machinery Society of Japan, Osaka, 1989, pp. 167–173.

15. S. Kawabata, and M. Niwa, A Proposal of the Standardized Measuring Condition for Mechanical Property of Apparel Fabrics, *Proc. Third Japan-Australia Symposium on Objective Measurement*, Textile machinery Society of Japan, Osaka, 1985, pp. 825–835.

11

Colorimetry for Textile Applications

PATRICK TAK FU CHONG   Spartan Mills, Spartanburg, South Carolina

I. INTRODUCTION

Since the introduction of CIE basic colorimetry by the Commission Internationale de l'Eclairage (CIE) in 1931, it has been the universal objective to utilize objective color measurement technology, whenever possible, for a wide variety of applications from color quality monitoring to color communication. The present chapter on colorimetry for textile applications is designed to provide the appropriate background information covering the fundamental concept of color science. A review of the various types of color measuring instrumentation and their selection is given. The proper color measuring procedure for textile materials is then introduced. Having established the necessary knowledge on color measurement technology, various practical applications in the textile industry are highlighted. Emphasis also is given to the major developments in the hardware and software of the color measuring system. The impact of such developments on the practical applications is addressed while the future prospects of the role of color measuring system in the direction of large-scale integration of color production system are highlighted. Finally, extensive references are provided in this chapter for those interested in pursuing an in-depth study of colorimetry as a textile characterization method.

II. BACKGROUND

Hardly a day goes by without the need for each of us to verbally describe at least one color. Color is not only used as a descriptive term, as most people do in their daily conversation, but also serves other important functions in commerce, industry, and science. The matching of color has been used for object identification in forensic science. The intensity of color has been used for the determination of chemical concentration in the chemical industry. Color has been used for the identification of tablets or capsules in the pharmaceutical industry. It also has been used for the assessment of the performance of photographic materials and processes. Furthermore, it also has been used to evaluate the color-rendering performance of the light sources. There are many other application areas, including textiles, food, cosmetics, coatings, plastic, metals, ceramics, paper, and the list goes on. In all these applications, it is important that an objective method be used to specify color accurately. This is analogous to the use of a micrometer to measure length, or the use of a weighing machine to measure the weight. Here, we need a color-measuring instrument to measure color.

Color communication, in the absence of objective color specification, is frequently confusing. This is because the appearance of color is subjected to influence simultaneously by at least three very different phenomena: the light source, the object, and the visual system. The Illuminating Engineering Society defines light as "visually evaluated radiant energy." Radiant energy comprises the whole gamut from cosmic rays to radio and power transmission. Without this wide electromagnetic radiation range, the human eye can only detect the visible spectrum range from about 380 nm to 780 nm. Changes in either the radiant quantity or its spectral distribution can alter the observed color. Examples of commonly used artificial lamps are incandescent lamps, fluorescent lamps, mercury halide lamps, and sodium lamps. On the other hand, daylight is the natural light for observing color. They all have different spectral power distributions. The nature of the object illuminated can modify the quantity and quality of the incident light through selective spectral absorption, transmission, reflection, and other kinds of interactions such as fluorescence. In the context of colorimetry, there are three main classes of objects: namely, the transparent object, the translucent object, and the reflecting object—each exhibiting different ways of modifying the incident light source. The final element in the perception of color is the visual system, that is, the physiological properties of the eye that detect the modified radiant energy from the object and sends signals to the brain and, finally, the psychological processes [1–3] of the brain interpreting the received signals into response, which we call color. The subject of color and colorimetry has been studied extensively and reported in the literature [4–7].

III. CIE COLORIMETRY

The Commission Internationale de l'Eclairage (CIE) is an international organization that promotes the advances of science, technology, and art in the fields of light and lighting. The corresponding English name is the International Commission on Illumination. At the sixth session of the CIE, held in Geneva in 1924, it was decided to set up a study group on colorimetry. At the seventeenth session of the CIE, held in New York in 1928, a working program was proposed to reach agreements on (1) colorimetric nomenclature, (2) a standard daylight for colorimetry, and (3) the "sensation curves" of the average human observer with normal color vision. At the eighth session of the CIE, held in England in 1931, major recommendations that laid the basis for colorimetry were made.

A. CIE Color Specification by Tristimulus Values

In 1931, the CIE established an objective method of color specification by tristimulus values [8,9]. In this method, light source is characterized by its spectral power distribution S(λ). Objects can be characterized by its spectral reflectance curves R(λ) or spectral transmittance curves T(λ). Figure 1 shows the spectral reflectance curves of several opaque colored materials. The color vision properties of the eye can be simulated by the use of the principle of trichromacy, which postulates the existence of three response functions of the human eye generating the signals sent to the brain. In this aspect, the CIE established a standard observer [9–12], which is expressed in the form of three sets of numerical data , $\bar{x}(\lambda)$, $\bar{y}(\lambda)$, and $\bar{z}(\lambda)$, representing the color-matching response of the average normal human observer under a standard state of adaptation and viewing conditions to the indi-

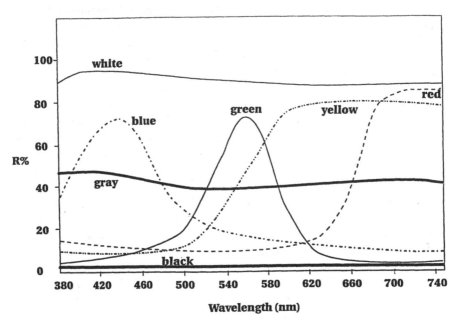

Figure 1   Spectral reflectance curves of several opaque materials.

vidual monochromatic spectrum colors. In this CIE colorimetry, three quantities called "tristimulus values" (X, Y, Z) are used to specify a color. The tristimulus values are obtained by multiplying, at each wavelength in the visible spectrum, the spectral power distribution $S(\lambda)$, the spectral reflectance factor $R(\lambda)$ in the case of reflecting object or the spectral transmittance $T(\lambda)$ in the case of transmitting object, and the CIE standard observers $\bar{x}(\lambda)$, or $\bar{y}(\lambda)$, or $\bar{z}(\lambda)$, and then summing the products over the visible wavelength range, as shown in the following equations:

$$X = k \sum_{\lambda} S(\lambda) \cdot R(\lambda) \cdot \bar{x}(\lambda) \, d\lambda \tag{1}$$

$$Y = k \sum_{\lambda} S(\lambda) \cdot R(\lambda) \cdot \bar{y}(\lambda) \, d\lambda \tag{2}$$

$$Z = k \sum_{\lambda} S(\lambda) \cdot R(\lambda) \cdot \bar{z}(\lambda) \, d\lambda \tag{3}$$

$$k = 100 \, / \sum_{\lambda} S(\lambda) \cdot \bar{y}(\lambda) \, d\lambda \tag{4}$$

where k is a normalizing factor, so that for a perfect white sample it has a tristimulus Y value of 100.

In the presence of numerous kinds of light sources, the CIE also has recommended sources with defined spectral power distributions called standard illuminants [9]. They are illuminant A, which simulates the incandescent lamp, and illuminants D, representing a range of the phases of daylight differentiated by correlated color temperatures. In addition, there are a variety of F-illuminants [9], which represent typical fluorescent lamps. Among these F-illuminants, illuminants F2, F7, and F11 should take priority over others when a few typical illuminants are to be selected. F2 is a typical cool white fluorescent lamp with a correlated color temperature of 4230 K and a CIE color rendering index of 64. F7 is a broad-band fluorescent lamp with a correlated color temperature of 6500 K and a CIE color rendering index of 90. F11 is a fluorescent lamp with three narrow bands, a correlated color temperature of 4000 K, and a CIE color rendering index of 83. Under the CIE system of colorimetry, the tristimulus values for an object would change if the spectral power distribution of the incident light source changes. Furthermore, two objects would be considered a match in color under a specified condition if they have the same tristimulus values. It is possible that such a pair of objects match under one kind of source but mismatch under another source of different spectral power distribution. This kind of match is known as metameric match. On the other hand, a pair of objects would match under any light source if they have the same spectral reflectance factors. This kind of match is called nonmetameric. CIE has recommended a Special Metamerism Index: Change in Illuminant, which provides a measure of the color difference between two metameric objects caused by substituting a test illuminant of different relative spectral composition for the reference illuminant [9].

B.   CIE 1976 (L* a* b*) Color Space

Subjectively, three variables commonly used to describe a color are hue, lightness, and chroma. Hue is the attribute corresponding to whether the object is red, orange, yellow, green, blue, or violet. Chroma is the attribute of a visual sensation as to the proportion of pure chromatic color. The lower the chroma of the color, the closer the color to neutral appearance. Thus a pastel tint has a low chroma, while a pure color is said to have high chroma. Lightness is the attribute of visual sensation associated with the luminous intensity of the object. Lightness can range from black to white for reflecting object and from black to perfectly clear and colorless for transparent object. The observed color difference for a pair of colors, therefore, constitutes the variations in hue, lightness, and chroma. Figure 2 illustrates the relationship of the color variables in a three-dimensional color space. Objectively, the CIE system of colorimetry also provides methods [9] for specifying the color difference between a pair of objects. The CIE 1976 (L* a* b*) Color Space [13] is a popular CIE color space to quantitatively interpret the differences of two colors in a three-dimensional color space using the L* axis, a* axis, and b* axis as illustrated in Figure 3. The alternate name for CIE 1976 (L* a* b*) color space is CIELAB color space. The L* axis runs through the center of the horizontal hue circle with 100 at the top representing white and 0 at the bottom representing black. At each horizontal plane cutting through the vertical L* axis, the a* axis crosses with the b* axis at the center where the L* axis runs through vertically. The a* axis shows red when positive (+) and green when negative (−). Sim-

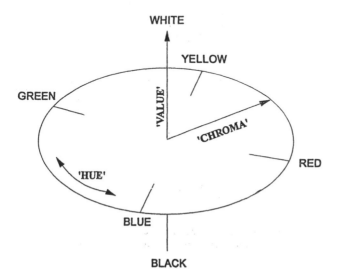

Figure 2   Relationship of hue, value, and chroma in three-dimensional space.

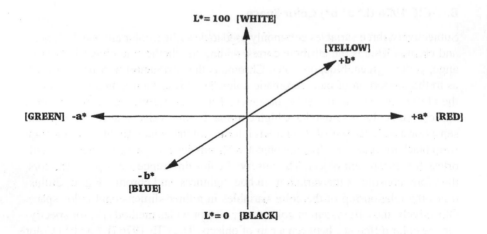

(a) 3 D CIELAB Color Space

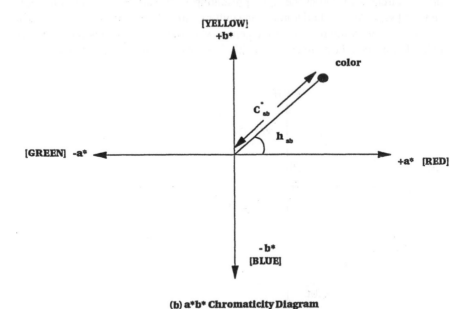

(b) a*b* Chromaticity Diagram
(Horizontal X-section of CIELAB Color Space)

Figure 3   CIE 1976 (L* a* b*) color space.

ilarly, the b* axis shows yellow when positive and blue when negative. Neutral color is indicated by a* and b* values being close or equal to zero. The chroma of the color increases, though not linearly, as the values, of a* or b* increase. The terms L*, a*, and b* are defined in the following equations.

$$L^* = 116(Y/Y_n)^{1/3} - 16 \qquad \text{if} \qquad Y/Y_n > 0.008856 \qquad (5)$$

or
$$L^* = 903.3(Y/Y_n)^{1/3} \qquad \text{if} \qquad Y/Y_n \leqslant 0.008856 \qquad (6)$$

$$a^* = 500[f(X/X_n) - f(Y/Y_n)] \qquad (7)$$

$$b^* = 200[f(Y/Y_n) - f(Z/Z_n)] \qquad (8)$$

where

$$f(X/X_n) = 7.787(X/X_n) + 16/116 \qquad \text{if} \qquad X/X_n \leq 0.008856$$

or
$$f(X/X_n) = (X/X_n)^{1/3} \qquad \text{if} \qquad X/X_n > 0.008856$$

$$f(Y/Y_n) = 7.787(Y/Y_n) + 16/116 \qquad \text{if} \qquad Y/Y_n \leq 0.008856$$

or
$$f(Y/Y_n) = (Y/Y_n)^{1/3} \qquad \text{if} \qquad Y/Y_n > 0.008856$$

$$f(Z/Z_n) = 7.787(Z/Z_n) + 16/116 \qquad \text{if} \qquad Z/Z_n \leq 0.008856$$

or
$$f(Z/Z_n) = (Z/Z_n)^{1/3} \qquad \text{if} \qquad Z/Z_n > 0.008856$$

For these equations, X_n, Y_n, and Z_n are the tristimulus values of the illuminant.

Thus the difference in L*, a*, and b* values between the reference color and the sample color may be interpreted in the following manner. Here, the convention is to subtract the reference values from the sample values.

$+\Delta L^*$: Sample color is lighter than the reference color
$-\Delta L^*$: Sample color is darker than the reference color
$+\Delta a^*$: Sample color is redder (or less green) than the reference color
$-\Delta a^*$: Sample color is greener (or less red) than the reference color
$+\Delta b^*$: Sample color is yellower (or less blue) than the reference color
$-\Delta b^*$: Sample color is bluer (or less yellow) than the reference color

The total color difference (ΔE_{ab}^*) between two colors is computed as the Euclidean distance between the points representing them in the CIELAB space:

$$\Delta E_{ab}^* = [(\Delta L^*)^2 + (\Delta a^*)^2 + (\Delta b^*)^2]^{1/2} \qquad (9)$$

On the other hand, when it is desired to express specifications in terms of the approximate correlates of lightness, chroma, and hue, the terms CIE 1976 lightness (L*), CIE 1976 a,b chroma (C_{ab}^*), and CIE 1976 a,b hue-angle (h_{ab}) should be used.

The CIE 1976 a,b chroma (C_{ab}^*) provides a direct correlate of the concept of chroma and is defined by the following equation.

$$C_{ab}^* = (a^{*2} + b^{*2})^{1/2} \qquad (10)$$

The CIE 1976 a,b hue-angle (h_{ab}) correlates the concept of hue. The 0° starting point is assigned to the horizontal +a* axis. The hue-angle increases counter-clockwise around the central vertical axis of the L* on the a*, b* diagram with 90° being +b* (yellow), 180° being −a* (green), and 270° being −b* (blue), and then 360° back to the +b* (red) axis again. This is illustrated in Figure 3, and the hue-angle is defined by the following equation:

$$h_{ab} = \arctan(b^*/a^*) \qquad (11)$$

Similarly, when it is desired to identify the components of color differences in terms of approximate correlates of lightness difference, chroma difference, and hue difference, the following terms should be used: CIE 1976 lightness difference (ΔL^*), CIE 1976 chroma difference (ΔC_{ab}^*), and CIE 1976 a,b hue difference (ΔH_{ab}^*).

The ΔL^* has been defined earlier, and ΔC_{ab}^* is the difference in C_{ab}^* values between the two colors. The ΔH_{ab}^* are defined by the following equation:

$$\Delta H_{ab}^* = [(\Delta E_{ab}^*)^2 - (\Delta L^*)^2 - (\Delta C_{ab}^*)^2]^{1/2} \qquad (12)$$

In 1991, R. Seve [14] derived the following alternative formula for computing ΔH_{ab}^*:

$$\Delta H_{ab}^* = 2(C_1^* \, C_2^*)^{1/2} \sin[(h_1 - h_2)/2] \qquad (13)$$

where subscripts 1 and 2 refer to color 1 and color 2.

Equation (13) is an improvement over Eq. (12) in that the quantity ΔH_{ab}^* is directly obtained with the same sign as the quantity $(h_1 - h_2)$ for convenient interpretation of color difference. Hence, a pair of colors that are colorimetrically matched has a calculated value of zero for the total color difference (ΔE_{ab}^*). This quantity increases as the mismatch increases. The splitting of the total color difference (i.e., ΔL^*, Δa^*, Δb^* or ΔL^*, ΔC_{ab}^*, ΔH_{ab}^*) also has been used for color quality monitoring of a coloration process [15]. For example, the color acceptability tolerance specifications may be expressed as $\pm \Delta L^*$, $\pm \Delta a^*$, $\pm \Delta b^*$ that may vary from one location to the other location in the CIELAB space. Such tolerance volume in the CIELAB space is therefore rectangular box shaped, which is not compatible to the many research results showing the tolerance volume being ellipsoidal. An alternative color space, 1976 L* u* v* (CIELUV), also was recommended in 1976 by CIE [9] but is preferred by those who work in the additive color mixing industry such as the color television manufacturers.

C. Advances in Color Difference Formulas

The CIELAB color difference formula has the limitation of being nonuniform color space; that is, equal distances in the CIELAB space do not represent equal visual color differences. Since 1976, many attempts have been made to improve this limitation [16–18] and have resulted in the following color difference formulas.

- JPC 79 formula [16]: This formula has been developed under the leadership of R. McDonald at J & P Coats, England.
- Datacolor formula: This formula has been developed under the leadership of E. Rohner of Datacolor AG (now Datacolor International), Switzerland. This color difference formula is proprietary and has not been published.
- M&S 89 formula: This formula has been developed by Marks & Spencer in collaboration with Instrumental Color Systems (now Datacolor International), England. This color difference formula is proprietary and has not been published.
- CMC (1:c) formula [17]: This formula has been developed by the Color Measurement Committee of the Society of Dyers and Colourists in England. It is an improvement of the JPC 79 formula.
- BFD (1:c) formula [19]: This formula has been developed under the leadership of B. Rigg and M. R. Luo at the University of Bradford, England.
- Berns and co-workers developed a new experimental data set that sampled color location and color difference direction to determine the mean color difference tolerance and the color difference variability of a population sample. These new results were used to fit simple empirical structures added to the CIELAB model [20].
- A team of researchers at Granada University has developed piecewise empirical models for differing areas of the chromaticity diagram to predict the coefficients of an ellipsoid formula [21].

Of these formulas, the CMC color difference formula has been adopted as the test method of the Society of Dyers and Colourists (United Kingdom) in 1984, the British Standard BS6923: 1988, and the American Association of Textile Chemists and Colorists Test Method 173–1989. In 1995, it has become an official standard of the International Standards Organization (ISO). The CMC color difference formula is an achievement of the Color Measurement Committee of the Society of Dyers and Colourists in the United Kingdom, and the Committee's initials have been adopted as the name of the formula. The CMC formula is widely used in the textile and apparel industries. The CMC color difference formula is given in the following equations:

$$\Delta E_{CMC(l,c)} = \{[\Delta L^*/(\ell \cdot S_L)] + [\Delta C_{ab}^*/(c \cdot S_c)]^2 + [\Delta H_{ab}^*/(S_H)]^2\}^{1/2} \tag{14}$$

where ΔL^*, ΔC_{ab}^*, and ΔH_{ab}^* have been defined by the CIELAB formula, and

$$S_L = 0.040975 \, L_s^* / (1 + 0.01765 \, L_s^*) \qquad \text{if} \qquad L_s^* \geq 16 \tag{15}$$

or

$$S_L = 0.511 \qquad \text{if} \qquad L_s^* < 16 \tag{16}$$

$$S_C = [0.0638 \, C_{ab,r}^*/(1 + 0.0131 \, C_{ab,r}^*)] + 0.638 \tag{17}$$

where the subscript r of $C_{ab,r}^*$ refers to the Reference Specimen value

$$S_H = (FT + 1 - F)S_C \tag{18}$$

where

$$F = \{ (C_{ab,r}^*)^4 / [(C_{ab,r}^*)^4 + 1900] \}^{1/2} \tag{19}$$

$$T = 0.36 + |0.4 \cdot \cos(35 + h_{ab})| \qquad \text{if} \qquad h_{ab} \geq 345° \quad \text{or} \quad \leq 164° \tag{20}$$

$$T = 0.56 + 0.2 \cos(168 + h_{ab}) \qquad \text{if} \qquad h_{ab} 164° \quad \text{or} \quad < 345° \tag{21}$$

and where ℓ and c are relative tolerances for lightness and chroma differences, respectively, and the value of c should always remain at 1.0. It has been demonstrated that changes in hue are generally much less tolerated than changes in lightness or chroma. Thus, the ℓ and c factors may be modified to place different emphasis on lightness and chroma, respectively, in relation to hue. The values of ℓ and c are generally set to 2.0 and 1.0 for acceptability application. Other values of ℓ may be required in cases where the surface characteristics dramatically differ from flat textiles.

The CMC formula has an autotolerancing feature that can define a reasonable volume of acceptance in terms of the S_L, S_C, and S_H values, based on the location of the standard color in the CIELAB space.

D. CIE TC1–29, Industrial Color-Difference Evaluation

In 1993, the CIE Technical Committee on Industrial Color-Difference Evaluation published a full draft of Recommendation on Industrial Color-Difference Evaluation [22]. In this draft, it recommended an extension of the CIE L* a* b* uniform color space and color-difference equations for industrial color-difference evaluation with added corrections for variation in perceived color difference resulting from variation in chroma level of the color standard. A set of base conditions is defined under which the recommended model is expected to perform well. The base conditions are defined by the illumination type, illuminance, observer, background field, viewing mode, sample size, sample separation, color difference magnitude, and sample structure. When conditions of use deviate significantly from the base conditions, the introduction of parametric factors may be used to correct for the effects of experimental or material variables. The complete color-difference model for industrial color-difference evaluation is termed the CIE 1994 (ΔL^* ΔC^*_{ab} ΔH^*_{ab}) color-difference model with the symbol, ΔE^*_{94}, and abbreviation CIE94. This formula is defined as follows:

$$\Delta E^*_{94} = \{[\Delta L^*/(K_L S_L)]^2 + [\Delta C^*_{ab}/(K_C S_C)]^2 + [\Delta H^*_{ab}/(K_H S_H)]^2\}^{1/2} \quad (22)$$

The total color difference, ΔE^*_{94}, is the distance between two color samples in lightness, chroma, and hue differences, ΔL^*, ΔC^*_{ab}, ΔH^*_{ab}, weighted by weighting functions S_L, S_C, and S_H, and parametric factors K_L, K_C, and K_H.

The weighting functions S_L, S_C, and S_H adjust the total color-difference equation to account for variation in perceived color-difference magnitude with variation in the color standard location in CIELAB space. The current best estimates of these weighting functions obtained by fitting with two visual color-difference perception data sets [19,20] are defined by the following formulas:

$$S_L = 1 \quad (23)$$
$$S_C = 1 + 0.045C_{ab}^* \quad (24)$$
$$S_H = 1 + 0.015C_{ab}^* \quad (25)$$

The parametric factors K_L, K_C, and K_H are correction terms for variation in perceived color-difference component sensitivity with variation in experimental conditions. Under the base conditions the parametric factors have assigned values of unity and have no effect on the total color difference. In the textile industry it is common to set the lightness parametric factor, K_L, to 2 while K_C and K_H are set to 1.0. The TCI-29 formulas retain the fundamental features of the CMC formula but modify the weighting functions, S_L, S_C, and S_H, based on the research results obtained by Berns and co-workers [20].

1. Reference Documents

- Publication CIE No. 15.2 (TC-1.3) Colorimetry, 2nd Edition, Wien 3. Bezirk, Kegelgasse 27/1, Austria, 1986.
- Publication CIE No. 17 (E-1.1) International Lighting Vocabulary, Wien 3. Bezirk, Kegelgasse 27/1, Austria.
- American Standard Test Method E308-95 for Computing the Colors of Objects by Using the CIE System, ASTM, 1916 Race St., Philadelphia, PA 19103, 1995.
- AATCC 173, *CMC: Calculation of Small Color Differences for Acceptability,* Technical Manual of the American Association of Textile Chemists and Colorists, AATCC, Research Triangle Park, NC 27709.
- CIE TC1-29 Industrial Color-difference Evaluation, Full Draft No. 2: Recommendation on Industrial Color-difference Evaluation, CIE, Wien 3. Bezirk, Kegelgasse 27/1, Austria, 1993.

IV. COLOR MEASURING SYSTEMS

Color measuring equipment is generally classified into spectrophotometer type and tristimulus colorimeter type. The former measures the spectral reflectance factors (and spectral transmittance for some equipments), while the latter measures the CIE tristimulus data directly.

A. Spectrophotometers

Fundamentally, the spectrophotometer is used to compare the radiant power leaving the object with that of a reference standard at each wavelength. The instrument itself consists of a light source whose emitted light is incident onto the objects and the reflected light is then passed into the monochromator. The monochromator disperses the incoming radiant energy spectrally and transmits it via a narrow band of wavelengths through the exit slit. The detector system receives the spectral radiant power reflected from the object and the standard in close succession and generates a ratio signal that is transmitted to the computer for analysis and display. The computer is interfaced with various components of the spectrophotometer and controls its operation. With the fundamental data of reflectance factors or transmittance, one can compute all kinds of useful colorimetric data for various kinds of practical applications. Figure 4 shows a simplified diagram of a spectrophotometer.

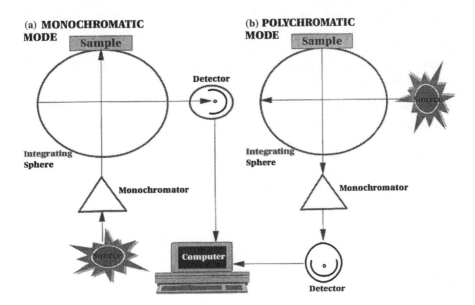

Figure 4   Simplied optical diagram of a spectrophotometer.

1. Reference Documents

- American Standard Test Method E1331 for Reflectance Factor and Color by Spectrophotometry Using Hemispherical Geometry, ASTM, 1916 Race St., Philadelphia, PA 19103.
- American Standard Test Method E1349 for Reflectance Factor and Color Using Bidirectional Geometry, ASTM, 1916 Race St., Philadelphia, PA 19103.

B. Tristimulus Colorimeter

A tristimulus colorimeter is an instrument with spectral response functions directly proportional to that of the CIE standard colorimetric observers. In this instrument, radiant power from the light source is incident onto the object. The reflected radiant power passes through one of the three tristimulus filters and falls onto the photodetector, causing it to give a response proportional to the corresponding tristimulus value of the object–source combination. These raw data are then transferred to a microprocessor for the computation of the absolute CIE tristimulus values. It is a useful tool for color quality monitoring of the production of a colored object. Most commercial tristimulus colorimeters are satisfactorily precise but may not agree with the tristimulus value obtained by spectrophotometry. However, there are many practical applications for which less accurate but precise instruments can still be useful, such as color quality monitoring. The tristimulus colorimeter is easy and quick to operate and is usually much cheaper than the spectrophotometric system. Figure 5 shows a simplified diagram of a tristimulus colorimeter.

1. Reference Document

- American Standard Test Method E1331 for Color and Color Difference Measurement by Tristimulus (Filter) Colorimetry, ASTM, 1916 Race St., Philadelphia, PA 19103.

C. Computer

At present all color measuring equipment is interfaced with computers to improve measurement speed and accuracy. A simple computer configuration includes a processor, monitor with graphic adapter, printer, magnetic disks, a suitable operating system, and the necessary application software. Additional features could include multiuser terminals and networking. The capacity of the computer ranges from microprocessor to minicomputer, depending on the application requirements. The computer serves at least four major functions.

1. Instrument Control

The computer controls the scanning of the color measuring equipment from wavelength to wavelength with the aid of a stepping motor. It also monitors the condi-

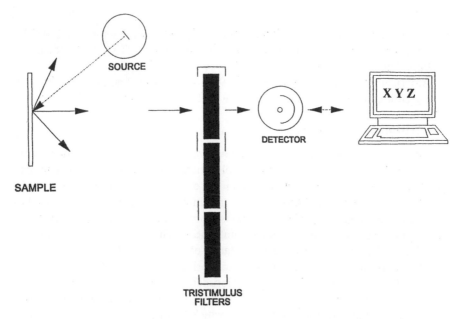

Figure 5   Tristimulus colorimeter.

tions of the equipment components such as the lamp stability. The scanning control, together with new instrument design, has improved the measurement speed dramatically in comparison with the older models.

2.   Data Transfer

The computer collects the measured data, as well as transferring the stored data for input into the specific application program for data processing.

3.   Data Storage

Many of the color measurement applications require the setup of a suitable database. Examples are the calibration information for computer color matching, acceptability data for color pass/fail, and spectral data for colorant identification. The prepared database is normally stored in the magnetic disks for later retrieval. Some users have found that the storage of the spectral data of color standards in the computer is preferred to the storage of the actual physical color standards, which may change over the storage period.

4.   Communication

Through the well-established regional or international computer network, all the useful measured information can be easily transferred to the interested party. Examples of such communications are laboratory to workshop, factory to factory, colorant supplier to users, and headquarters to overseas manufacturing plants.

D. Selection of Color Measuring Instruments

With the advances in computer technology and the revolution in equipment design, various color measuring systems exist in the market. Different applications in different environments require different instrumentations. The following items should be considered in the selection of color measuring systems.

1. Color measuring instruments
2. Computer
3. Technical support

1. Color Measuring Instruments

The two major types of color measuring equipments, the spectrophotometric system and the tristimulus colorimeter, measure different fundamental data. The former measures the spectral data while the latter measures the tristimulus values (usually for a specific illuminant) directly. Hence, if an application is only for color monitoring purposes with respect to color assessments for a specific illuminant, a tristimulus colorimeter is suitable. On the other hand, for computer color matching and for most colorant solution (liquid) evaluation tasks, the spectrophotometric system is a must, as the method involved requires the spectral data for implementation. In addition, a spectrophotometric system can handle all the tasks performed by a tristimulus colorimeter, although it is usually more expensive. If the system is used for color communication, then consideration should be given to whether the information would be subsequently used for colormonitoring purposes or for applications that require spectral data, such as in the case of colorant formulation.

(*a*) *Spectrophotometer.*   In evaluating the spectrophotometer, it would be useful to examine the following features.

1. *Spectral Range.*   The spectral energy distribution of the light source, the transmission characteristics of the monochromator, the intermediate optics, and the spectral responsivity of the detector should be designed for the working wavelength range. Normally, the spectral range from 400 to 700 nm with minimum data available at 10-nm intervals is practically sufficient. On the other hand, it would be preferable to have the spectral range from 380 to 780 nm with data available at 5-nm intervals in order to conform with the CIE recommendations [9]. In some special applications, the spectral range would have to be expanded up to 1100 nm for the measurement of camouflage materials for military application and down to 300 nm for colorant quantitative and qualitative analysis. Colorant analysis is usually done in solution medium, and hence the spectrophotometer should also be equipped with a transmission sample compartment, together with good resolution performance.

2. *Equipment Mode.*   For color measurement, the instrument is normally set up in the polychromatic mode, such that the light source is incident onto the sample first and the reflected or transmitted light is then monochromated. If the equipment is set up in the monocromatic mode (i.e., the light source is monochromated first and the selected monochromatic light is then incident onto the sample), the measured result would be the same for the nonfluorescent sample in principle but would be incorrect for the fluorescent sample [23]. Even in the polychromatic mode, it is important that the spectral distribution of the source system should be the same as that of the CIE Standard Illuminant used for calculating the tristimulus values when measuring the fluorescent samples [23]. Some equipment types provide an option of either mode for special application such as in the evaluation of true reflectance data for the fluorescent sample for colorant formulation [24]. Figure 4 shows a simplified diagram of polychromatic and monochromatic modes.

3. *Illumination and Viewing Geometries.*   The CIE has recommended four geometries defining the direction of the incident light and the direction of detecting the reflected light [9]. This can be further grouped into two major types.

In the *bidirection* type, the sample is illuminated by a beam at one angle and the reflected beam is detected at another angle such as the 45/0 or 0/45 geometries. The first number designates the illuminating incident angle while the second number designates the angle viewed by the detector. Such geometry is suitable for measuring samples that have a smooth surface, such as paint, plastic, and ink samples. In this measurement, the surface or the mirror reflection is always excluded if the sample surface is flat.

For samples with a nonsmooth surface, such as textile fabric, the so-called circumferential 45/0 (or 0/circumferential 45) geometry is preferred. Circumferential 45/0 geometry refers to the illumination of the sample at an angle of 45 degree in multiple directions around the viewing axis normal to the sample surface.

In the *sphere* type, the sample is placed at one of the port openings of a sphere, coated white internally. The sample is illuminated by diffuse light at all angles from the internal sphere wall, and the reflected light is viewed by the detector at or near the normal to the sample surface. This geometry is designated by the CIE as D/0, indicating diffuse illumination and normal viewing. Alternatively, the sample can be illuminated at or near the normal and viewed diffusely, that is, 0/D geometry. The D/0 or 0/D sphere type geometries can be used to measure samples with a relatively nonsmooth sample surface such as textile samples, with good repeatability. For flat-surface samples, the D/8 or 8/D (the number 8 designates 8 degrees from the normal) sphere type geometries provide an optional measurement of including or excluding the mirror reflection by placing a white specular component or a black trap along the 8-degree direction. Normally, for color appearance evaluation, the mirror reflection is excluded during measurement. For quantitative analysis, such as for colorant formulation application, the mirror reflection is included. Figure 6 shows the CIE recommended illuminating and viewing geometries.

4. *Other Considerations.*   Other regular features have to be considered as well. These include the measurement accuracy; the measurement repeatability and reproducibility; the speed and ease of operation; the availability of special accessories such as special sample holders for powder, ultraviolet (UV), and infrared (IR) cutoff filters; and small area of view for small sample measurement. Such considerations have been reported by the Inter-Society Color Council [25].

(*b*)   *Tristimulus Colorimeter.*   As the fundamental quantities measured by the tristimulus colorimeter are tristimulus values, the spectral range of the instrument

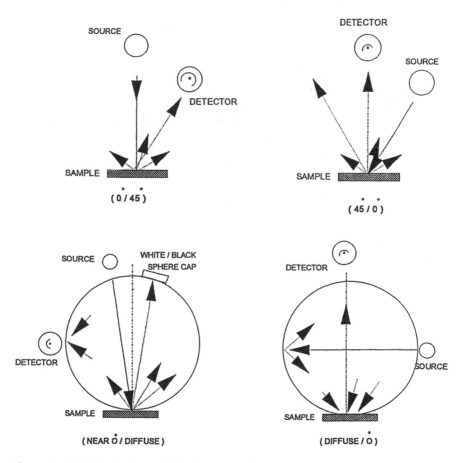

Figure 6   CIE illuminating and viewing geometries.

is limited to the visible region. It is a standard practice that the instrument is set up with the sample being illuminated directly by the light source for the reason explained earlier. Other points discussed in Section IV.D.1 should also be examined.

2. Computer

There are basically two major types of computer systems for the color measuring equipment in the market. They are the single-user type and the multi-user type. Selection of either type depends on the present and the future requirements. The multiuser type includes a minicomputer or multiple personal computer in a networked fashion, while the single-user type is usually a personal computer. Because of strong competition in personal computer sales in conjunction with rapid advances in technology, the personal computer type color measuring system is particularly attractive and popular.

3. Technical Support

A good color measuring system supplier should provide a quick response to users needs through the provision of a range of services. These include the installation of the system, basic and updating training programs, hardware and software maintenance, system development, and problem-solving consultancy service. In some cases, the services provided are inadequate when the place of origin of the goods is outside the users' territory. In this situation, one may need to turn to a third party for assistance, which is described in Section VII.

E. Developments in Color Measuring Instruments

The major development for both the tristimulus type and spectrophotometric type of color measuring instruments is the reduction in physical size to a portable format with an approximate size and weight of 0.1 ft^3 and 3 lb, respectively. Such a portable format is made possible by the use of much smaller components, such as a camera-type xenon flash lamp as the illuminating source, fiber optics for light transmission, and tiny spectral filter and silicon photodiode arrays for reflected or transmitted light reception. The data processing device is also being made compact by using a microprocessor, liquid crystal display, and built-in bar-code reader. The use of these components also improves the measurement and computation speed. The portable format is attractive in measuring color objects that cannot be conveniently moved, such as of automobiles or building architecture, and for colorimetric measurements in a location not possible with a conventional tabletop color measuring instrument. In using these portable instruments at the production sites where the environment is generally hostile and dusty, the users have to be careful in measuring textile samples whose color appearance is sensitive to moisture content and heat. The relatively small sample port of the portable type instruments may affect the precision performance in measuring textile samples with enhanced surface texture. Also, major advances in terms of precision and flexibility are happening with the conventional benchtop spectrophotometer [26] as summarized in the following section

1. Illuminating System

The instrument illuminating system is generally designed to promote precision in measuring thermochromic and fluorescent samples. The heating of the sample by the illuminating light source is minimized via the use of a sample exposure shutter or a cooler pulsed Xenon source. In addition, the fast measurement also minimizes the sample exposure time. Calibration and control of the ultraviolet (UV) content of the illuminating source is made possible by electronically controlled UV cutoff filters for accurate measurement of optically whitened textiles and to evaluate textiles colored with chromatic fluorescent dyes. In some instruments, the light source is filtered to simulate D65 illuminant to enhance the correlation of measurement results with the visual assessment of fluorescent samples under daylight illumination. The intensity of the illuminating source is also kept stable and high to enhance accurate readings of dark and saturated colors.

2. Sample Compartment

Both the D/0 type and circumferential 45/0 type illuminating and viewing geometries remain popular for the measurement of textile samples. Some of the instrument setup can be performed automatically. This includes the selection of large area of view versus small area of view, specular component included or excluded in the case of D/0 geometry, and UV cutoff filter included or excluded. The instrument is generally equipped with a transmission compartment for transmission measurement of transparent or translucent samples, such as liquid or thin films.

3. Monochromator System

The essential device for light dispersion has been gradually shifted from prism monochromator to the grating type monochromator. The grating version also has been switched from the ruled grating type to the holographic grating type. The holographic grating makes use of the phenomenon of light-wave interference, photographic recording, and chemical etching methods so as to rule a grating in an optical material with relatively less stray light than the mechanically ruled grating.

Another method for selectively producing narrow wavelength regions of the spectrum is the use of interference filters. This restricts the radiant energy transmitted to the relatively narrow spectral transmission band of the filter. An earlier type of such a device is discrete; that is, each filter is responsible for each narrow spectral band region and readings at intermediate wavelength regions are not available. An alternative version is the continuous variable interference wedge, whose spectral transmission varies as a continuous function of the physical location of the incident energy across the surface of the wedge. The bandwidth of the interference filter is relatively larger than that of the prism or grating type mono-chromator.

4. Detector System

The detector is the light-sensitive device that registers the quantity of light being reflected or transmitted from the sample. In a single-beam instrument, the detector must remain stable between the time of detecting the light from the standard and the time for that from the sample. With a double-beam instrument the time between readings of reference and sample beams is reduced to fractions of a second, and the detector must therefore have a frequency response compatible with that used in alternating between the two optical paths.

The traditional light-detecting device is mainly the photomultiplier. This has been replaced by the tiny solid-state silicon diode photodetector array, which is positioned accurately in the dispersed spectrum so that each detector responds to a specific small band of wavelengths. This has made possible the measurement of the entire spectral data simultaneously, thus increasing the measurement speed drastically. With such a design, the measurement time for the entire visible range can be reduced to less than 1 sec, as opposed to the conventional sequential type with a measurement time of about 10 sec or more. However, such a design converts the equipment to an abridged spectrophotometer. In other words, the spectral information cannot be measured continuously throughout the spectral range and the number of measured discrete spectral data depends on the number of the silicon detectors. Instruments with as many as 76 detector elements are available in order to obtain spectral data at 5-nm intervals in the visible range. The CIE recommends that the spectral data be taken at wavelength intervals as small as possible for tristimulus integration for better accuracy [9]. Generally speaking, most applications can be carried out satisfactorily with the abridged spectrophotometer. On the other hand, the unavailable spectral information can be predicted by mathematical interpolation based on the discrete spectral data. The use of silicon detectors, coupled with the suitable light source and monochromator, can also expand the spectral range to 1100 nm for the near-infrared measurement of camouflage materials.

Recently, instrument manufacturers have created a new line of color measuring spectrophotometers called compact benchtop spectrophotometers to suit those users who do not need portability but require reasonable price, precision, and reliability [27]. The term "compact" refers to the size of the instrument, somewhere between the portable size and that of the conventional benchtop instrument. These compact benchtop spectrophotometers are of interest to large users where multiple instruments are required for a variety of applications.

The development of the color measuring sensors, coupled with the revolution in computer power and versatility as well as their integration, has led to significant advances in the performance of the color measuring systems in terms of speed, accuracy and precision. The gains in system performance have had a great impact on their areas of application.

V. COLOR MEASURING PROCEDURE

A proper procedure to perform the measurement of colors is important for useful, practical applications. A normal measurement procedure involves the following steps.

A. Instrument Setup

The spectrophotometer has to be set up in the proper mode prior to color measurement, according to the instrument manufacturer's recommendations. For example, in the case of a sphere-type reflectance spectrophotometer, the possible list of parameters to be set up includes the spectral range, the polychromatic/ monochromatic mode, the specular component in/out, the sample port size, and the selection of filters as detailed in Section IV.D.1.

B. Instrument Calibration

After the instrument has been properly set up, it has to be calibrated for its photometric scale with respect to the 100% line and the 0% line using a white standard and a black standard, respectively, in the case of reflectance spectrophotometer. In the case of the transmission spectrophotometer, a clear solution for the solution sample (or air for the color filter sample) is used to set up the 100% reference line, and the 0% reference line is normally established by blocking the illumination beam with an opaque sample. The calibration procedure establishes a set of correction factors at each wavelength and is applied to the subsequent spectral measurements to obtain absolute spectral data.

C. Instrument Verification

The performance of the spectrophotometer in terms of precision and accuracy can be checked by the measurement of color standards with calibrated spectral data. Precision refers to how repeatable the measurements are for the same sample over a period of time, while accuracy refers to how close the measured reading of a sample is to its absolute true reading. Examples of transmitting color standards include a series of filters (2101–2105) supplied by the National Institute of Standards and Technology (NIST). In addition, a didymium filter is useful for wavelength accuracy checking. Examples of reflective color standards include a set of 12 ceramic color tiles supplied by the British Ceramic Research Association. These standards may also be used to check the instrument precision (i.e., repeatability) by comparing the repeat measurements on a short-term or long-term basis. However, some of the standards are temperature sensitive and hence both the room and the sample temperature have to be conditioned to a standard temperature prior to testing.

D.   Sample Preparation and Measurement

The importance of the sample itself in providing reliable color measurement data should not be overlooked. There are a number of factors that may affect the measurement precision and accuracy. The following outlines the major items to be observed during the measurement of textile samples.

1.   Sample Temperature and Moisture Content

The temperature and moisture content of a textile sample could change its color appearance significantly and hence its measurements. It is therefore important to condition all textile samples in a room or chamber with controlled humidity and temperature for a suitable period prior to color measurement.

2.   Sample Format

A good technique of sample presentation for measurement is to ensure an identical format of presenting all the samples to be intercompared at the instrument sample port for color measurement.

(*a*)   *Sample Opacity.*   The textile fabric is usually folded to complete opacity to avoid background influence during color measurement. Thus it is important that all samples to be intercompared be folded to the same number of layers. If the measurement of the sample backed with a white background is equivalent to the measurement of the same sample backed with a black background, the sample thickness has reached complete opacity. Yarn samples should be wound onto a rigid card uniformly with identical layers. Loose fibers should be placed into a transparent cup holder with identical thickness under identical pressure. The transparent bottom of the cup holder is then presented to the sample port of the instrument for color measurement.

(*b*)   *Sample Planarity.*   Color measuring instruments are generally designed for measurement of flat samples to be placed at the sample port. If the sample extends inside the port or is displaced away from the port, different measured readings may result. If the textile sample flatness is difficult to achieve, due to surface texture, measurement behind glass will help. However, the measurement results must be corrected for effects of the cover glass, such as the Fresnel reflection.

3.   Thermochromic and Photochromic Property

Some textiles have colorants that are sensitive to heat and light. Color change on exposure to heat and light is called thermochromism and photochromism, respectively. Both the thermochromic effect and the photochromic effect can be eliminated or reduced by minimizing the time of sample exposure to the illuminating source during measurement. In addition, the thermochromism effect may be further reduced by using an illuminating source of low infrared content such as a flash xenon lamp in conjunction with an infrared cutoff filter.

4. Fluorescence

Some textiles are colored with fluorescent colorants whose spectral emissions are sensitive to the spectral power distribution of the instrument's illuminating source system [23]. The prerequisite to measure fluorescent samples is to set up the spectrophotometer in the polychromatic mode as described in Section IV.D.1.b. The measurement repeatability of the same instrument and the measurement reproducibility of instruments of the same model depend heavily on the stability of the spectral power distribution of the illuminating source system. It is therefore desirable to measure all the fluorescent samples to be intercompared at about the same time and at the same instrument. Furthermore, the spectral power distribution of the instrument source system and the illuminating source for visual assessment should be compatible in order to obtain consistent results between instrumental and visual assessments.

In general, it is always a good practice to perform measurement averaging of multiple measurements at different locations and orientations of the test sample to achieve repeatable measurement.

5. Reference Documents

- *Color Technology in the Textile Industry,* American Association of Textile Chemists & Colorists, P.O. Box 12215, Research Triangle Park, NC 27709. (See chapters on "The Calibration of a Spectrophotometer for Color Measurement" by Henry Hemmendinger and "Preparation and Mounting Textile Sample for Color Measurement" by R. L. Connelly.
- Society of Automotive Engineers, Warrendale, PA, Test Method J1545, "Instrumental Color Difference Measurement for Exterior Finishes, Textiles, and Colored Trim."

VI. TEXTILE APPLICATIONS

The primary applications of color measuring systems in the textile and textile-related industries into the following four major areas: color matching, color quality monitoring, colorant solution evaluation, and color communication.

A. Color Matching

In today's competitive world, one of the important problems of the manufacturing industry that uses color technology is how to arrive at a perfect color match with minimum of cost of dyeing and minimum time expenditure using a mixture of a few colorants. Traditionally, the procedure followed in the dyeing plant is usually a trial and error method, which gives a metameric match. During the process, one must be aware of the performance, cost, and availability of the particular colorant. The greatest drawback of this method is that no colorimetric record is kept in the trials and it takes considerable time for these trials. The method also depends on the experience of the color matcher. There is no measurement of color at any stage, and only visual assessment is done.

Computer color matching (CCM) can reduce the cost of production by:

1. Saving time in developing shade accurately
2. Providing a large number of alternative combinations of dyes for achieving the colorant formulas matching the target
3. Choosing the colorant formula for a specific requirement, such as minimum cost or minimum metamerism
4. Integration with complementary production systems for accurate and efficient data communication
5. Other associated applications, such as checking on the strength of the incoming dyes, or formulation with waste colorants

The basis of CCM is largely built upon the theory postulated by Kubelka and Munk [28] in 1931. The theory describes the scattering and absorption of radiant energy in a turbid medium in terms of reflectance, defining the quantities of radiant energy absorption and scattering by the coefficients K and S, respectively. Equation (26) defines the relationship between the reflectance factor and the Kubelka-Munk coefficients at a certain wavelength in its simplified form.

$$K/S\,(\lambda) = [1 - R\,(\lambda)]^2/2R\,(\lambda) \tag{26}$$

The quantity K/S is related more or less linearly to the concentration of the colorant in the substrate medium and is therefore very useful in predicting the colorant formulations in conjunction with the CIE colorimetric system, to match a given color standard [29].

Although the fundamental concept of match prediction was laid down in the early 1930s, the first commercial computer color matching device was not available until 1958. This was the colorant mixture computer (COMIC), developed by Davidson and Hemmindinger [30]. It did not gain extensive popularity, largely because of its speed and flexibility, as the computer was an analog version. In the 1960s, digital computers became available, and most leading colorant makers installed their own systems of CCM to service their customers. These systems included the instrumental match prediction (IMP) system of the Imperial Chemical Industry [31] in 1963, the computer color matching (CCM) system of American Cyanamid [32, 33] in 1963, the automatic recipe formulation and optimalization (ARFO) system of Sandoz in 1964, the programmed match prediction technique (PROMPT) of Du Pont in 1965, and the computer color matching system of Ciba-

Geigy [34]. However, these systems usually had the disadvantage of poor accuracy, as the fundamental colorant calibration data were not made by the user, and the colorant formulations generated were restricted to one colorant maker. Furthermore, degree of metamerism was not indicated for certain systems. In the late 1960s, time-sharing CCM systems became available in which users could develop their own databanks. These included the systems developed by General Electric, IBM, Beckman Instruments, and the Applied Color System. So far, all these CCM systems were abridged; that is, data measured on the color measuring equipment cannot be transferred to the computer directly. In the late 1970s, CCM and other practical color measurement applications gained wide popularity because of the availability of relatively low-cost mini-computers, which are interfaced directly to the color measuring equipment. At this time, many users can afford to have such an integrated in-house CCM system with improved speed and accuracy.

In the early 1980s, CCM reached a new stage with the introduction of IBM or IBM-compatible personal computer (PC). The interfacing of the color measuring sensor with the PC meant a significant decrease in the cost of the CCM system such that the CCM system was no longer a privilege of medium-sized to large dyehouses but could also be afforded by smaller dyehouses, especially in developing countries, or for those companies that have already owned PC. Because of the open architecture of the personal computer and the affordable price, the system has become more versatile, and there are thousands of third-party softwares and accessories prepared for all kinds of applications in virtually any aspect. At the same time, the CCM system has become much more compact in size. On the other hand, the drawbacks of such systems are the much smaller central processing unit (CPU) memory and storage capacity, much slower speed, and the poorer performance in a network environment in comparison with the minicomputer version. This has rendered the PC-type CCM system more or less a personal or stand-alone system in the 1980s, whereas the minicomputer type is the multi-workstation system with a much stronger performance in network environment.

In the 1990s, the advances in information technology has greatly enhanced the performance of PC in terms of speed, memory capacity, storage capacity, and connectivity, as well as the PC network. The use of PC has dominated most color measuring system applications.

As the computer technology advanced, innovations in commercial CCM software occurred. The most evident of these is that the software is now written in a much more user-friendly manner, usually in a menu- or window-driven format with plenty of help messages and colorful graphic interpretations that were not available in the 1970s and early 1980s. Other innovations are:

1. Storage and retrieval of color standards along with relevant useful formula and process information.
2. Input of standards can be achieved by a variety of means including measurement, manual input, and electronic data transfer.

3.   Assignment of performance factors for individual dyes and substrate for the correction of strength and exhaustion variation.
4.   Creation of suitable dyeing groups of compatible dyestuff, substrate, and dyeing process for formulation.
5.   Varieties of graphical presentations and simulated color images for assessment of the measured calibration dyeings and the predicted or corrected formulas.
6.   Automatic queued match predictions of combinations of variable numbers of dyes per formula based on the preassigned standards, dye candidates, and tolerances.
7.   Manual formulation or correction with support of graphical presentations.
8.   Formulation for the use of surplus dyes or materials.
9.   Correction can be achieved by using original dyes or new dyes or their combinations.
10. Special color matching program, such as for blending of various colored fibers for matching.

In fact, the use of CCM is so popular that such services are available at the store level in some countries. In the United States, you can walk into a paint store with a color standard requesting the store to prepare cans of paint to match the standard. Within a reasonable waiting period, a formulation based on the measurement of the color standard is predicted and the necessary amounts of the paint ingredients are automatically dispensed.

Although the technique of CCM has been practiced over 30 years, there are still a number of limitations [35], with the major ones being the poor accuracy of CCM for fiber-blends coloration [36] and for coloration with fluorescent colorants [37, 38]. However, Gibson [39] reported that the use of a neural network approach for colorant formulation shows positive results with fluorescent dyes.

B.   Color Quality Monitoring

1.   Color Pass/Fail Systems

Color pass/fail systems screen the color of the products against preset tolerances in color requirement. It is especially important in the case of large purchases. Such a preset tolerance can guarantee exactly how closely the color requirement will be met. An off shade could mean unacceptable products. Traditionally the decision as to whether a batch color is close enough to the standard being matched has been made subjectively. Due to inter- and intraobserver variability and other influential factors, subjective assessment cannot be accurate even if the observer is very experienced. The CIE system of colorimetry is normally used as a basis to carry out the color pass/fail assessment. In this system, the color difference formula is used to set the size of the tolerance for acceptability by applying lower maximum value(s) of the total color difference, chroma difference, lightness difference, hue difference as described in Section III, or any selected combinations of these variables [40–42]. In the 1990s, the use of single-color difference value for shade acceptability judgment has gained recognition in the textile industry. Reports on the performance of the use of single color difference for shade acceptability judgment have been published [43–45].

The use of color measuring systems for color quality monitoring has been extended from an offline basis to an online basis. The colored material is measured by a color measuring sensor at the production line to monitor the shade uniformity and acceptability on a real-time basis [46]. Important features of such monitoring device are the fast measurement speed, noncontact capability, good depth of focus, and large area of view. More recently, there has been development in the monitoring of color quality of textile prints by means of the colorimetric CCD camera [47].

2.   Color Sorting Systems

Color sorting systems are designed to identify the parts that can be put together in a finished product without noticeable (or unacceptable) color differences among the parts. A typical example is the application of such a system in the garment factory. Here, it is important that the various patterns that constitute the entire garment should not have noticeable color variation for a solid-shade garment. This is primarily because the various dye lots, from which the patterns are cut, have some color variations. Hence, it is necessary for the garment or dyeing factories to carry out color sorting of the dye lots prior to the pattern cutting process.

Like the color pass/fail system, the color sorting system is built on the basis of the CIE colorimetric system. The principle is to subdivide an acceptable volume of a color space, with reference to a standard color, into individual smaller volumes in which all colors located in each of these units are compatible in color and can be merged together without any unacceptable color differences. These units are usually identified with a shade numbering system based on the relative position of the individual unit from the central unit housing the standard color in a color space. Thus the individual colors inside a basic unit would be assigned with the same shade number. A popular shape of the basic unit for the shade sorting system is rectangular. One of the popular shade numbering systems is the Simon method, known as the "555" system [48]. In this system, each color is given a three-digit numeric shade sort code. Using the CIE $L^* C^*_{ab} h_{ab}$ color space as an example, the first digit is an indication of the lightness of the color as compared to the standard color. If the color is lighter than the standard this digit will be above 5, and below 5 if it is darker. If the color is more saturated than the standard color the second digit will be above 5 and below 5 if it is duller than the standard. Similarly, the third digit in the shade sort code indicates the hue variation from the standard. For example, a color having the same saturation and hue except just one step lighter than the standard color would be identified with a shade sort code of 655. The dimensions of the individual units have to be varied for different color regions of the color space, and various guidelines have been provided in the literature [40, 41, 49]. Alternative shapes of the individual units in terms of rhombic dodecahedron and truncated octahedron for improved shade sorting performance in fewer groupings have been reported by McLaren [50].

The primary disadvantage of the 555 shade sorting method is that the borderline colors occupying the corners of the rectangular block are much farther removed from the center of the block than are the borderline colors that occupy the center of the faces, which may produce anomalies in shade sorting [51]. A new shade sorting technique known as Clemson color clustering (CCC) has been devised by Aspland et al. [51] to overcome this disadvantage with less shade-sorted groups. However, CCC sorting is carried out without reference to a standard. Thus the nature of the color difference of the individual color from the standard color cannot be deduced from the CCC shade sort codes. A similar color clustering technique known as the Scotsort, designed to overcome this difficulty partially by means of a primary cluster, is reported by Wardman et al. [52]. Since these color clustering techniques do not sort the colors with reference to the standard color completely, the shade sort codes for a production lot have no relationship to the shade sort codes for another production lot even though both lots are using the same standard color for acceptability judgments. To avoid this problem, it is necessary to merge the colorimetric data of the two production lots and sort the data again.

The need for color sorting is obvious and is particularly useful to those industries whose products are made up by parts in different locations.

3. Colorant Strength Evaluation

The relative concentration of colorant compared to that of the corresponding standard colorant is routinely assessed by the colorant manufacturer during the standardization of colorants in order to maintain a high consistent quality. It is also being used by colorant users during the quality evaluation of the new shipments in order that the performance of colorants be maintained during coloration. Determinations of relative colorant strength from reflectance measurements are usually based on the Kubelka-Munk function as defined by Eq. (26). As the function is to a large extent linearly related to the colorant concentration in a substrate, the ratio of Kubelka-Munk functions of the sample and the standard at equal prepared concentration can indicate the relative colorant strength. Standard procedures for such evaluation have been worked out by some dyestuff manufacturers [53] and the Inter-Society Color Council (ISCC) [54]. A comparative assessment of the performance of various colorant strength formulae was reported by R. Hirschler [55] at the 1992 25th anniversary conference of the International Color Association (AIC) at Princeton, NJ.

4. Whiteness Evaluation

White is a color of freshness, purity, and cleanliness. It has been used as an indicator of qualities such as freedom from contamination. The determination of the degree of whiteness has been an interesting subject for many years. In principle, it can be measured by the amount of departure from the "perfect white" position in a three-dimensional color space. However, agreement on the "perfect white" has not been reached because of a number of problems. The major problem is that strong preference in the concept of whiteness is governed by trade, nationality, habit, and product. This problem is further enhanced by the introduction of fluorescent whitening agents, the conditions of observation, and the measurement accuracy [23]. As a result, no single formula for whiteness is universally applicable. The principles for deriving whiteness formula have been described by Ganz [56–58]. In 1981, the CIE recommended field trials of a new whiteness formula [9]. The CIE whiteness formula was adopted by the American Association of Textile Chemists and Coloristis in 1989 as AATCC Method 110–1989 [59].

(*a*) *Reference Documents*
- American Standard Test Method E313 for Indexes of Whiteness and Yellowness of Near-White, Opaque Materials, ASTM, 1916 Race St., Philadelphia, PA 19103.
- Publication CIE No. 15.2 (TC-1.3) Colorimetry, 2nd edition, Wien 3. Bezirk, Kegelgasse 27/1, Austria, 1986.
- AATCC 110–1989, *Whiteness of Textiles*, Technical Manual of the American Association of Textile Chemists and Colorists, AATCC, Research Triangle Park, NC 27709.

5. Yellowness Evaluation

The preferential absorption of white light in the short-wavelength region (380–440 nm) by the material usually causes an appearance of yellowness. Interest has developed in determining the degree of yellowness as it is considered to be associated with soiling, scorching, and product degradation by exposure to light, atmospheric gases, and other chemicals. A number of yellowness scales have been developed over the years [60–62].

(*a*) *Reference Document*
- American Standard Test Method E313 for Indexes of Whiteness and Yellowness of Near-White, Opaque Materials, ASTM, 1916 Race St., Philadelphia, PA 19103.

6. Metamerism Evaluation

Metamerism refers to a pair of visual stimuli that in the human eye give rise to identical colors but that have different spectral energy distribution, as described in Section III.A. Visual stimuli could be the light source entering directly into the eye or modified reflected light into the eye from the objects. In the latter case, the visual stimuli could be modified by changing the spectral power distribution of the light source. Either change causes an identical match or mismatch of the objects. This phenomenon is commonly called source metamerism. It is important to predict the degree of mismatch for a set of color-matched products when the light source changes, using a source metamerism index in computer color matching. One can select the least metameric recipe from a large number of alternative recipes generated by the system if one can evaluate the size of metamerism for each predicted recipe. The CIE has recommended a special index of metamerism for change of illuminant in 1972 [63]. Other means of computing metameric indices were assessed by Badcock [64] and Choudhury et al. [65].

a. Reference Document
- Publication CIE No. 15.2 (TC-1.3), Colorimetry, 2nd edition, Wien 3. Bezirk, Kegelgasse 27/1, Austria, 1986.

7. Color Fastness Assessment

Most methods in color fastness assessment of textile materials involved treating the dyed material in a standardized manner and then comparing the treated textile and the original untreated textile (with respect to the color change) visually versus a gray scale that carries a series of pairs of color chips with increasing color difference magnitude. There are two kinds of gray scales, one for the "staining" test and the other for "change of shade." The obvious disadvantage of determining the color fastness rating by means of visual assessment is the poor reproducibility from observer to observer. Methods for instrumental assessment of staining and change of shade have been developed by various professional bodies. At the International Standards Organization (ISO) technical committee (TC38/SCI) meeting in Bad Soden, Germany, in 1987, a German proposal for instrumental assessment of staining was accepted [66]. At the ISO TC 38/SCI meeting in Williamsburg, VA in 1989, the Swiss proposal on instrumental assessment of change in shade was accepted [66].

(a) Reference Document
- B. Rigg, Instrumental methods in fastness testing, *Journal of the Society of Dyers & Colourists,* 107 (7/8):244–246 (1991).

8. Luster Evaluation

Some textile fabrics are finished with a lustrous appearance via a calendering process. The luster is the gloss appearance associated with the contrast between the specularly reflecting area of fabric and the surrounding diffusely reflecting area [67]. Hunter has developed a formula [67] to express this relationship.

$$\text{Luster} = 100(1 - R_d/R_s) \tag{27}$$

where R_d is the diffuse reflectance factor and R_s is the specular reflectance factor.

C. Colorant Solution Evaluation

A spectrophotometer equipped with transmission measurement may also be used to evaluate a colorant solution for a number of applications. This includes the determination of solubility and solution stability of water-soluble dyes [68, 69], the evaluation of relative colorant strength based on solution measurement [70], the monitoring of dye exhaustion characteristics [71], and the evaluation of formalde-hyde content [72]. These evaluations are largely based on Beer's law, which states that the measured quantity of absorbance is directly proportional to the concen-tration of the absorbing species present in a solution [73]. The quantity of ab-sorbance is equal to the logarithm of the inverse of the transmittance. This law has been very useful in various quantitative analysis and the investigation of dyeing mechanisms [74]. On the other hand, transmission measurement of colorant solu-tion has also been used for qualitative analysis of organic pigment [75].

With increasing concern in regard to the effluent color of the textile finishing industry, quantitative techniques, based on transmission measurement, have been devised by the American Dye Manufacturers' Institute [76] in the United States and the National Rivers Authority [77] in the United Kingdom to monitor the color of wastewater effluent as an indicator of water quality.

D.   Color Communication

Various ways, with differing accuracy, have been devised to communicate color. These includes the use of general color names, the method of designating colors developed by the Inter-Society Color Council and the National Bureau of Stan-dards [78], the use of color order systems with a systematic collections of color standards sampling the color space such as the Munsell notations [79], and the CIE system [9]. Of these methods, there is no doubt that the CIE system provides the highest precision. Furthermore, one can easily use the CIE color specifications for quick distant communication via an international telecommunication system. One drawback of the CIE color specification system is the absence of the real physical color accompanying the numeric specifications. Otherwise, it would be useful as a color development tool by designers for styling applications for example. Such a drawback has been, to some extent, overcome by the development of calibrated color display system where it is possible to generate a variety of image colors with CIE colorimetric specifications by controlling the red, blue, and green guns of the cathode ray tube in an appropriate fashion [80–85]. Calibrated color display sys-tems have been utilized in computer-aided color manipulation (CCMAN) systems to aid color selection and visualization in the design creation process as well as for color communication with external systems. However, the success of this method requires further research due to a variety of technical and visual observation prob-lems [81, 86]. The phenomenon of metamerism complicates the colorimetric cal-ibration of CAD/CCMAN systems including the color monitors, scanners, and printers. The colors of the original and its reproduction at the monitors or printers are usually predicted to match only under one illuminant. In addition, color re-production performance is further limited by the mismatch of color gamuts among the monitors, printers, and scanners. CIE Technical Committee 1-27 has published guidelines for evaluation of color appearance models for reflection print and self-luminous display image comparisons with respect to color reproduction [87].

VII. SOURCES OF TECHNICAL INFORMATION ON COLOR SCIENCE

The science and technology of color have been changing and advancing rapidly. Research on various aspects is being carried out at various places. It is sometimes difficult for a beginner to locate the necessary information as well as the bodies that provide the required technical assistance. The following is an attempt to provide the sources where technical information or assistance is available.

A. International Associations

There are two major international associations that promote the study, advancement, and exchange of information on color science. They are the International Commission on Illumination (CIE) and the International Color Association (AIC) [88]. CIE is particularly active in recommending technical standards and working procedures.

B. National Associations

On a national basis, most countries have their own national association in developing the science and technology of color. Most national associations are also members of the AIC and CIE. Examples are the Inter-Society Color Council (United States), the Canadian Society for Color in Art, Industry and Science, the Color Group (Great Britain), the Hungarian National Color Committee (Hungary), the Color Science Association of Japan, the Hong Kong Illumination Committee, AIC-Verbindungsausschuss (F.R. Germany), Centre d'Information de la Couleur (France), Associazione Ottica Italiana, and Pro Colore (Switzerland).

C. Technical Assistance

Technical assistance on the consultation, measurement, and training programs is available from three major sources: technical institutes, educational institutes, and instrument manufacturers. The technical institutes are usually supported by the local governments. Examples are the National Institute of Standards and Technology (United States), National Physical Laboratories (United Kingdom), National Research Council (Canada), Bundesanstalt fur Materialprufung (F.R. Germany), Electro-Technical Laboratory (Japan), and National Office of Measures (Hungary). Many educational institutes are particularly active in research and the provision of training programs. These include the Rochester Institute of Technology

(Unites States), Clemson University (United States), Hong Kong Polytechnic, and University of Derby (United Kingdom). Equipment manufacturers also provide technical assistance such as training programs, which are, as expected, usually product oriented. Examples of color measuring equipment manufacturers are HunterLab, Kollmorgan, BYK-Gardner, and X-Rite (all United States), and Datacolor International (Switzerland) and Minolta (Japan). On the other hand, the readers should not neglect the abundant valuable publications available in the public domain that report the latest developments in color research and application. Examples of these publications can be found in the Reference section.

VIII. FUTURE PROSPECTS

Since the introduction of the CIE basic colorimetry in 1931, the developments in color measuring systems have been enormous. Substantial advancements have been achieved in both the hardware and software applications. During this growth period, other coloration-related systems have also been developed:

- Computer-aided design (CAD) system. A system to aid designers to create/manipulate design with digital information.
- Computer-aided color manipulation (CCMAN) system. A system, complementary to the CAD system, to aid color selection and visualization in the design creation process with CIE colorimetric data for color communication with external systems. The CAD and the CCMAN systems are sometimes merged into a single system.
- Computer-aided colorant dispensing (CCD) system. A system for dispensing the required amount of the individual colorants and chemical auxiliaries into the color production system. The colorant formula is usually originated from the computer color matching (CCM) system.
- Computer-aided process control (CPC) system. A system for monitoring and controlling the process variables of the coloration process.
- Computer-aided color monitoring (CCMON) system. A system for monitoring the color quality of the color production.

To some extent, substantial integration of these systems has already been achieved in the laboratory and production environment. In the next decade, technology development will be in the direction of total integration for the purposes of improving accuracy, precision, and efficiency, as well as quick response. The color of the design obtained from CAD/CCMAN will be transmitted to the CCM, where the colorant formulations will be predicted, and the selected recipe will pass on to the CCD for dispensing the required amounts of colorants and chemicals into the production system, where the CPC with the aid of CCMON will monitor and control the entire production process. Such an integrated system can communicate with another similar system or just a CAD/CCMAN device at another location via

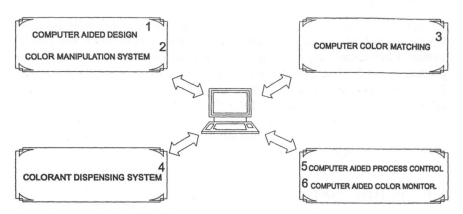

Figure 7 An integrated color production system:
(1) a system to aid product design;
(2) a system to aid color selection and manipulation with CIE colorimetric specifications for color communication;
(3) a system to assist colorant formulations;
(4) a system to dispense the required amount of each colorant into the production system;
(5) a sytem to control the operation of the production system; and
(6) a system to monitor the color quality of the product.

network. The national and international information superhighways will play a major role in linking the appropriate systems together to be used by all elements of the textile product chain to enhance productivity and competitiveness. Figure 7 shows the concept of a large-scale integrated color production system that can interface to other similar systems.

IX. CONCLUSION

In less than half a century, color measuring devices have evolved from a standalone unit to a computer-interfaced system and are moving in the direction of total integration. At the same time, the application environment has also evolved from the research laboratory to the manufacturing and retail sectors. The target is to let the usefulness of such devices reach the general public in their daily lives. The evolution in color science has been greatly assisted by the developments in and use of computers.

ACKNOWLEDGMENTS

The author would like to express his appreciation to the Society of Dyers and Colourists and to Textile Asia for permission to utilize some of the published papers written by the author as a foundation framework for the present chapter.

REFERENCES

1. K. Motokawa, *Physiology of Color and Pattern Vision*, Springer-Verlag, New York, 1970.
2. Visual psychophysics, *Handbook of Sensory Physiology,* Vol. VII/4 (D. Jameson and L. M. Hurvich, eds.), Springer-Verlag, New York, 1970.
3. K. H. Ruddock, *Contemp. Phys. 12*:229 (1971).
4. F. W. Billmeyer, Jr., and M. Saltzman, *Principles of Color Technology*, 2nd ed., Wiley-Interscience, New York, 1981.
5. D. B. Judd and G. Wyszecki, *Color in Business, Science and Industry*, 3rd ed., John Wiley & Sons, New York, 1975.
6. G. Wyszecki and W. S. Stiles, *Color Science*, 2nd ed. John Wiley & Sons, New York, 1982.
7. *Journal of Color Research and Application* (F. W. Billmeyer, Funder Editor), John Wiley & Sons, New York.
8. *Proceedings of the Eighth Session (Cambridge, England, 1931)*, International Commission on Illumination, Wien 3. Bezirk, Kegelgasse 27/1, Austria.
9. CIE Publication No. 15.2 (TC-1.3), Colorimetry, Official Recommendation, International Commission on Illumination, Wien 3. Bezirk, Kegelgasse 27/1, Austria, 1986, 2nd ed.
10. D. B. Judd, *J. Opt. Soc. Am. 23*:359 (1933).
11. W. D. Wright, *Trans. Opt. Soc. London 30*:141 (1928–1929).
12. J. Guild, *Phil. Trans. Roy. Soc. London A230*:149 (1931).
13. A. Robertson, *J. Color Res. Appl. 2*:7 (1977).
14. R. Seve, New formula for the computation of CIE 1976 hue difference, *J. Color Res. Appl. 16*:217 (1991).
15. S. M. Jaeckel and C. D. Ward, *J. Soc. Dyers Color. 92*:353 (1976).
16. R. McDonald, *J. Soc. Dyers Color. 96:*418, 486 (1980).
17. F. J. J. Clarke, R. McDonald, and B. Rigg, Modification to the JPC79 color difference formula, *J. Soc. Dyers Color. 100:*128 (1984). Errata, *J. Soc. Dyers Color. 100*:281 (1984).
18. K. Witt, Proceedings of 5th Congress of the International Color Association, 1985, p. 50.
19. M. R. Luo and B. Rigg, BFD(1:c) color difference formula (Part 1 & Part 2), *J. Soc. Dyers Color. 103*:86, 126 (1987).
20. R. S. Roy, D. H. Alman, L. Reniff, G. D. Snyder, and M. R. Balonon-Rosen, Visual determination of suprathreshold color-difference tolerances using probit analysis, *J. Color Res. Appl. 16*:297 (1991).
21. J. A. Garcia, J. Romero, L. J. del Barco, and E. Hita, Improved formula for evaluating color-differential thresholds, *Appl. Opt. 31*:6292 (1992).
22. CIE TC1-29 Industrial Color-difference Evaluation, Full Draft No. 2 : Recommendation on Industrial Color-Difference Evaluation, CIE, Wien 3. Bezirk, Kegelgasse 27/1, Austria, 1993. *J. Color Res. Appl. 18*:137 (1993).
23. T. F. Chong and F. W. Billmeyer, Jr., Proceedings of 19th CIE Session, Kyoto, Japan, 1979, p. 167–171.
24. F. T. Simon, *J. Color Res. Appl. 1*:5 (1972).

25. Committee report, Color Measuring Instruments: A Guide to Their Selection, Inter-Society Color Council, Problem Subcommittee 24, 1971.
26. Four state-of-the-art spectrophotometers, *J. Soc. Dyers Color. 107*:240 (1991).
27. Competitively priced colour control systems, *International Dyer Aug.*:43–44 (1994).
28. P. Kubelka and K. Munk, *Zeitschrift fuer Technische Physik 12*:593 (1931).
29. E. Allen, *J. Opt. Soc. Am. 56*:1256 (1966).
30. H. R. Davidson, H. Hemmendinger, and J. L. R. Landry. *J. Soc. Dyers Color. 79*:577 (1963).
31. J. V. Alderson et al., *J. Soc. Dyers Color. 17*:657 (1961) and *79*:723 (1963).
32. U. Gugerli, Proceedings of th International Symposium on Color, Luzern, 1965.
33. U. Gugerli, United States Patent 3,368,864, 1968E.
34. R. C. Allison, *Textile Industries Oct.*:166 (1969).
35. A. Brockes, *Text. Chem. Color. 5*:98 (1974).
36. T. F. Chong, Proceedings of the Annual World Conference of Textile Institute (UK), Hong Kong, 1984, pp. 384–411.
37. J. S. Bonham, *J. Color Res. Appl. 11*:223 (1986).
38. M. H. Brill, Meeting Report: 1994 ISCC Williamsburg Conference on Colorimetry of Fluorescent Materials, *J. Color Res. Appl. 19*:313 (1994).
39. R. Gibson, Proceedings of the AATCC 1992 International Conference, 1992, pp. 154–159.
40. C. B. Smith, *Text. Chem. Color. 17*:13 (1985).
41. J. R. Aspland, and J. P. Jarvis, *Text. Chem. Color. 18*:27 (1986).
42. Color Measurement Committee (SDC), The use of colour-difference measurement in quality control, *J. Soc. Dyers Color. June*:204 (1974).
43. J. Rieker and E. Fuchs, *Textilveredlung 22*:361 (1987).
44. C. T. Witt, Proceeding of the 44th Annual Conference of the American Society for Quality Control, Memphis, TN, 1994, Vol. 22.
45. K. Witt, *J. Color Res. Appl. 19*:273 (1994).
46. R. Willis, *Text. Chem. Color. 24*:19 (1992).
47. M. Jarvis, Proceedings of AATCC 1994 International Conference, Charlotte, NC, 1994.
48. F. T. Simon, *Am. Dyestuff Rep. 73*:17 (1984).
49. R. Harold, *Text. Chem. Color. 19*:23 (1987).
50. K. McLaren, Shade sorting, *The Colour Science of Dyes & Pigments*, 2nd ed., Adam Hilger Ltd., Bristol, UK, 1986, p. 150.
51. J. R. Aspland, C. W. Jarvis, and P. R. Jarvis, *Text. Chem. Color. 19*:21 (1987).
52. R. H. Wardman, P. J. Weedall, and D. A. Lavelle, *J. Soc. Dyers Color. 108*:74 (1992).
53. W. Baumann et al., *J. Soc. Dyers Color. 103*:100 (1987).
54. ISCC Problem Subcommittee 25 (Dyes), *Text. Chem. Color. 6*:104 (1974).
55. R. Hirschler et al., Proceeding of the AIC (25th Anniversary) and ISCC (61st Annual Meeting) at Princeton, NJ, 21–24 June 1992.
56. E. Ganz, *Appl. Opt. 15*:2039 (1976).
57. E. Ganz, *Appl. Opt. 18*:2963 (1979).
58. E. Ganz et al., *J. Opt. Soc. Am. 20*:1395 (1981).
59. AATCC 110-1989, Whiteness of textiles, *Technical Manual of the American Association of Textile Chemists and Colorists*, 1994, AATCC, Research Triangle Park, NC.

60. ASTM Designation: D 1925, American Society for Testing and Materials, Philadelphia, PA, 1991, p. 77.
61. F. W. Billmeyer, Jr., *Mater. Res. Stand.* 6:295 (1966).
62. R. S. Hunter, *J. Opt. Soc. Am.* 50:44 (1960).
63. Supplement No. 1 to Publication CIE No. 15, Colorimetry (E.-1.3.1) 1971, Wien 3. Bezirk, Kegelgasse 27/1, Austria.
64. T. Badcock, *J. Soc. Dyers Color.* 108:31 (1992).
65. A. K. R. Choudhury and S. M. Chatterjee, *Rev. Prog. Color. Soc. Dyers Color.* 22:42 (1992).
66. B. Rigg, *J. Soc. Dyers Color.* 107:244 (1991).
67. R. S. Hunter, *The Measurement of Appearance*, 2nd ed. John Wiley & Sons, New York.
68. A. Berger et al., *J. Soc. Dyers Color.* 103:138 (1987).
69. A. Berger et al., *J. Soc. Dyers Color.* 103:140 (1987).
70. P. Brossman et al., *J. Soc. Dyers Color.* 103:38 (1987).
71. M. Leferber, K. Beck, C. B. Smith, and R. McGregor, *Text. Chem. Color.* 26:30 (1994).
72. AATCC Testing Method 112-1982, *Technical Manual of the American Association of Textile Chemists and Colorists*, 1994, AATCC, Research Triangle Park, NC.
73. A. Beer, *Ann. Phys. Chem.* 86:78 (1852).
74. J. Park and J. Shore, *J. Soc. Dyers Color.* 102:330 (1986).
75. F. W. Billmeyer, Jr., M. Saltzman, and R. Kumar, *J. Color Res. Appl.* 7:327 (1982).
76. American Dye Manufacturers' Institute Color Index, American Dye Manufacturers' Institute (ADMI).
77. J. Pierce, *J. Soc. Dyers Color.* 110:131 (1994).
78. K. L. Kelly, and D. B. Judd, *The ISCC-NBS Color Names Dictionary and the Universal Color Language* (The ISCC-NBS Method of Designating Colors and a Dictionary of Color Names), NBS Circular 553, November 1, 1955, 7th printing, 1976.
79. A. H. Munsell, *A Color Notation,* 12th ed., Munsell Color Co., Inc., Baltimore, MD.
80. Meeting papers of 1986 Interim Meeting of the International Color Association, Toronto, Canada, 19–20 June 1986. Papers are reprinted in *J. Color Res. Appl.* 11(Suppl.) (1986).
81. R. McDonald, *Text. Chem. Color.* 24:11 (1992).
82. T. F. Chong, C. K. Yeung, and S. K. Ku, *J. Text. Asia XXIV*:62 (1993).
83. C. J. Hawkyard and D. P. Oulton, *J. Soc. Dyers Color.* 107:309 (1991).
84. *Color in Electronic Displays* (H. Widdel and D. L. Post, eds.), Plenum Press, New York, 1992.
85. R. Berns, R. J. Motta, and M. E. Gorzynski, *J. Color Res. Appl.* 18:299 (1993).
86. Comparison of color images presented in different media, Proceedings 1992 Vol. 2, Technical Association of the Graphics Arts and Inter-Society Color Council, Williamsburg, VA, 1992.
87. CIE Technical Committee 1-27, CIE guidelines for coordinated research on evaluation of colour appearance models for reflection print and self-luminous display image comparisons, *J. Color Res. Appl.* 19:48 (1994).
88. G. Tonnquist, Proceedings of 3rd Congress of the International Color Association, Troy, NY, 1977, p. 13–32, Adam Hilger Ltd., Bristol, England.

12

Assessment of Chemical Barrier Properties

JEFFREY O. STULL   International Personnel Protection, Inc., Austin, Texas

I.   INTRODUCTION AND BACKGROUND

Various forms of chemical resistance testing are used to assess the barrier properties of textile materials when used in applications requiring barrier performance. In most cases, information from this testing is used to support decisions for the development, selection, and use of products that require some level of chemical resistance. One of the foremost applications of textile materials involving the need for chemical resistance is for their use in protective clothing (Fig. 1). As a consequence, many of the chemical resistance evaluation methods have evolved from the chemical protective clothing industry. An important part of this industry has been the establishment of standard test methods that ultimately guide product innovation and claims. While many test techniques have been developed specifically for evaluating protective clothing, they may easily be applied to any use of textile material where chemical resistance determinations are needed. This chapter has been written to provide an understanding of the test methods available for evaluating the chemical resistance of textile materials and is intended to help those with testing needs to correctly choose test methods and interpret test results for chemical barrier performance.

A.   Chemical Barrier Materials

Barrier performance-based products require unique materials that are capable of preventing chemicals in a particular state and concentration from passing through the materials used in their construction. For the most part, textiles by themselves are generally only able to resist solid (particulate) penetration, and sometimes liquid penetration depending on the nature of the liquid challenge and surface characteristics of the material. Most chemical barrier materials for liquids

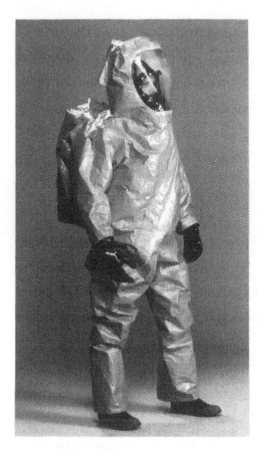

Figure 1 Barrier materials are an integral part of protective clothing. Shown here is a proprietary plastic laminate on a nonwoven substrate as part of encapsulating suit for total protection of the wearer. (Courtesy of Kappler, Inc.)

and gases are composed of a film or coating in combination with a substrate fabric. The film or coating may be on one side or both sides of the fabric. In general, the film or coating provides the barrier properties of the material, while the substrate fabric mainly provides material strength and support.

Films and coatings comprise a number of materials, either elastomeric or plastic in polymer composition. Examples of elastomeric films include Neoprene (chloroprene), nitrile rubber, butyl rubber, chlorobutyl rubber, Viton (fluoropolymer), and various combinations of these polymers. Plastics and thermoplastics are increasingly being used for barrier products owing to their relatively "good" chemical resistance. Traditionally, polyvinyl chloride, polyurethane, and polyethylene have been used. However, the diversity of these plastics is expanding as manufacturers produce a variety of laminate materials that include different polymer layers with varying levels of chemical resistance in order to provide overall chemical resistance to a broader range of chemicals, such as in Saranex and a proprietary Teflon film [1].

Another type of materials being used in barrier products is microporous films, which are engineered with pores ranging in size from 0.01 to 10 μm. These materials generally offer "breathability" in terms of water vapor transmission and sometimes air permeability (see Fig. 2). These materials are designed to offer liquid or particulate penetration resistance, but because of their structure, they cannot provide an effective barrier to vapor penetration by most chemicals [2].

Substrate fabrics may be of woven or nonwoven types. Nylon, polyester (Dacron), Nomex, and fiberglass fabrics are common examples of woven supporting fabrics used in chemical protective garments. Nonwoven materials include polyester, polypropylene, and spun-bonded polyethylene. The substrate fabrics are either laminated to the plastic or rubber film/sheet under heat and pressure or coated with a solution of the plastic/rubber material.

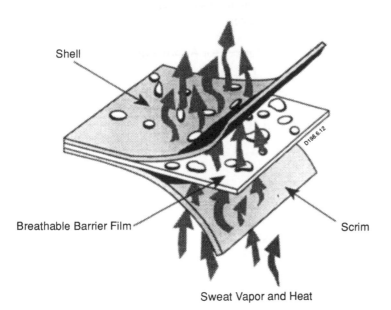

Shell

Breathable Barrier Film

Scrim

Sweat Vapor and Heat

Figure 2 Some protective barrier materials incorporate a microporous film. These materials are designed to prevent liquid penetration while allowing air and water vapor to pass through the material, providing greater comfort to the wearer. (Courtesy of W. L. Gore & Associates.)

Textiles may also be combined with adsorbent materials that are designed to prevent the passage of aerosols or chemical vapors. These types of fabrics have been traditionally used for chemical warfare agent protection within the military, but their breathable characteristics make these types of fabrics suitable for applications where low levels of air chemical contamination may be encountered [3].

B.   Standards Pertaining to Chemical Barrier Performance

Though a number of driving forces are responsible for large changes in chemical barrier technology, one of the primary reasons for these changes can be traced to the introduction of new testing standards. In the United States, consensus group standards established by the American Society for Testing and Materials (ASTM) and related industry trade group standards have been predominantly used for evaluation of product chemical barrier performance. Many of these have been developed by ASTM Committee F-23 on Protective Clothing, Committee D-11 on Rubber, Committee D-13 on Textiles, and Committee D-20 on Plastics. Outside the United States, the most significant standards development groups are the European Standardization Committee (CEN), which is in the process of developing standards for the European Community, and the International Standards Organization (ISO). A list of standards relevant to this chapter, with their current edition, is provided in Appendix A of this chapter.

C.   Overview of Chemical Resistance Test Approaches

Chemical resistance test approaches for chemical barrier textile-based products may be segregated into small-scale, material-based tests and full item evaluations. Material test approaches can be classified into three types, which describe how chemicals may interact with materials:

- Degradation resistance
- Penetration resistance
- Permeation resistance

A number of different procedures exist for the measurement of each phenomenon, depending on the type of chemical challenge and the level of sophistication for performing the tests. Of the material testing approaches, both penetration and permeation resistance testing allow assessment for the barrier qualities of a protective clothing material, whereas degradation resistance does not. Penetration testing may involve chemical particulates, liquids, or vapors (gases). This chapter specifically addresses liquid and vapor challenges. Similarly, since various applications involve chemical barrier materials, overall product testing is not described in this chapter.

The individual procedures available for measuring chemical degradation, penetration, and permeation resistance are described in the following sections.

II. DEGRADATION RESISTANCE

Degradation is defined by ASTM's F-23 Committee as the "change in a material's physical properties as the result of chemical exposure." Physical properties may include material weight, dimensions, tensile strength, hardness, or any characteristic that relates to a material's performance when used in a particular application. As such, the test is used to determine the effects of specific chemicals on materials. In some cases chemical effects may be dramatic, showing clear incompatibility of the material with the chemical. Figure 3 shows a specimen of a protective clothing material before and after its exposure to a selected chemical, illustrating a severe case of material chemical degradation. In other cases, chemical degradation effects may be very subtle.

Various groups have examined different approaches for measuring the chemical degradation resistance of barrier materials, but no single generalized test method has been developed by consensus organizations within the United States, Europe, or internationally [4]. Nevertheless, various techniques are commonly used for rubber and plastic materials within different barrier material industries. These procedures and their utility in evaluating chemical barrier materials are discussed next.

A.   Specific Testing Approaches

While not specific to any particular product, a few test methods have been developed for evaluating chemical resistance of different materials. These are:

* ASTM D471, *Test Method for Rubber Property-Effect of Liquids*
* ASTM D543, *Test Method for Resistance of Plastics to Chemical Reagents*
* ASTM D3132, *Test Method for Solubility Range of Resins and Polymers*

Figure 3   Plastic-coated nonwoven fabric before (left) and after (right) exposure to sulfur trioxide. (Courtesy of TRI/Environmental, Inc.)

However, a few specific test methods have been developed for protective clothing:

- ASTM F1407, *Test Method for Resistance of Chemical Protective Materials to Liquid Permeation-Permeation Cup Method.* This test as developed by ASTM's F23 Committee also allows for determining chemical degradation of protective clothing materials; however, this method is primarily intended to provide a simple technique for measuring chemical permeation of protective clothing.
- ISO 2025, *Lined industrial rubber boots with general purpose oil resistance.* This test involves a footwear specification that include measuring footwear material degradation to oils.
- The National Aeronautics and Space Administration (NASA) has developed a method to determine degradation of fabrics used in propellant handlers ensembles [5].

These techniques have been grouped by the type of approach used in the following sections. Table 1 summarizes the key characteristics, differences, and applications of each approach for measuring the chemical degradation of protective clothing materials.

1. Degradation Tests Using Immersion-Based Techniques

ASTM D471 and D543 establish standardized procedures for measuring specific properties of material specimens before and after immersion in the selected liquid(s) for a specified period of time at a particular test temperature. Test results are reported as the percentage change in the property of interest. ASTM D471 provides techniques for comparing the effect of selected chemicals on rubber or rubber-like materials, and is also intended for use with coated fabrics. ASTM

Table 1 Comparison of Chemical Degradation Test Methods

Test method	Type of contact	Contact period	Determination	Sample handling
ASTM D471	Both immersion and one-sided	22–760 hr	Weight, volume, or other physical property change	Acetone rinsing followed by blotting
ASTM D543	Immersion only	Up to 7 days	Weight, volume, or other physical property change	Solvent rinsing depending on chemical followed by blotting
ASTM D 3132	Immersion only	24 hr	Visible condition of sample	None
ASTM F1407	One-sided	Unspecified	Weight Change	Blotting
NASA MTB-175-88	One-sided	1 hr	Weight change, visible condition of sample	Blotting

D543 covers testing of plastic materials, including cast and laminated products, and sheet materials for resistance to chemical reagents. In each test, a minimum of three specimens are used whose shape and size are dependent on the form of the material being evaluated and the tests to be performed. An appropriately sized vessel, usually glass, is used for immersing the material specimens in the selected chemical(s). Testing with volatile chemicals typically requires either replenishment of liquid or a reflux chamber above the vessel to prevent evaporation.

The two sets of procedures prescribe a variety of exposure conditions and recommended physical properties. In general, the test methods can be applied to any type of liquid chemical challenge. ASTM D471 cites a number of ASTM oils, reference fuels, service fluids, and reagent-grade water. ASTM D543 lists 50 different standard reagents, which include representative inorganic and organic chemicals. In ASTM D471, 17 different test temperatures ranging from $-75°C$ to $250°C$, and five different immersion periods (22 to 760 h) are recommended. ASTM D543 suggests a 7-day exposure at either 50 or 70°C. For determining chemical degradation resistance, ASTM D471 specifies procedures for measuring changes in mass, volume, tensile strength, elongation, and hardness of rubber material, and breaking strength, burst strength, tear strength, and coating adhesion for coated fabrics. Measurement of material specimen mass and dimensions is recommended in ASTM D543, while other properties may be selected that are appropriate for the material's application.

Both test methods indicate that the selected exposure conditions and physical properties measured should be representative of the material's use. For protective apparel material testing, this will usually mean specifying significantly shorter test durations and ambient temperature exposures. Since the methods are intended for comparing materials against similar chemical challenges, no criteria are given for determining acceptable performance.

2. Degradation Tests Using One-Sided Exposure Techniques

Section 12 of ASTM D471 provides a procedure for evaluating the effects of chemicals when the exposure is one-sided. This technique is particularly useful for the evaluation of protective clothing materials, particularly those involving coated fabrics, laminates, and any nonhomogeneous material. In this procedure, the material specimen is clamped into a test cell (Figure 4) that allows liquid chemical contact on its normal external (outer) surface. Usually changes in mass are measured for this testing approach, since the size of the material specimens is limited.

A similar procedure was developed by the ASTM F23 committee [6] but was never established as a standard test method. This test method involved the same approach as in ASTM D471 (Section 12) but was specific to evaluation of protective clothing materials. The proposed test method used a chemical degradation test fixture that was a simple "sandwich" configuration consisting of three identical one-chamber glass pipe cells held between top and bottom polyethylene boards by a series of wing nuts. Material specimens were positioned between the bottom

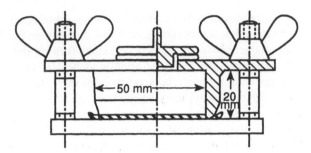

Figure 4 Degradation test cell design for one-sided exposure as specified in ASTM D471.

board and the glass pipe cells. Liquid chemical was then poured into the cell to initiate contact with the material specimens. Following the end of the exposure period, the material specimens were blotted dry and then evaluated for the property of interest. Mass, thickness, and elongation were originally prescribed as physical property measurements in the technique, but the committee later debated the need for evaluating material elongation since different types of substrates significantly affected the ability of the test to compare supported and unsupported materials. An interlaboratory test program for validating the technique also indicated serious problems in method reproducibility and difficulty in conducting the test when severe chemical degradation occurred.

The ASTM F23 Committee more recently developed a procedure that, while intended for measuring chemical permeation resistance, serves as a useful technique for evaluating chemical degradation resistance of protective clothing materials [7,8]. ASTM F1407 employs a lightweight test cell in which the material specimen is clamped between a Teflon-coated metal cup filled with the selected chemical and a metal ring (flange). The entire cup assembly is inverted and allowed to rest on protruding metal pins, which hold the test cell off the table surface. In the mode of permeation testing, the weight of the entire assembly is monitored; however, for use as a degradation test, the test cell serves as a convenient means for evaluating changes in material mass and thickness. Visible observations are also recorded as part of the testing protocol. Figure 5 shows a photograph of the permeation test cup specified by this method.

Each of the test methods described thus far for surface contact is provided for the testing of liquids. Protective clothing specimen chemical degradation can also be evaluated for chemical gas or vapor exposures. During one study for screening chemical resistance of protective suit [9], a special test cell was designed for evaluating material performance against gases. This cell was configured to provide a leak-free seal with the material specimen and to allow the flow of the gas into and out of the test cell. Degradation results for three protective clothing materials are shown in Table 2 for both liquids and gases.

Figure 5 Permeation test cup and field test kit conforming to ASTM F1407, which may also be used for measuring barrier material chemical degradation resistance. (Courtesy of TRI/Environmental, Inc.)

NASA has developed and uses an internal procedure (MTB-175-88) for examining material responses to liquid propellants [5]. In one part of the test, 0.5 ml of the test chemical is placed on the middle of a 2-inch-square material specimen. Temperature rise of the material sample is monitored using a sheathed thermocouple, with subsequent observation of visible changes in the sample's appearance such as burning, smoking, frothing, bubbling, charring, solubilizing, fracturing, and swelling. The second part of the test involves applying a larger amount of test chemical to material specimens for the purpose of measuring changes in thickness and tensile strength.

An extension of the NASA procedure involves applying differential scanning calorimetry and thermogravimetric analysis as another means for determining material chemical degradation. Bryan and Hampton [10] examined the effects of liquid nitrogen tetroxide and monomethyl hydrazine on the breaking strength of a chlorobutyl rubber coated Nomex material. They found that the time of exposure could be determined by differential scanning calorimetry and that a correlation existed between the breaking strength of the fabric and an endothermic event occurring during the analysis of the fabric.

3. Degradation Testing Using Solubility-Based Evaluations

Henriksen [11] first proposed the application of solubility parameter measurements in the selection of chemical protective clothing barriers. His recommenda-

Table 2 One-Sided Immersion Degradation Resistance Data for Selected Materials and Chemicals

Chemical	Viton/chlorobutyl laminate			Chlorinated polyethylene			FEP/Surlyn laminate		
	Percent weight change	Percent elongation change	Visual observation	Percent weight change	Percent elongation change	Visual observation	Percent weight change	Percent elongation change	Visual observation
Acetaldehyde	10	0		24	11		0	−5	
Acrylonitrile	9	0	Delaminated	35	Failed		−1	0	
Benzene	2	0		60	Failed		0	0	
Chloroform	4	0		72	Failed		0	0	
Dichloropropane	3	0		120	Failed		−2	0	
Ethyl acrylate	17	0	Curled	160	Failed		0	0	
Ethylene oxide	2	0		13	11		0	0	
Hydrogen fluoride	4	0	Discolored	2	11		4	0	
Nitric acid	9	0	Discolored	8	−6	Discolored	−1	0	

Note: FEP Fluorinated ethylene propylene. Percent elongation based on elongation measured using ASTM D412 for exposed and unexposed samples; "failed" results indicate materials not tested due to weight changes over 25%.
Source: Adapted from Ref. 9.

tion of use of three-dimensional solubility parameters was based on the early work of Hansen [12], who successfully demonstrated that chemical effects could be correlated with the solubility parameters of individual chemicals for resins and other homogeneous materials. Hansen postulated that the chemical's energy of evaporation, ΔE, is the sum of the energies arising from dispersion forces, ΔE_d, polar forces, ΔE_p, and hydrogen bonding forces, ΔE_h. Dividing ΔE by molar volume (V_m) gives the cohesive energy density for each solvent:

$$\delta = (\frac{\Delta E}{V_m})^{1/2} \tag{1}$$

The three-dimensional solubility parameter is defined as a vector of magnitude δ, with components, δ_d, δ_p, and δ_h, derived from the energies resulting from the three types of molecular forces. Thus, the solubility parameter of a given substance can be visualized as a fixed point in three-dimensional space. Hansen [13] conjectured that the closer the solubility parameters of two substances lie within the three dimensional system, the greater their affinity and similarity of response to other substances. Hansen found that a sphere could be defined in solubility parameter space for each polymer such that when exposed to solvents with solubility parameters lying within the sphere, the polymer would interact (i.e., dissolve, swell, etc.), whereas those solvents lying outside had relatively little effect. Figure 6 shows this schematically with two-dimensional projections on Cartesian planes.

Estimation of a three-dimensional solubility parameter "sphere" has become one technique to evaluate the chemical degradation resistance over a wide range of substances. Holcomb [14] advanced a technique for determining the solubility parameters of polymer substances that involved observing and measuring changes in a material when exposed to a large battery of chemicals representing ranges in each of three solubility parameter dimensions. The principle of this technique follows immersion testing where small specimens of a material are placed in each of the solvents and changes in visual appearance, mass, or volume are determined. These procedures are embodied in ASTM D3132, where a battery of solvents is used to show effects on the material by virtue of the measurements or observations that define the solubility space for the polymer. Bentz and Billing [15] and Perkins and Tippit [16] both applied this technique to a number of polymer substances. Instead of using the 90 different solvents and solvent mixtures specified in ASTM D3132, they used the smaller 55 "neat" chemicals listed in Table 3 and originally suggested by Holcomb [14].

Use of solubility parameter-based techniques for measuring the chemical degradation of protective clothing materials allows the researcher to quickly and comprehensively characterize the effects of chemicals on a given material. In essence, when a solubility sphere can be defined for the test material, the technique can be used as a predictive model to characterize performance against other solvents (those not in the test battery) without testing. Nevertheless, the method

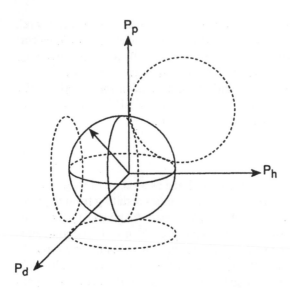

Figure 6 Hansen three-dimensional solubility plot. The sphere indicates the solubility space. The circles represent solubility areas projected by the sphere in two dimensions. (From Ref. 16.)

Table 3 List of Chemicals Used in Degradation Tests to Determine Material Solubility Parameters

Number	Chemical	S_p	δD	δP	δH
1	Acetic anhydride	10.9	7.8	5.7	5.0
2	Acetone	9.8	7.6	5.1	3.4
3	Acetonitrile	11.6	7.5	6.3	6.3
4	Acetophenone	10.6	9.6	4.2	1.8
5	Acrylonitrile	12.1	8.0	8.5	3.3
6	Aniline	12.0	9.8	3.6	6.0
7	Benzaldehyde	10.5	9.5	3.6	2.6
8	Benzene	9.2	9.2	0.0	0.3
9	1,3-Butanediol	14.2	8.1	4.9	10.5
10	1-Butanol	11.3	7.8	2.8	7.7
11	2-Butoxyethanol	10.2	7.8	2.5	6.0
12	Carbon disulfide	10.7	10.7	0.0	0.3
13	Carbon tetrachloride	8.7	8.7	0.0	0.3
14	Chloroform	9.3	8.7	1.5	2.8
15	Cyclohexanol	11.0	8.5	2.0	6.6
16	Cyclohexanone	9.6	8.7	3.1	2.5

Table 3 Continued

Number	Chemical	S_p	δD	δP	δH
17	Dichloromethane	9.9	8.9	3.1	3.0
18	Diethylenetriamine	12.6	8.2	6.5	7.0
19	1,4-Dioxane	10.0	9.3	0.9	3.6
20	Dimethyl formamide	11.6	8.3	6.7	4.5
21	Dimethyl phthalate	10.8	8.3	6.5	2.4
22	Epichlorohydrin	10.7	9.3	5.0	1.8
23	Ethanol	13.0	7.7	4.3	9.5
24	1-Ethanolamine	15.4	8.4	7.6	10.4
25	2-Ethoxyethanol	11.5	7.9	4.5	7.0
26	2-Ethoxyethanol acetate	9.6	7.8	2.3	5.2
27	Ethyl acetate	9.0	7.4	2.6	4.5
28	Ethylene carbonate	14.5	9.5	10.6	2.5
29	Ethylene chloride	10.1	9.3	3.3	2.0
30	Ethylene glycol	16.1	8.3	5.4	12.7
31	Formamide	17.9	8.4	12.8	9.3
32	2-Furaldehyde	11.9	9.1	7.3	2.5
33	Furfuryl alcohol	11.9	8.5	3.7	7.4
34	Glycerol	17.6	8.5	5.9	14.3
35	*n*-Heptane	7.5	7.5	0.0	0.0
36	*n*-Hexane	7.3	7.3	0.0	0.0
37	Ethylene cyanohydrin	15.1	8.4	9.2	8.6
38	*iso*-Octane	7.0	7.0	0.0	0.0
39	Methanol	14.5	7.4	6.0	10.9
40	Methyl ethyl ketone	9.2	7.7	4.4	2.5
41	4-Methyl-2-pentanone	8.4	7.6	3.0	2.0
42	1-Methyl-2-Pyrrolidinone	11.2	8.8	6.0	3.5
43	Methyl sulfoxide	13.0	9.0	8.0	5.0
44	Nitrobenzene	11.7	9.8	6.0	2.0
45	Nitromethane	12.8	8.2	9.0	3.8
46	2-Nitropropane	10.9	7.8	7.0	3.0
47	Octanol	9.9	7.9	1.6	5.8
48	3-Penanone	8.9	7.7	3.7	2.3
49	Propanol	12.1	7.8	3.5	8.5
50	Pyridine	10.7	9.9	3.2	2.2
51	2-Pyrrolinone	13.9	9.5	8.5	5.5
52	Toluene	8.9	8.8	0.7	1.0
53	1,1,1-Trichlooethane	8.6	8.3	2.1	1.0
54	Triethylene glycol	13.0	7.8	5.1	9.1
55	Xylene	8.8	8.7	0.5	1.5

Source: Adapted from Ref. 14.

works well only with homogeneous substances, usually in the absence of any sup-porting substrate. Less success has been achieved in characterizing multilayer ma-terials or laminates, since individually layers may be affected differently [15,17,18].

B.   Application of Test Data

Chemical degradation by itself cannot fully demonstrate product barrier perfor-mance against chemicals. This form of chemical resistance testing does not ascer-tain the barrier properties of materials. While a material that shows substantial effects when exposed to a chemical can be ruled out as a protective membrane, it remains uncertain whether materials that show no observable or measurable effect provide a barrier against the test chemical. For this reason, chemical degradation data are typically used as a screening technique to eliminate a material from con-sideration for further chemical resistance testing (i.e., penetration or permeation resistance) [6,19].

To better understand the application of chemical degradation resistance data, it is instructive to know how this data is now being used within the protective clothing industry and how it could be used in a comprehensive material evaluation program.

1.   Current Protective Clothing Industry Practices

Within the chemical protective clothing and related apparel industries, a great deal of chemical degradation resistance data has been generated and is presented in var-ious product literature. Unfortunately, most presentations of material chemical degradation resistance are based on qualitative ratings such as "excellent," "good," "fair," "poor," and "not recommended." Ratings of this type provide lit-tle information to the end user, particularly when the basis of the ratings are not explained or cannot be related to a particular application.

Manufacturers that use degradation data often base their ratings on either ob-served visual changes or weight gain. In a few cases, tensile or breaking strength differences are also used to qualify material chemical degradation. Ratings are then based on arbitrarily set levels of degradation. One manufacturer uses the fol-lowing scale for its degradation rating system:

Rating	Percent weight change
Excellent	0–10%
Good	11–20%
Fair	21–30%
Poor	31–50%
Not recommended	Over 50%

This scale and probably all other rating systems used in the market have little bearing on the selection and use of protective clothing materials. Their presentation simply helps the end user to compare chemical effects on different materials. Since there is no one standard test method used in the industry for measuring chemical degradation resistance of protective clothing materials, chemical effect information cannot be compared from one source to the next because different test approaches and criteria are used [19].

The majority of chemical degradation resistance data is reported in the glove industry. This is because most gloves are made from elastomeric materials. As a class of materials, elastomers, when compared to plastics, show greater affinity for chemical adsorption and swelling [20]. Therefore, elastomeric materials are generally more susceptible to measurable chemical effects. This is particularly true today, because the majority of garment materials are composed of different plastic layers that have few observable degradation effects.

2.   Recommended Use of Degradation Testing

Chemical degradation resistance testing is best used to qualify material candidates for subsequent barrier forms of chemical resistance testing. Materials that show significant signs of chemical degradation as determined by relevant criteria can be eliminated from further consideration in the required application. The premise for using degradation testing in this fashion is to configure the test and choose criteria that reflect how the material will be used and what barrier performance should be demonstrated. Establishment of a degradation testing protocol should include the following decisions:

- How long should the exposure be conducted?
- Should the exposure be one-sided or by complete immersion?
- What material properties should be measured?
- How should material specimens be handled following exposure?
- What criteria should be used for accepting or rejecting a material for a given chemical?

In general, the length of chemical exposure should be as long as the maximum duration for which chemical contact can occur or for the same period of time being used for the barrier test procedure selected. In some cases, longer exposure periods are used because the longer chemical contact can accentuate degradation effects that may be difficult to observe or measure.

Immersion offers the easiest approach for measuring chemical degradation resistance but may not be suitable for some materials. As described for ASTM D471 and D543 test methods, immersion-based degradation testing is best applied to homogeneous materials. Nevertheless, the ease of placing material specimens into a container filled with chemical can also be applied to more complex material to determine if separation of layers or substrate occurs as the result of the chemical exposure. The principle argument against immersion-based testing is that many products are typically exposed on their external surfaces only. This means that the substrate and different layers inside the material matrix are not likely to be exposed to chemicals unless the external surface is breached in some manner. For this reason, one-sided exposures are considered more realistic for the evaluation of protective clothing chemical degradation. On the other hand, one-sided degradation testing is more difficult to perform, usually limits the size of the material specimen for subsequent physical property testing, and may not provide any more information about the materials performance following chemical exposure.

Physical properties should be selected to measure a material characteristic of interest. For the use of degradation testing as a means for screening material chemical resistance, observable changes in specimen appearance and weight gain are suggested. While disintegration of a test specimen is an easily recorded event, other differences in the visual appearance of test specimens may not be easily discerned. When chemicals are darkly colored, it may be impossible to carefully examine the material. Furthermore, it is difficult to achieve consistent material performance determinations using visual observations as the sole basis for rating material chemical degradation resistance. Nevertheless, the use of visual observations offers the easiest approach for conducting degradation testing with a minimum of specimen handling. When used, operator comments should be confined to certain observations. Examples include:

- Discoloration (if the material is not darkly colored)
- Curling
- Swelling
- Delamination (for multilayer or fabric supported materials)
- Disintegration

Depending on both the nature of the chemical and material being evaluated, other "standardized" observation categories can also be created.

Measurement of weight change provides a simple, quantitative approach for assessing material degradation resistance. Material specimens may either gain or lose weight, depending on how the chemical interacts with the material. Weight gain is caused by adsorption or absorption of a chemical on or in the material specimen and is often evidenced by swelling. Specimen weight loss is primarily due to some disintegration of the material or removal of particular components, such as plasticizers and other additives. In some cases, both phenomena can occur and result in very little weight change, even though significant material effects have occurred. Weight change is always indicated as the percent change based on the sample's original weight.

Chemical degradation resistance testing involving the measurement of weight change is often affected by the techniques used for specimen handling. This handling can remove chemical from the material's surface. In addition, removal of the material specimen from the chemical allows evaporation of volatile chemicals

both off the surface and from inside the material matrix. The design of a degradation testing procedure must include uniform specimen handling techniques to
avoid systematic errors. Each test method defines slightly different procedures:

- ASTM D471 specifies quickly dipping the specimen in acetone, blotting with a
  lint-free blotter paper, and then placing it in a tared, stoppered weighing bottle.
- ASTM D543 requires similar handling but indicates that the rinsing should be
  done with water, acetone, or not at all, depending on the exposure chemical.
- ASTM F1407 uses blotting only (without any sample rinsing).

The choice of specimen handling procedures should be based on the nature of the
chemical being testing and may need to be different for different chemicals, particularly when both low-vapor-pressure, nonvolatile and high-vapor-pressure,
volatile chemicals are tested. The former class of chemicals should require some
blotting, with the time between removal from the chemical and weighing not as
critical. Handling of specimens that have been exposed to high-vapor-pressure,
volatile chemicals will most likely dictate no specimen rinse and a uniform period
between specimen removal and weighing.

Development of screening criteria should be based on experience with the
material(s) being evaluated and the barrier test chosen for assessment of chemical
resistance. For example, small weight changes in some materials, such as Teflon
or polyethylene laminates, may provide a clue of rapid permeation. But for other
types of materials, these changes may not be significant for their permeation resistance. Whether the material specimen includes substrate fabric can also be a factor in setting acceptance and rejection criteria. Some fabrics can readily absorb test
chemicals, particularly when immersion-based testing is used. If degradation testing is performed for screening materials prior to penetration resistance testing,
then only severe material changes that could lead to penetration of liquids through
the material of interest need be considered.

The last consideration for using chemical degradation resistance testing is
economic. The costs of the tests for degradation screening combined with subsequent barrier tests should not exceed the cost of barrier testing alone with all material and chemical combinations.

The other use for chemical degradation resistance testing is to identify potential modes of a product's failure. This form of degradation testing is most suitable to nonplanar components such as seams, closure, gaskets, or any other
accessories used on the product. The chemical performance of these items can be
evaluated by degradation tests that combine chemical exposure with an appropriate physical property test. Some examples include:

- Evaluation of rigid-formed materials for environmental stress cracking (see
  ASTM D1975)
- Assessment of transparent material clarity following chemical exposure using
  light transmittance and haze testing (see ASTM D1003)

- Measurement of changes in seam or closure strength (see ASTM D751 for rubber materials or coated fabrics, D1683 for textiles, and F88 for plastic materials)
- Operability of zippers (see ASTM D2062)
- Hardness of product gaskets or interface materials (see ASTM D2240)

Overall, chemical degradation resistance testing can be a useful means for evaluating chemical barrier materials. However, it is important that the limitations of this testing be recognized and that decisions for barrier product selection and use never be based solely on chemical degradation resistance data.

III. PENETRATION RESISTANCE

In the U.S. protective clothing industry, the ASTM F-23 Committee has defined penetration as "the flow of chemical through closures, porous materials, seams, and pinholes and other imperfection in a protective clothing material on a non-molecular basis." This definition is intended to accommodate both liquids and gases, but all U.S., European, and international test methods focus on liquid penetration. Liquid suspended in air as aerosols and solid particulates can also penetrate protective clothing materials, but the discussion of penetration resistance in this chapter relates to liquids exclusively.

Much of the liquid penetration resistance testing pertains to water as the challenge. The ASTM D-11 Committee on Rubber and Rubber Products (which includes coated fabrics) defines water repellency, waterproofness, and water resistance for coated fabrics as follows:

- Water repellency—the property of being resistant to wetting by water
- Waterproofness—the property of impenetrability by liquid water
- Water resistance—the property of retarding both penetration and wetting by liquid water

Liquid repellency and penetration resistance are related, since wettability of the fabric affects the ability of the liquid to penetrate.

For porous fabrics, a liquid of surface tension γ will penetrate given sufficient applied pressure, p, when its pores are of diameter D, according the relationship known as Darcy's law:

$$D = k\frac{4\gamma\cos\theta}{p} \tag{2}$$

where

θ = contact angle of liquid with the material
k = shape factor for the material pores

For nonporous fabrics, particularly coated fabrics or laminate materials, liquid penetration may still take place as the result of degradation. Given a sufficient period of contact, chemicals may cause deterioration of the barrier film to allow pathways for liquid to penetrate. In this sense, penetration testing allows an assessment of both material barrier performance to liquid chemicals and chemical degradation resistance.

There are two fundamentally different approaches used in liquid penetration resistance test methodologies: (1) "runoff-based" methods, and (2) hydrostatic-based methods. *Runoff-based techniques* involve contact of the liquid chemicals with the material by the force of gravity over a specified distance. The driving force for penetration is the weight of the liquid and the length of contact with the material specimen. Usually the material specimen is supported at an incline, allowing the chemical to run off, hence the name for this class of penetration tests. *Hydrostatic-based techniques* involve the pressurization of liquid behind or underneath the material specimen. It is this hydrostatic force that is the principal driver for liquid penetration. Though the term "hydrostatic" is used to describe this class of test methods, one of the tests in this class can accommodate a wide range of liquid chemicals.

Both classes of test methods are described next; however, specific discussion focuses on those test methods most generally used for assessing protective clothing chemical penetration resistance.

A.   Runoff-Based Test Methods

Runoff-based tests are characterized by three features:

- Impact of the liquid from a stationary source onto a material specimen
- Orientation of the material specimen at an incline with respect to the point of liquid contact
- Use of a blotter material underneath the material specimen to absorb penetrating liquid

Runoff-based tests differ in the distance separating the liquid source from the point of contact with the material specimen, the type of nozzle through which liquid is delivered, the amount of liquid and the rate at which it is delivered, the angle of the incline, and the type of test measurements made.

1.   Specific Runoff-Based Test Methods Available

There are a number of liquid penetration tests that are based on runoff techniques. These include:

American Association of Textile Chemist and Colorist (AATCC) Test Methods
- AATCC 42, Water Resistance: Impact Penetration Test
- AATCC 118, Oil Repellency: Hydrocarbon Resistance Test

European Community (CEN) Test Method
• EN 368, Protective clothing—Protection against liquid chemicals—Test
  Method: Resistance of materials to penetration by liquids
International Standards Organization (ISO) Test Method
• ISO 6530, Protective clothing—Protection against liquid chemicals—Deter-
  mination of resistance of materials to penetration by liquids
U.S. Federal Government Test Methods (FTMS)
• FTMS 191A,5520—Water Resistance of Cloth; Drop Penetration Method
• FTMS 191A,5522—Water Resistance of Cloth; Water Impact Penetration
  Method
• FTMS 191A,5524—Water Resistance of Cloth; Rain Penetration Method

Table 4 provides a comparison of the key characteristics of each of the referenced
test methods. It should be noted that some methods from different organizations
are very similar to each other. These methods are also cross-referenced in Table 4.
   As the titles for several methods denote, the majority of these methods are intended for use with water as the liquid challenge only. Physically, many of the
methods are suitable for testing with other liquids; however, the containment as-

Table 4 Characteristics of Runoff-Based Penetration Tests

Test Method	Type of Delivery	Liquid Amount and Rate	Sample Orientation	Measurements
FTMS 191A,5520	Polystyrene plate with 31 0.4-mm ID capillary holes 1.73 m above sample	Determined by test end point	45 Degrees, clamped onto perforated disk	Time to collect 10 ml water (from penetrating sample)
FTMS 191A, 5522	Spray nozzle at end of funnel with 19 0.89-mm holes 610 mm above sample	500 ml	45 Degrees, under 0.45 kg tension force	Weight of penetrating water
FTMS 191A,5524	Spray nozzle at end of funnel with 12 1-mm holes 305 mm above sample	300 sec at selected pressure head	Horizontal	Weight of penetrating water
EN 368	Single, 0.8-mm bore hypodermic needle 100 mm above sample	10 ml at 1 ml/sec	45 Degrees, over blotter in semicircular "gutter"	Index of repellency; index of penetration
ISO 6530	Same as EN 368			

pects of these test methods vary and some are clearly inappropriate for use with hazardous chemicals.

The majority of test methods listed in Table 4 involve delivering relatively large quantities of water onto a sample and measuring the amount of water absorbed in a blotter paper placed underneath the material specimen. This approach is characteristic of AATCC 42, FTMS 191A,5522, and FTMS 191A,5524. Figure 7 shows a picture of a spray impact tested used in both AATCC 42 and FTMS 191A,5522. One method, FTMS 191A,5520, is used for materials where a significant amount of water is expected to penetrate since the time to obtain 10 ml of water is used as the test endpoint. The large quantities of water specified and the lack of containment in the design of the apparatus make these test methods unsuitable for other liquids.

AATCC 118 was designed to measure fabric resistance to oil. However, this test is a repellency type test, where surface appearance of the material specimen is rated after its exposure to selected hydrocarbons. This test is similar to several other tests in the literature involving water where either percent absorption of liquid by the fabric is measured or the pattern of wetness on the underside of the material specimen is rated on percent coverage.

Two of the test methods are essentially identical and are designated for use with various liquid chemicals. Both EN 368 and ISO 6530, the so-called "gutter test," use a system where the liquid chemical is delivered by a single, small-bore nozzle onto the material specimen at a distance of 100 mm (see Fig. 8). The material is supported in a rigid transparent gutter, which is covered with a protective film and blotter material set at a 45-degree angle with respect to the horizontal plane. A small beaker is used to collect liquid running off the sample. The two results reported in these tests are the indices of penetration and repellency. The "index of penetration" is the proportion of liquid deposited in the blotter paper:

$$\text{Index of penetration (P)} = \frac{M_p \times 100}{M_t} \tag{3}$$

where

M_p = mass of test liquid deposited on the absorbent paper/protective film combination

M_t = mass of the test liquid discharged onto the test material specimen

The "index of repellency" is the proportion of liquid deposited in the blotter paper:

$$\text{Index of repellency (R)} = \frac{M_r \times 100}{M_t} \tag{4}$$

where

M_r = mass of test liquid collected in the beaker

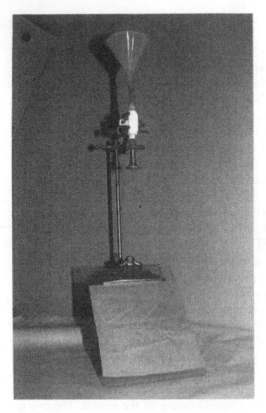

Figure 7   Spray impact tester used for testing in accordance with AATCC 42 and FTMS 191A,5222. (Courtesy of TRI/Environmental, Inc.)

A mass balance of the liquid also allows calculation for liquid retained in the material specimen.

2.   Application of Runoff-Based Test Methods

Not all of the tests described so far can be considered "true" liquid penetration tests. Penetration with these procedures can only be characterized when some assessment or measurement of liquid passing through the material specimen is made. Typically this is done by examination of the blotter material, either visually or gravimetrically.

Runoff tests are generally used on textile materials that have surface finishes designed to prevent penetration of liquid splashes. Many of these tests easily accommodate uncoated or nonlaminated materials, since the driving force for liquid penetration is relatively low (when compared to hydrostatic-based test methods). As a consequence, runoff tests may be infrequently specified for chemical barrier-

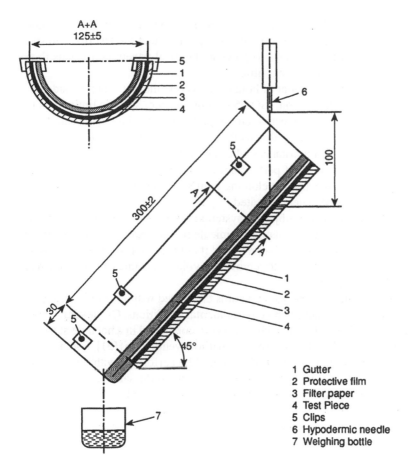

Figure 8 Specifications for CEN and ISO "gutter test."

based clothing. The European Community, while developing a range of chemical protective clothing standards, uses EN 368 in none but its lowest clothing classification (for partial body coverings). ISO 6530 is proposed as a test method for fire-fighting protective clothing in terms of evaluating composite materials against chemicals encountered in fire fighting. In the United States, only a few material manufacturers use runoff-based tests for characterizing the chemical penetration resistance of their fabrics. When used in this fashion, the fabrics tested are either uncoated or have thin simple plastic films.

3. Specific Use of ISO 6530

Table 5 provides runoff data on three different fabrics for selected chemicals using ISO 6530. One fabric has special surface finishes while another has a thin polyethylene coating. A third material involves a microporous film laminated to a nonwoven fabric. From this testing, it is readily apparent that the test distinguished fabrics relying on surface finishes to prevent chemical penetration versus those that are coated or laminated with a film.

The relatively small amount of liquid involved in the test is not considered a strong challenge. For this reason, ISO 6530 contains very specific limitations for its use in testing chemical protective clothing: "Clothing which has been developed from materials selected by this method of test (i.e., ISO 6530) should only be used in well-defined circumstances when an evaluation of the finished item has indicated an acceptable level of performance." In other words, ISO recommends that the test be used only when the clothing item's overall integrity for preventing liquid penetration has been demonstrated.

Use of ISO 6530 is also subject to systemic errors. As with degradation tests described earlier, testing with volatile chemicals requires special handling procedures to minimize evaporation of solvent and its impact on test results. Likewise, test operators must be careful not to remove nonvolatile chemical through the handling of the blotter material.

The "open" nature of the test apparatus combined with its gravimetric basis may also be strongly influenced by environmental conditions. One testing laboratory has reported different results when tests are conducted in a hood versus those that are conducted on the laboratory bench (without ventilation) [21]. For these reasons, penetration testing using ISO 6530 should be performed with uniform handling procedures and in a controlled environment that is the same from test to test.

Some in the protective clothing industry do not consider any of the runoff tests as legitimate liquid penetration tests since these methods fail to demonstrate "liquidproof" performance for protective clothing material performance. In this

Table 5 Penetration Data for Selected Materials and Chemicals Using ISO 6530

Material	Chemical	Index of Penetration	Index of Repellency
Woven cloth	Ethanol	16.3	37.5
	1.6% Orthene™ in water	24.4	48.2
	13.3% Carbaryl in water	15.5	64.0
Coated nonwoven fabric	Ethanol	0.9	84.4
	1.6% Orthene™ in water	0.0	95.9
	13.3% Carbaryl in water	0.0	92.9
Microporous film laminated nonwoven fabric	Ethanol	0.6	88.4
	1.6% Orthene™ in water	0.0	95.4
	13.3% Carbaryl in water	0.02	92.5

context, liquidproof performance is often defined as the ability of the material to prevent liquid penetration under conditions representative of use. These researchers contend that runoff-based or repellency penetration tests are an evaluation of the surface wettability characteristics for material finishes, and thus not true barrier-oriented techniques.

B.   Hydrostatic-Based Test Methods

Another class of penetration test methods involves those based on hydrostatic techniques. In this testing approach, liquid is contacted with the material specimen, with at least some portion of the test period having the liquid under pressure. Different devices or test cells are available for providing this type of liquid contact with the material specimen, in essence representing the differences among representative test methods. Like runoff-based test methods, the majority of the industry tests are designed for use with water. Many of the devices described in this section cannot be used with other liquids or may even be damaged if anything but water is used in the respective tests.

1.   Specific Hydrostatic-Based Test Methods Available

There are a number of liquid penetration tests based on runoff techniques. These include:

American Association of Textile Chemist and Colorist (AATCC) Test Method
• AATCC 127, Water Resistance: Hydrostatic Pressure Test
American Society for Testing and Materials (ASTM) Test Methods
• ASTM D751, Methods for Testing Coated Fabrics, Hydrostatic Test
• ASTM D3393, Specification for Coated Fabrics—Waterproofness
• ASTM F903, Resistance of Protective Clothing Materials to Penetration by Liquids
International Standards Organization (ISO) Test Method
• ISO 8096, Rubber- or plastics coated fabrics water-resistant clothing—Specification—Part 1: PVC-coated fabrics; Part 2: Polyurethane- and silicone elastomer-coated fabrics
U.S. Federal Government Test Methods (FTMS)
• FTMS 191A,5512—Water Resistance of Coated Cloth; High Range, Hydrostatic Pressure Method
• FTMS 191A,5514—Water Resistance of Coated Cloth; Low Range, Hydrostatic Pressure Method
• FTMS 191A,5516—Water Resistance of Cloth; Water Permeability, Hydrostatic Pressure Method

Two different types of testing machines prevail for measuring hydrostatic resistance. AATCC 127, ISO 8096, FTMS 191A,5514, and FTMS 191A,5516 all use similar devices, where water is introduced above the clamped material specimen at a pressure controlled by water in a rising column. A mirror is affixed below the specimen to allow the test operator to view the underside of the specimen for the appearance of water droplets. Both the pressure and length of exposure are to be specified for the particular application. AATCC 127 and FTMS 191A,5514 define water penetration as the pressure when a drop or drops appear at three different places of the test area (on the specimen). When a specific hydrostatic head is specified, test results are reported as pass or fail. FTMS 191A,5516 also permits measurement of the amount of water penetration, collected from a funnel and drain hose underneath the material specimen (in lieu of the observation mirror).

ASTM D751, ASTM D3393, and FTMS 191A,5512 use a motor-driven hydrostatic tester (pictured in Fig. 9). Water contacts the underside of the material specimen, which is clamped into a circular opening. Increasing hydraulic pressure is applied to the clamped material specimen at a specified rate until leakage occurs. The pressure at which this leakage occurs is noted and reported as the test result.

Of the already listed tests, only ASTM F903 was developed for testing liquids other than water [22]. In this test method, a 70-mm-square material specimen is exposed on one side to the test chemical for a specified period of time using a special penetration test cell (see Figs. 10 and 11). The test cell is positioned vertically to allow easy viewing of the material specimen. During the chemical exposure, a pressure head may be applied to the liquid for part of the test period. Penetration is detected visually and sometimes with the aid of dyes or fluorescent light. The test is generally pass/fail; that is, if penetration is detected within the test period, the material fails. Observations of material condition following chemical exposure are also usually provided. Different test specifications exist for the amount of chemical contact time and pressurization.

2. Application of Hydrostatic-Based Test Methods

Penetration resistance using hydrostatic-based test methods can accommodate different types of material and clothing test specimens, including:

- Plastic or rubber films
- Coated fabrics
- Textiles
- Microporous films
- Clothing seam samples
- Clothing closure samples

For these types of material specimens there are different modes of failure. Continuous film or film coated fabrics generally only fail due to:

- Imperfections in the material, such as cuts or pinholes
- Deterioration (degradation) of the film, providing an avenue for liquid penetration

Figure 9 Mullen hydrostatic tester. (Courtesy of TRI/Environmental, Inc.)

Figure 10 ASTM F903 penetration test cell. (Courtesy of TRI/Environmental Inc.)

The latter type of failure often depends on the thickness of the film or coating as well as the contact time and amount of pressure applied to the specimen.

Textiles and microporous film products provide another set of possible failure mechanisms. Textiles may be considered as liquid barriers when they have been treated with water/chemical-repellent finishes. The ease of liquid penetration

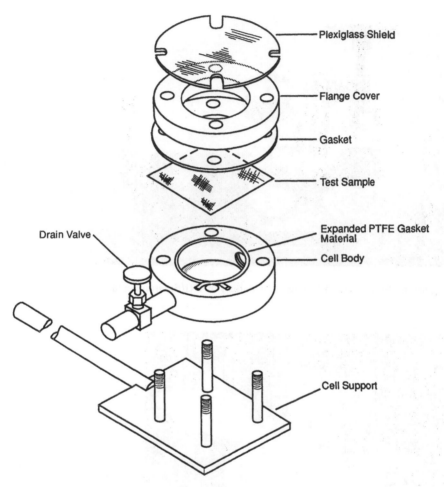

Figure 11   Exploded view of ASTM F903 penetration test cell.

is therefore more a function of repellant finish quality and the surface tension of the liquid being tested. Table 6 provides the surface tension of the liquid chemicals in ASTM F1001. Also, penetration may still be the result of material degradation while in contact with the chemical.

Microporous film products represent a unique class of test materials since by design they afford transmission of vapors but prevent liquid penetration. These materials therefore require careful observations since significant vapor penetration may occur. Like textiles, surface tension may be a factor, though most microporous films have pore sizes that preclude penetration of most common liquids at relative low pressures (less than 12 kPa).

The integrity of seams, closures, and other clothing material interfaces is easily evaluated using penetration resistance testing [23]. Their uneven sample profiles must be accommodated through special gaskets or sealing techniques. For zipper closures, a groove can be in the test cell to provide a better seal on the protruding teeth portion of the zipper. In assessing penetration resistance for these items, failure may occur because of:

- Penetration of liquid through stitching holes in seams
- Solvation of seam adhesives
- Degradation of seam tapes or other seam components
- Degradation of materials joining seam, causing lifting of seam tapes or destruction of seam integrity
- Physical leakage of closures

Berardinelli and Cottingham [23] demonstrated the utility of this test on a number of material, seam, and closure samples as shown in Table 7. Understanding how clothing specimens may fail provides insight into identifying protection offered by the overall clothing item. The qualitative nature of the penetration test requires that test operators be familiar with failure modes so that they can correctly assess whether liquid penetration has or has not occurred.

3. Specific Use of ASTM F903

Penetration testing per ASTM F903 provides a test for assessing the barrier performance of materials against liquid chemicals [24]. Though measuring specimen weight change is not required, this testing can also serve to measure material

Table 6 Surface Tensions for ASTM F1001 Liquids

Chemical	Surface Tension (dyn/cm)
Acetonitrile	26.6
Carbon disulfide	31.6
Dichloromethane	27.2
Dimethylformamide	36.2
Ethyl acetate	23.4
Hexane	17.9
Methanol	22.1
Nitrobenzene	43.3
Sodium hydroxide	103.0
Sulfuric acid	55.1
Tetrachloroethylene	31.8
Tetrahydrofuran	26.5
Toluene	27.9

Table 7 Penetration Resistance Data on Selected Materials, Seams, Closures, and Chemicals

Material	Component	Chemical	Result, 5 min	Result, 10 min
Saranex-Tyvek	Fabric	Water	Pass	Pass
		Isooctane	Pass	Pass
		MEK	Pass	Pass
		TCE	Pass	Pass
	Stitched seam	Water	Fail	Fail
	Taped seam	Water	Pass	Pass
		Isooctane	Pass	Fail
		MEK	Fail	Fail
	Zipper with flap	Water	Pass	Fail
		Isooctane	Fail	Fail
PVC coated cotton	Fabric	Water	Pass	Pass
		Isooctane	Pass	Fail
		MEK	Pass	Pass
		TCE	Pass	Pass
	Double-sewn, armored seam	Water	Pass	Pass
		Isooctane	Pass	Fail
		MEK	Pass	Pass
		TCE	Fail	Fail

Note: Based on overall 15-min exposure with first 5 min at ambient pressure and second 10 min at 13.8 kPa (2 psig) using ASTM F903. MEK, methyl ethyl ketone; TCE, trichloroethylene.
Source: Adapted from Ref. 23.

degradation since visual observations are required. Degradation of the material, in turn, may be a primary route for penetration by some chemicals. The difficulty in penetration testing lies in making a clear-cut determination of liquid penetration. Many high-vapor-pressure, low-surface-tension solvents spread thinly over the material and evaporate quickly. Therefore, actual liquid penetration may be difficult to observe even when enhanced by using dyes (Sudan III is recommended for most organic solvents). Still, the test serves as a good indicator of material performance against liquid contact or splashes. Since test length and pressurization periods depend on the selected procedure within the method, pass/fail determinations are discernible and easily define acceptable material–chemical combinations.

While ASTM F903 establishes clear, detailed procedures for measuring material penetration resistance, experience in laboratories dictate careful attention to several areas, which include:

- Problems with visually interpreting penetration
- The influence of contact time and pressurization on liquid penetration

(*a*) *Interpretation of Liquid Penetration.*   As stated earlier, penetration is determined by the visual observations of the test operator. As such, this determination lends itself to the bias and experience of the individual test operator. In most cases, liquid penetration is obvious with the appearance of fine droplets of liquid over the entire specimen surface or from a specific surface defect (Figs. 12 and 13). Alternatively, if a material flaw exists such as a pinhole, liquid may appear at a separate point on the test specimen. The visual detection of penetration may not always be an easy determination. This is based on several potential problems in which liquid penetration may seem to occur but is actually the result of:

- A poor seal between test cell and material
- Wicking of material textiles that may be present on both sides of the material being tested

Figure 12   Overall ASTM F903 penetration test showing three test cells operated in parallel. (Courtesy of TRI/Environmental, Inc.)

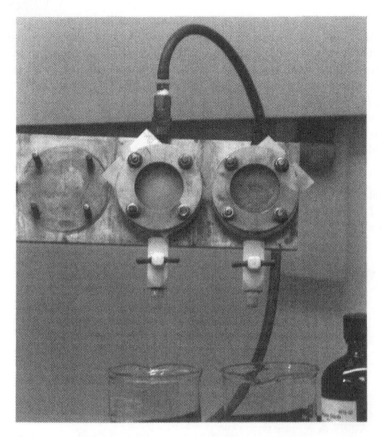

Figure 13 The penetration test cell on the right shows a pinpoint failure in the barrier material as evidenced by discoloration of the material surface. (Courtesy of TRI/Environmental, Inc.)

- Penetration of liquid through some layers of the material specimen giving the appearance that the liquid is at the surface
- Vapor permeation through the sample with condensation of liquid on the Plexiglas shield of material surface

These situations would constitute "false positives." False negatives may occur when penetration occurs but goes undetected by the test operator. This can happen with high-vapor-pressure, low-surface-tension chemicals that spread thinly over the viewing side of the test specimen and evaporate quickly.

A number of practices can be implemented in penetration test programs that eliminate problems in detecting liquid penetration [25]:

- The test method already encourages the use of dyes to enhance visual detection of low-surface-tension, volatile chemicals. Sudan III works well with organic chemicals; red food coloring can be used for inorganic solutions, acids, and bases.
- Small droplets of test chemical can be placed on the inside (viewing) surface of the test material to determine how the chemical might appear when it does penetrate.
- Expanded polytetrafluoroethylene (PTFE) gasket material can be used in lieu of the standard rubber gaskets specified by the method to achieve better specimen sealing.
- For film products with textile fabrics on one or both sides of the material, specimen edges may be dipped in paraffin or wax to seal edges and prevent wicking.
- Blotting paper can be used to help determine if liquid does actually appear on the specimen viewing surface.

Leakage from a poor seal between the test specimen and cell may be evident by close observation of the viewing side of the material during the test. If leakage first occurs at the peripheral edge of the cell, this may be due to a poor seal or damage to the specimen due to compression of the sample at the sealing surface. This phenomenon is most likely to occur when the chemical is first added or at the beginning of any pressurization period. Failures of this nature should require repeating the test but can only be ascertained by careful observation of the test specimen during the test.

ASTM F903 provides the following criteria for interpreting liquid penetration:

- When a droplet of liquid appears or discoloration of the viewing side of the sample, or both; or
- Appearance of liquid or discoloration of the viewing side of the sample due to chemical permeation.

The practices just described help to determine the case of liquid penetration under the first criterion, but additional judgment is necessary when the penetration is the result of significant vapor permeation or penetration. The passage of chemical vapor through the material specimen may appear as condensation on the Plexiglas shield or on the material itself. Occlusion of the Plexiglas shield often inhibits observation of the sample. The Plexiglas shield itself is intended more as a safety feature than as a principal part of the test cell. Since the purpose of penetration testing is to determine material barrier effectiveness to liquids, only the appearance of liquid on the sample should be used for judging failure of the material. This mode of failure is best determined by applying blotting paper to the specimen viewing surface at the end of the test period.

Table 8   ASTM F903 Penetration Test Variations

Procedure	Initial contact period (minutes at 0 kPa)	Pressurization Period (minutes/pressure)	Subsequent contact period (minutes at 0 kPa)
A	5	10/1	None
B	5	10/2	None
C	5	1/2	54
D	60	None	None

(*b*)   *Effect of Contact Time and Pressurization.*   The most recent edition of
ASTM F903 (1995) incorporates four types of contact time and pressure exposure formats (see Table 8). The original protocol consisted of exposing the material to the liquid for a five minute period at ambient pressure, followed by
a 10-min period at 13.8 kPa (2 psig). This exposure condition was selected as a
test pressure to simulate the force on a protective garment of a liquid coming out
a burst pipe at an approximate distance of 3 m. A lower pressure of 6.9 kPa
(1 psig) was adopted later on because many materials would "balloon" away
from the test cell as pressure was applied. Some differences in material performance due to degradation effects have been noted as shown in Table 9 [22]. The
new contact time/pressure formats in ASTM F903 were included to accommodate practices being used by the National Fire Protection Association in their
requirements for chemical protective suit material and component penetration
resistance [26].

The effects of pressure and contact time are important on the outcome of the
test, and test specifiers should realize that differences in material performance can
be expected when different contact time/pressure formats are chosen. How con-

Table 9   Effect of Contact Time and Pressure on Penetration of Selected Material–Chemical Combinations

Material	Chemical	Penetration test exposure protocol	Penetration time (min)
PVC/Nylon	Dimethylformamide	A	None
		B	None
		C	40
		D	50
Microporous film/ nonwoven laminate	Hexane	A	None
		B	5
		C	5
		D	None

tact time and pressure affect the test is dependent on the potential modes of failure as follows:

- Increased contact time allows for increased degradation of material samples.
- Increased pressure is more likely to enable detection of material imperfections in films, penetration of low-surface-tension liquids in textiles or breathable films, and location of construction or design flows in suit components (seams and closures).

Many unsupported film samples cannot be tested at high pressure since they burst when the pressure is applied. Severe "ballooning" of these test specimens in testing may also give questionable test results due to the relatively large forces placing the material in tension. In these cases, the true barrier properties of the material to liquid penetration are not tested. For this reason, the optional use of a screen having more than 50% open area has been specified in the latest form of the test method.

IV. PERMEATION RESISTANCE

Permeation involves a process in which chemicals move through a material at a molecular level. In the permeation process, chemical is first adsorbed on the external or exposure side of the material, then diffuses through the material, and finally desorbs from the other surface (interior or nonexposure side) as shown in Figure 14. As a consequence, the permeation process involves two fundamental modes of interaction with the material: the solubility of the chemical(s) in the

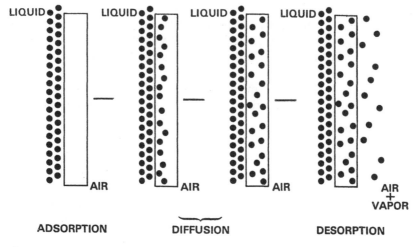

Figure 14 Illustration of the permeation process.

material, and the propensity for the chemical(s) to diffuse through the material. In some cases, chemical solubility may be of little importance, such as in the transmission of an inert gas through a thin polymer film. However, many chemicals do interact with the barrier materials, and material–chemical solubility is a significant factor for chemical permeation resistance [27].

Material permeation resistance is generally characterized using two test results: breakthrough time and permeation rate. Breakthrough time is the time when a chemical is first detected on the "interior" side of the material. As is discussed later, its determination is strongly dependent on how the test is configured and the sensitivity of the detector. Permeation rate is a measure of the mass flux through a unit area of material for a unit time. Permeation rate is most commonly expressed in units of micrograms per square centimeter per minute (μg/cm^2 min). For a given material–chemical combination, the steady-state or maximum observed permeation rates are reported.

Understanding permeation results requires some knowledge of the theory and factors that define material–chemical permeation behavior as well as the different approaches that can be used for its measurement.

A. Permeation Theory

The permeation process can be modeled using the combination of theory for solubility and theory for diffusion. From solubility theory, two substances are soluble if, upon mixing, the free energy of the mixture is less than the sum of the free energies of the two pure substances. The free energy of mixing is defined as

$$\Delta G_m = \Delta H_m - T\,\Delta S_m \tag{5}$$

where ΔG_m is the free energy of mixing, ΔH_m is the enthalpy of mixing, T is the temperature, and ΔS_m is the entropy of mixing. The significance of this equation is that enthalpy mixing term (ΔH_m) must be relatively small in order to have a negative difference in the free energy. The larger the negative difference, the better the solubility.

As previously explained, the molecular forces holding a liquid together include dispersion, polarity, and hydrogen bonding as governed by the relationship

$$C^2 = D^2 + P^2 + H^2 \tag{6}$$

where C is the cohesive energy density, D is the dispersion parameter, P is the polar parameter, and H is the hydrogen bonding parameter [28]. The square root of the cohesive energy density is called the one-dimensional, or total, solubility parameter. The three energy components taken as a vector are commonly referred to as the three-dimensional solubility parameter. Holcombe [14], Henriksen [11], Perkins and Tippet [16], and Bentz and Billing [15] have applied solubility theory to the prediction of chemical permeation through homogeneous materials. Use of solubility theory alone to characterize permeation behavior of more complex materials has been less successful [29].

Diffusion is the random movement of molecules such that, given enough time, the distribution of molecules tends toward even concentration over space [20]. The mathematical equation that describes diffusion is Fick's law:

$$J = -D \frac{\partial C}{\partial X} \tag{7}$$

where

$\quad$ J = rate of transfer per unit area
$\quad$ C = concentration of the diffusion substance
$\quad$ x = distance into the material
$\quad$ D = proportionality constant, called the diffusion coefficient

This equation assumes that rate of transfer through a unit area of material is proportional to the concentration gradient measured normal to the material.

The fundamental, one-dimensional, differential equation of diffusion, which can be derived from Fick's law by considering diffusion into and out of a volume element, is

$$\frac{dC}{dt} = D \frac{\partial^2 C}{\partial x^2} \tag{8}$$

Equations (7) and (8) can be solved with various initial and boundary conditions to obtain explicit relationships for the permeation rate and cumulative mass permeated as a function of time [31]. The simplest set of conditions and assumptions include:

- At time zero, there is no chemical in the material.
- Upon exposure to the chemical, the external surface of the material immediately equilibrates with the chemical. This means that the surface of the material is at its saturation concentration, or solubility, S.
- The concentration of the chemical on the interior side of the material is maintained at essentially zero.
- The material–chemical solubility and diffusion coefficient remain constant (concentration and time independent).
- The material does not swell with chemical.

Under these conditions, the amount of chemical that permeates a material of thickness x at any time t is

$$M = Sx \left\{ \left(\frac{Dt}{x^2} \right) - \left(\frac{1}{6} \right) - 2 \Sigma \left[\frac{(-1)^n}{(n\pi)^2} \right] \exp \left[-(n\pi)^2 \left(\frac{Dt}{x^2} \right) \right] \right\} \tag{9}$$

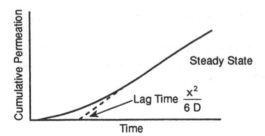

Figure 15   Theoretical plot of chemical concentration versus time for the permeation process. (From Ref. 38.)

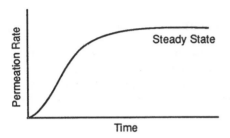

Figure 16   Theoretical plot of permeation rate versus time for the permeation process. (From Ref. 38.)

and the rate of permeation is

$$J = \left(\frac{SD}{x}\right)\left\{1 + 2\sum(-1)^n \exp\left[-(n\pi)^2\left(\frac{Dt}{x^2}\right)\right]\right\} \tag{10}$$

Figure 15 shows the plot of cumulative permeation as a function of time using Eq. (9) whereas Figure 16 represents a plot of permeation rate versus time based on Eq. (10). The curve in Figure 16 is the first derivative of the curve in Figure 15. On Figure 15, the intersection of the slope for steady-state permeation with the x axis defines the lag time (T_l). This extrapolation can be mathematically determined, using the same conditions and assumptions, as follows:

$$T_1 = \frac{x^2}{6D} \tag{11}$$

Breakthrough time is not shown in Figures 15 and 16 because it is a function of several experimental parameters. Its determination is based not only on the test method used, but on the specific parameters of each test.

B. Permeation Resistance Test Methods

The measurement of chemical permeation resistance is specified in different standard test methods offered by ASTM, CEN, and ISO:

American Society for Testing and Materials (ASTM) Test Methods
- ASTM E814, Test Method for Rubber Property—Vapor Transmission of Volatile Liquids
- ASTM D1434, Test Method for Determining Gas Permeability Characteristics of Plastic Film and Sheeting
- ASTM D3985, Test Method for Oxygen Gas Transmission Rate Through Plastic Film and Sheeting using a Coulometric Sensor
- ASTM E96, Test Methods for Water Transmission of Materials
- ASTM F739, Test Method for Resistance of Protective Clothing Materials to Permeation by Liquids and Gases Under Conditions of Continuous Contact
- ASTM F1383, Test Method for Resistance of Protective Clothing Materials to Permeation by Liquids and Gases Under Conditions of Intermittent Contact
- ASTM F1407, Test Method for Resistance of Chemical Protective Materials to Liquid Permeation—Permeation Cup Method

European Community (CEN) Test Method
- EN 369, Protective clothing—Protection against liquid chemicals—Test method: Resistance of materials to permeation by liquids.
- EN 374-3, Protective gloves against chemicals and microorganisms—Part 3: Determination of resistance to permeation by chemicals.

International Standards Organization (ISO) Test Method
- ISO 2556, Plastics—Determination of the gas transmission rate of films and thin sheets under atmospheric pressure—Manometric method
- ISO 6529, Protective clothing—Protection against liquid chemicals—Determination of resistance of air-impermeable materials to permeation by liquids

These test methods may be segregated into two classes: vapor transmission tests and chemical permeation test methods. ASTM D481, ASTM D1434, ASTM D3985, ASTM E96, and ISO 2556 each represent gas or vapor transmission tests. Each of the techniques represented by these standards involve contact of the material's external surface with a gas or chemical vapor. In contrast, ASTM F739, ASTM F1383, ASTM F1407, EN 369, EN 374-3, and ISO 6529 are chemical permeation tests involving either liquid or gaseous chemical contact with the material and assessment of permeation as affected by both chemical solubility and diffusion through the test material.

1. Vapor Transmission Tests

While vapor transmission may involve permeation, it may also be the result of vapor penetration. As a consequence, this testing is applied not only to continuous films but also to microporous films and uncoated textiles. The vapor transmission process is primarily governed by the diffusion of the gas or vapor through the material. Solubility effects are relatively small and in many cases insignificant.

Different techniques are used for the actual measurement of gas transmission rates in each of the preceding tests. ASTM D481 measures the vapor transmission of volatile liquids by placing the liquid in a test jar and allowing the vapor above the liquid to diffuse through a material that is mounted in the top of the jar. Loss of liquid is measured gravimetrically, and vapor transmission rates are reported in milligrams per square meter per second ($mg/m^2/sec$). ASTM E96 provides similar procedures for assessment of water vapor transmission, replacing the liquid with water, but also offers an alternative procedure involving desiccant inside the sealed jar and measurement of desiccant weight gain under carefully controlled temperature and humidity conditions. Both of these methods rely on the vapor pressure of the selected liquid and the relative concentration of the vapor exterior to the test cell as the driving forces for vapor transmission.

ASTM D1434, ASTM D3985, and ISO 2556 involve techniques where a difference is established between the partial pressures of the gas on either side of the membrane (material), usually by applying a slight vacuum on the interior side of the material. Each of these tests is intended for pure gases only and is typically applied in the assessment of packaging material permeation resistance. ASTM D1434 provides procedures for both manometric and volumetric determination of gas transmission. The gas transmission cell for manometric gas transmission testing is shown in Figure 17. ISO 2556 is equivalent to the pressure-based technique in ASTM D1434. ASTM D3985 is intended specifically for measuring oxygen gas transmission and entails the use of a coulometric sensor instead of differences in volume or pressure. Each of these standard test methods provide for measurement of gas transmission rate, permeance, and permeability. Permeance is the ratio of the gas transmission rate to the difference in partial pressure in the two sides of the film. Permeability is the product of the permeance and the film's thickness. The latter measurement is intended for homogeneous materials only.

2. Chemical Permeation Tests

ASTM F739 was first established in 1981 as the first standard test method for measuring material permeation resistance to liquid chemicals [32]. ASTM F739, EN 369, EN 374-3, and ISO 6529 provide standardized procedures for measuring the resistance of protective clothing to permeation by chemicals using continuous contact of the chemical with the material's exterior surface. ASTM F1383 is a variation of ASTM F739 that involves testing under conditions of intermittent chemical contact. ASTM F1407 represents a simplified form of testing where

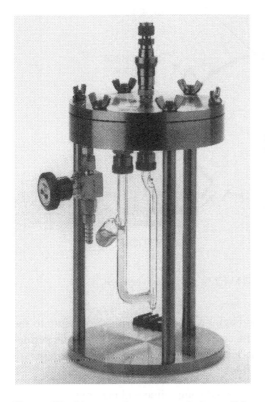

Figure 17 Manometric gas transmission cell for measuring gas transmission rates in accordance with ASTM D1434. (Courtesy of Custom Scientific Instruments).

permeation is determined gravimetrically. Based on its limited sensitivity, this method is primarily used as a screening test or field method.

In each of the tests (except ASTM F1407), a similarly designed test cell is used for mounting the material specimen. The test cell consists of two hemispherical halves divided by the material specimen. One half of the test cell serves as the "challenge" side where chemical is placed for contacting the material chamber. The other half is used as the "collection" side, which is sampled for the presence of chemical permeating through the material specimen. Figure 18 shows a conceptual illustration of the permeation test cell.

The basic procedure in each test is to charge chemical into the challenge side of the test cell and to measure the concentration of test chemical in the collection side of the test cell as a function of time. Of principal interest in permeation

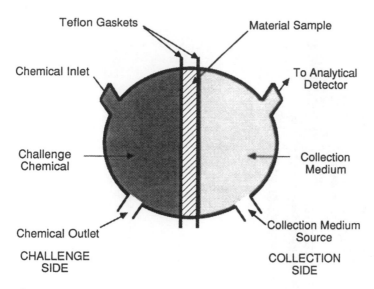

Figure 18   Conceptual diagram of a ASTM F739 permeation test cell.

testing are the elapsed time from the beginning of the chemical exposure to the first detection of the chemical (i.e., the so-called breakthrough time), the permeation rate, and the cumulative amount of chemical permeated. The results reported are dependent on the test method chosen:

- ASTM F739 requires reporting of breakthrough time and maximum or steady state permeation rate.
- ASTM F1383 specifies reporting breakthrough time and cumulative permeation.
- ASTM F1407 permits reporting either cumulative permeation or breakthrough time and permeation (maximum or steady state).
- EN 369 and ISO 6529 require reporting of breakthrough time with the total cumulative mass permeated at 30 and 60 min.

Other significant differences exist between the already listed test methods as described in the following section. Table 10 provides a comparison of key characteristics for each of the different permeation test methods.

C. Parameters Affecting Permeation Resistance Testing

Although the permeation test procedure is simple in concept and generalized procedures are specified by each of the test methods already listed, a number of significant variations exist in the manner in which permeation testing can be conducted. These variables include:

Table 10 Differences Among Permeation Test Methods

Test method	Chemicals permitted	Type of contact	Collection medium flow rate(s)	Minimum test sensitivity ($\mu g/cm^2$ min)	Test results reported
ASTM F739	Liquids and gases	Continuous	50–150 ml/min	0.1	Breakthrough time Permeation rate
ASTM F1383	Liquids and gases	Intermittent	50–150 ml/min	0.1	Breakthrough time Cumulative permeation
ASTM F1407	Liquids only	Continuous	Not applicable	$\simeq$20.0*	Breakthrough time Permeation rate Cumulative permeation
EN 374-3	Liquids and gases	Continuous	50–150 ml/min	1.0	Breakthrough time Permeation rate
EN369	Liquids only	Continuous	520 ml/min (gas) 260 ml/min (liquid)	1.0	Breakthrough time Cumulative permeation at 30 and 60 min
ISO 6529	Liquids only	Continuous	520 ml/min (gas) 260 ml/min (liquid)	1.0	Breakthrough time Cumulative permeation at 30 and 60 min

*Depends on analytical balance, exposed specimen surface area, and time interval between measurements.

- The general configuration of the test apparatus
- How the chemical contacts the material specimen in the test cell
- The type of collection medium used and frequency of sampling
- The type of detector and detection strategy used
- The test temperature
- The effect of multicomponent solutions

In addition, a separate class of materials—porous textile materials containing adsorbent particles—must be tested in totally different manner.

The variety of available test techniques and conditions allows several different approaches for conducting permeation testing and can provide different results for testing the same material and chemical combination.

1. Test Apparatus Configuration

The overall test apparatus includes the test cell, chemical delivery system, and collection/detector system. This apparatus should be configured to meet testing needs and accommodate the characteristics of the chemical(s) being tested. Test cells are generally specified by the test method, but alternative designs are available and may be necessary for testing with specific chemicals or chemical mixtures. Likewise, the chemical delivery and collection/detection systems are dependent on the nature of the chemical and the requirements for running the test. Their selection is discussed later in this section. The way that each part of the apparatus is operated comprises the test apparatus configuration. There are two basic modes for configuring permeation test systems: closed-loop or open-loop. Aspects of test apparatus configurations are described next.

(*a*) *Permeation Test Cells.*   ASTM F739 and ASTM F1383 specify a 51-mm (2.0-in) diameter test cell, which is constructed of two sections of straight glass pipe blown to meet the design shown in Figure 19. The two glass sections are joined by flanges on both sides, which are bolted together to provide a seal between the PTFE gaskets, glass sections, and material specimen. Collection and challenge sides of the test cell are configured to handle the types of chemicals being tested. When a liquid challenge is used, a single, relatively large inlet can be used, whereas gas challenges require both an inlet and outlet for charging gas into and out of the challenge chamber. Typically, collection sides of the test cell also contain an inlet and an outlet. When liquid is used as the collection medium, a stirring rod may be placed in one of these ports to affect uniform mixing of the collection medium with permeating chemical. Gas collection media require an inlet for charging gas into the test cell and an outlet that permits sampling near the material surface.

Both of these tests permit alternative test cells, as long as their equivalency can be demonstrated. Some laboratories use a small version of the test cell that

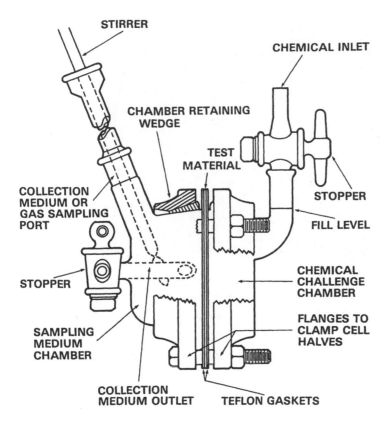

Figure 19 Specification ASTM F739 standard permeation test cell.

has a 25.4-mm (1.0-in) diameter (see Fig. 20). These smaller permeation cells use less chemical and minimize disposal problems, particularly for testing with hazardous chemicals. Billing and Bentz [33] and Henry [34] both investigated differences between the standard ASTM and smaller permeation test cells. Billing and Bentz [33] reported slightly longer breakthrough times for the 25-mm test cell but slightly better precision. Henry [34] concluded that there was no statistical difference in breakthrough times and permeation rates between the two permeation cells. The evaluation by Berardinelli et al. [35] of a different kind of glass permeation cell having a 30-mm exposure diameter compared favorably with the ASTM cell for determination of breakthrough time but yielded less precision in calculated permeation rates. Patton et al. [36] examined a proprietary stainless steel-constructed microcell to determine its equivalency with the ASTM standard cell. They found difficulties in traditional approaches to relate test performance

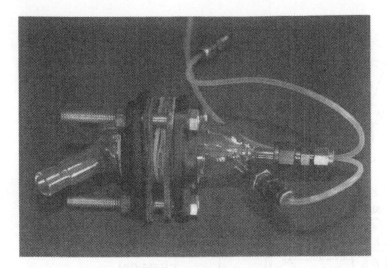

Figure 20 Small version of ASTM F739 permeation test cell having 25mm inner diameter exposure area. (Courtesy of TRI/Environmental, Inc.)

and recommended using nonparametric statistics to evaluate the equivalency of permeation test cells.

While EN 374-3 specifies a similar cell as used in ASTM F739, EN 369 and ISO 6529 specify a 25-mm diameter exposure area with a cylindrical shape that is to be constructed of an "inert" material. Brass is recommended for gaseous collection media, while PTFE or glass is suggested for liquid collection medium. Unlike the ASTM test cell, the challenge side has a loose cover for adding the liquid challenge. This design of the test cell does not permit testing with chemical gases or vapors. Testing by one researcher has shown no significant differences, based on cell design alone, between CEN/ISO and ASTM test methods [37].

The test cell used for ASTM F1407 was mentioned as part of the discussion on degradation testing. This test cell is a shallow cup, usually fashioned out of metal coated with a PTFE film. An outer ring secures the material specimen onto the cup flange by a series of thumb screws. The side opposite the challenge side is open to the atmosphere as the test cell is inverted and placed either on a stand or elevated off of the testing surface [7,8] as depicted in Figure 21.

b. Closed- Versus Open-Loop Permeation Systems
In closed-loop permeation systems, the volume of collection fluid is maintained throughout the test. This volume may be contained fully within the collection chamber or it may be circulated through the chamber, into a nonintrusive detector,

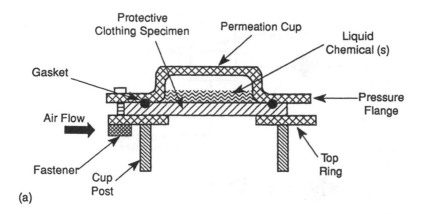

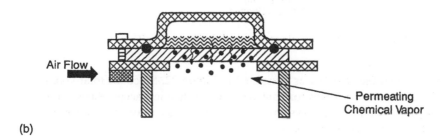

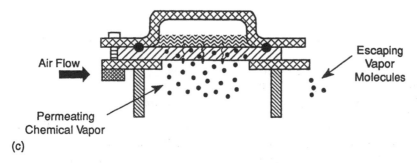

Figure 21 ASTM F1407 permeation test cup: (a) initial setup, (b) chemicals begin to permeate material, and (c) steady-state permeation.

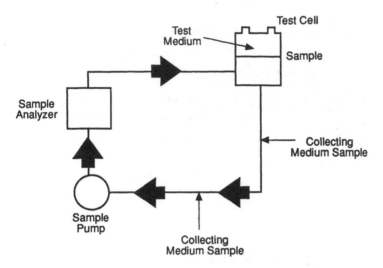

Figure 22   Closed-loop permeation test system configuration. (From Ref. 35.)

and back into the chamber as illustrated in Figure 22. Since the total volume of collection medium remains constant, permeating chemical accumulates within the collection medium. In this system, the permeation rate must account for this accumulation of permeant as follows:

$$\text{Rate} = \frac{(C_n - C_{n-1})}{(t_n - t_{n-1})} \times \frac{V}{A} \qquad (12)$$

where

$(C_n - C_{n-1})$ = change in concentration of the challenge chemical between sampling periods
$(t_n - t_{n-1})$ = time between sampling periods
V = volume of collection medium
A = exposed area of the material specimen

In the open-loop permeation systems, a gas or liquid collection medium is passed through the collection side out of the test cell to the detection system (see Fig. 23). This collection medium stream can be evaluated discretely or continuously depending on the detector selected. Therefore, collection of permeant is specific to the sample taken (over a discrete time period), and permeation rates can be directly calculated as a factor of the collection medium permeant concentration (C) and flow rate (F):

$$\text{Rate} = \frac{C \times F}{A} \qquad (13)$$

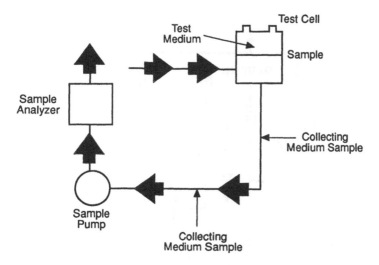

Figure 23 Open-loop permeation test system configuration. (From Ref. 35.)

The choice of a closed- or open-loop system is most often determined by the properties of the chemical and the available detector. Some chemicals such as inorganic substances often require closed-loop systems, particularly if ion-specific electrodes are used that have recovery time constraints. Open-loop testing is preferred for many volatile organic chemicals because these systems can be easily automated as shown in Figures 24 and 25.

Both Henry [34] and Berardinelli et al. [35] examined differences in permeation test results for closed- and open-loop systems. For the small number of material–chemical combinations investigated, both researchers could not find appreciable differences in the collection medium in either breakthrough time or permeation rate. Schwope et al. [38] looked at intrinsic differences between closed- and open-loop systems through a modeling approach based on the use of Fickian equations and assumed values for the diffusion coefficient, solubility parameter, and thickness of a material–chemical combination. In this treatment, they were able to show the effect of several system variables on permeation breakthrough time, including material thickness (assuming homogeneity), specimen exposed surface area, and detector analytical sensitivity.

2. Methods for Chemical Contact

With the exception of ASTM F1383, each of the listed permeation test methods specify testing with neat chemicals under conditions of continuous exposure. Both EN 369 and ISO 6529 accommodate liquid chemicals only. In liquid exposures, the chemical or chemical mixture of interest is placed directly in the challenge por-

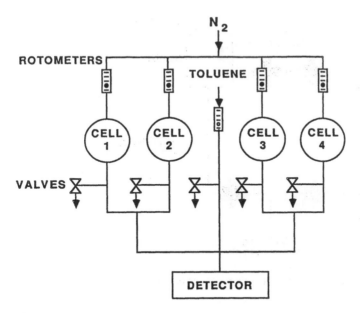

Figure 24   Design of an automated permeation test cell for testing of three samples and "blank" simultaneously. Each test cell is periodically monitored for permeating chemical by pressure switching of collection medium (dry nitrogen) flows to detector. Monitoring of "blank" test cell and toluene standard provide baseline and test system calibration, respectively.

tion of the test cell and left in contact with the material specimen for the selected duration of the test.

ASTM F739 and ASTM F1383 also permit testing with gases, using the modifications to the test cell as described earlier and the test system configuration pictured in Figure 26. When testing gases at 100% concentrations, time zero in the test is established by passing five volumes of the gas through the challenge chamber within a 1-min period. Following this initial period, the flow of gases is reduced to a rate that ensures the concentration of gas in the chamber does not change with time. Special considerations are needed for testing of gases to ensure integrity of the test cell and proper disposal of the effluent challenge gas [39,40].

Permeation testing may also be conducted against vapors of liquid chemicals per ASTM F739. These tests require a high level of temperature control to achieve consistency in the vapor concentration of the chemical and a different orientation of the test cell [39]. Most tests are performed with the chemical at its saturated vapor pressure at the test temperature using a test system configuration as shown in Figure 27. The test cell must be maintained in a horizontal position such that the air space above the liquid becomes saturated with vapor and the liquid does not

Figure 25 Automated permeation system. (Courtesy of TRI/Environmental, Inc.)

contact the material specimen. Unfortunately, tests conducted in this manner may incur a relatively long transition time based on volatility of the test chemical. Table 11 show representative permeation data for two different material–chemical combinations as vapors.

Some research has also been reported for conducting permeation tests with a solid. Lara and Drolet [41] describe a modified test cell were a gel containing nitroglycerin was placed on the material's external surface for permeation testing.

Intermittent forms of chemical contact akin to splash-like exposures are prescribed in ASTM F1383. In this test method, the time of material specimen exposure to chemical is varied in a periodic fashion. Chemical is charged into the challenge side of the test cell and then removed after a specified time. This type of exposure may be repeated in a cyclic fashion. ASTM F1383 suggests three different exposure conditions as shown in Table 12.

The use of intermittent exposure conditions gives rise to permeation curves with a cyclic appearance (see Fig. 28). As a consequence, breakthrough time with cumulative permeation is reported in lieu of permeation rate for these tests. Schwope et al. [42] illustrated this behavior for a number of material–chemical

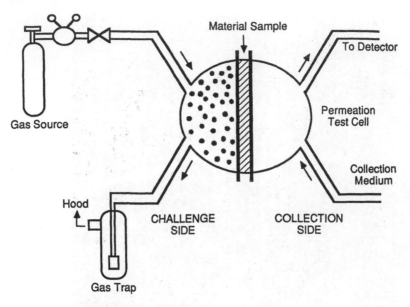

Figure 26 Configuration of permeation test system for evaluating gaseous chemical challenges.

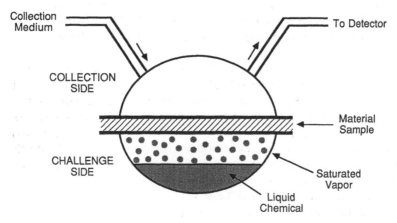

Figure 27 Configuration of permeation test system for evaluating chemical vapor challenges.

Table 11 Permeation Data for Chemical Vapors for Selected Material–Chemical Combinations

Material–chemical combination	Chemical challenge	Breakthrough time (min)	Permeation rate (μ/cm^2 min)
Ethylene dichloride	Saturated vapor at 27°C	4	>25,000
against PVC glove	10 ppm in nitrogen	4	350
Dichloromethane	Liquid	16	470
against Viton–butyl	Saturated vapor at 27°C	28	280
suit material	100 ppm in air	No BT	Not applicable

Note: No BT, No breakthrough observed in a 3 h period for testing per ASTM F739.
Source: Adapted from Ref. 39.

Table 12 ASTM F1383 Recommended Exposure Conditions

Condition	Contact time (min)	Purge time (min)	Number of cycles
A	1	10	10
B	5	10	7
C	15	60	3

combinations and found the cumulative permeation to be proportional to the relative exposure time. Man et al. [43] compared permeation breakthrough times of protective clothing materials against specific chemicals using liquid contact, liquid splashes, and vapors. Their findings showed significant differences between the different exposure conditions for some combinations of materials and chemicals, but lesser changes in breakthrough time for other material–chemical sets. They postulated that the different wetting characteristics of the test materials contributed to this phenomenon indicating those materials that easily wet by a chemical may show similar permeation for liquid splash exposures as with continuous liquid exposure.

3. Types of Collection Media and Frequency of Sampling
Collection media should be chosen to reflect the permeating chemical being evaluated with its particular characteristics that affect acceptable levels of detection. Sampling of the collection medium is dependent on the chosen detector.

(a) Types of Collection Media. The collection medium must have a high capacity for the permeating chemical(s), allow ready mixing, be readily analyzed for the chemical(s) of interest, and have no effect on the clothing material being tested [44]. A collection medium's capacity refers to the relative amount of chemical that can be collected. Collection media with low capacities will hinder the detection of

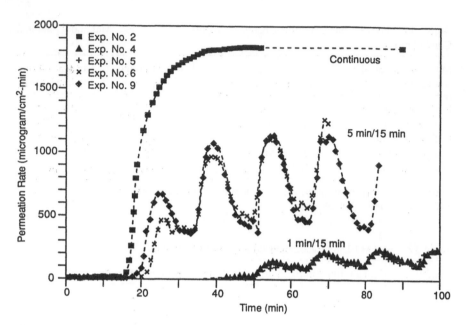

Figure 28 Cyclic permeation observed during permeation test involving intermittent contact. (Adapted from Ref. 42.)

permeation, showing lower than actual permeation rates. Schwope et al. [38] recommended that the concentration of the permeant in the collection medium at the clothing/collection medium interface and in the bulk of the collection medium be maintained below 20% of the solubility of the permeant in the collection medium when the challenge is a neat chemical.

Air, nitrogen, helium, and water are common collection media. In general, these collection media have no effect on the clothing material and are amenable to most analytical techniques. Fricker and Hardy [45] developed a test method involving a saline collection medium to simulate sweat on skin.

In cases where the test chemical has a relatively low vapor pressure, gaseous collection media may have inadequate capacity for the permeant requiring a different choice of collection medium. Similar concerns arise for test chemicals having low water solubility when water is the collection medium. These situations may be addressed by circulating large volumes of fresh gas or water through the collection chamber; however, this practice will dilute permeant concentration in the collection medium and reduce test sensitivity.

Some test chemicals exhibit both low water solubility and vapor pressure. Chemicals with these characteristics generally include higher molecular weight chemicals such as polynuclear aromatics, polychlorinated biphenyls (PCBs), and some pesticides. One approach for conducting permeation tests with these chemicals has been to use solid collection media [42,46]. This technique involves placing a solid, highly absorbent film directly against the material specimen. Ehntholt [46] designed a special test cell successfully using a silicone rubber material for collection of pesticides. Unfortunately, this technique has also been reported to be very labor-intensive, involves multiple replicates to determine breakthrough times, and is subject to cross-contamination [44]. In addition, swelling of the test material can prevent uniform contact between the specimen and the solid collection medium. An alternative approach advocated by Pinnette and Stull [44] and Swearengen et al. [47] has been the use of a liquid splash collection. In these approaches, a solvent medium is briefly contacted with the material specimen on the collection side and the extract evaluated for the chemical(s) of interest. If such an approach is followed, it first must be demonstrated that the solvent does not affect the barrier properties of the test material. For example, the absorbance and back diffusion of the solvent from the collection medium into the clothing material could swell or soften the material and thereby promote more rapid permeation of the challenge chemical. A third approach was used by Spence [48] for permeation testing with halogenated pyridines. His method employed a technique for concentrating the permeant in the collection medium by use of a trap built with the detector gas chromatograph.

(*b*) *Agitation of Collection Media.* Agitation of the collection medium is recommended to ensure that it is homogeneous for sampling and analytical purposes, and to prevent or minimize concentration boundary layers of permeant at the interface of the clothing material and the collection medium. It may or may not be needed, depending on the medium chosen, the capacity of the medium for the test chemical(s), and the nature of the chemical challenge. ASTM F739 specifies a range of flow rates from 50 to 150 ml/min (ISO 6529 specifies a rate of 520 ml/min for gaseous collection media and 260 m/min for liquid collection media). Mixing is particularly important for closed-system testing. In open-loop systems, flow of the collection medium is usually considered acceptable for test systems where the collection medium has high capacities. Agitation of the collection medium is recommended when the concentration of the permeant is above 20% of the partition equilibrium at the clothing/collection medium interface [38]. Testing with multicomponent chemical mixtures may also require agitation to maintain a constant concentration of the challenge at the mixture/material specimen boundary.

(*c*) *Frequency of Sampling.* Permeant sampling of the collection medium may be either continuous or discrete. Some permeation test system configurations permit continuous sampling of permeant in the collection media. In these systems, the permeant concentration or permeation rate can be constantly determined directly, providing the permeation curve for the test system. For closed-loop permeation systems, nondestructive analytical methods must be used for continuous sampling.

Other system configurations or test parameters dictate discrete sampling at periodic intervals dependent on the analytical detector chosen and its ability to recover prior to a new analysis. Alternatively, discrete samples may be taken for later analysis that affect the total number of separate analyses to be performed. Discrete sampling schemes may also be necessary to accommodate test systems that involve multiple test cells operated in parallel. Discrete samples influence the determination of breakthrough time, since breakthrough time must be reported at the time of the previous interval when permeant is found in a sample. For example, if discrete samples are taken at 5, 10, and 15 min and a permeant is detected in the 10-min sample, the permeation breakthrough time is 5 min in following the procedure established in ASTM F739. More frequent sampling may show the permeation breakthrough time to lie sometime between 5 and 10 min.

4. Detection Strategies

The method for monitoring the collection medium for the permeating chemical is selected by the tester. In selecting an analytical detection method, the tester must consider its sensitivity and selectivity as well as its compatibility with the collection medium. Breakthrough time is totally dependent on the sensitivity of the detector. Breakthrough time has been analytically defined as a detector response twice the background level of the system. Background levels can be determined using a "blank" test cell, one that contains material, but not challenge chemical.

(a) *Detector Sensitivity.*   The analytical method by itself may be very sensitive, but the sensitivity of the permeation test can be orders of magnitude less. For example, consider two tests performed with the same analytical instrument and detection limit but with different flow rates of collection media. The same permeation rate will produce a lower concentration of permeating chemical in the stream with the higher flow rate. Breakthrough will be detected at a later time for this test. In the extreme, if the flow rate were very high relative to the permeation rate, it is possible that breakthrough would go undetected. Figure 29 shows how different analytical sensitivities affect breakthrough time.

The detection of chemical breakthrough is highly dependent on the sensitivity of the analytical method. For this reason, ASTM F739 requires that the analytical sensitivity must be reported along with breakthrough time. Reporting of analytical sensitivity alone, however, is insufficient to allow interpretation of test results. The analytical sensitivity of the detector may have little or no relevance to test method sensitivity, which is defined by the analytical sensitivity, the surface area of the clothing material sample, and the collection medium flow rate (open-loop systems) or volume (closed-loop systems).

In a simplistic analysis, the sensitivity of a gravimetric system, such as one described in ASTM F1407, shows the effect of these variables. In this case, the sensitivity of the balance, exposed surface area, and time interval between mea-

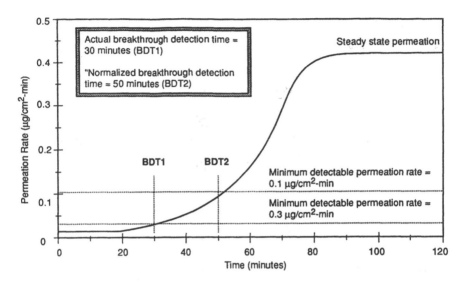

Figure 29 Impact of analytical sensitivity on interpretation of breakthrough time in permeation testing (actual versus normalized breakthrough detection time).

surements directly provide a means for determining the limited sensitivity of the test method using the following equation:

$$\text{Minimum detectable rate} = \frac{W}{(A \times t)} \tag{14}$$

where W is the balance sensitivity in mass unit, A is the material surface exposure area, and t is the time between test measurements. Mickelsen et al. [49] showed good correlation with a gravimetric-based permeation test technique [50] as compared with gas chromatography using an infrared detector.

When analytical test measurements are made as part of the test procedure, more sophisticated procedures must be used to determine the sensitivity of the entire test system. ASTM F739 specifies a technique for measuring minimum detection limits (or minimum detectable permeation rates) based on a modified form of the test method itself. Verschoor et al. [51] developed a technique where a specialized permeation test cell has a collection side with an additional port, and aluminum foil replaces the material sample (see Fig. 30). No chemical is placed on the challenge side. The test chemical is injected near the surface of foil at a known rate using a syringe pump, and the detector response is determined and compared with a known detector calibration gas response. The concentration of chemical injected is successively increased to identify the lower limit of detection. The ratio of the detector response of the test chemical with the detector response of the

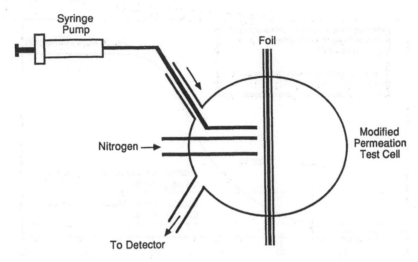

Figure 30   Design of specialized permeation test cell for measuring system detection limits.

calibration gas can then be used in subsequent testing with the selected chemical. This approach requires determining the test sensitivity for each chemical. Figure 31 shows how a blank cell, calibration gas (toluene), and specialized test cell can be integrated into an automated system for measuring minimum detection limits and permeation rates.

(*b*)   *Types of Detectors.*   Various detectors have been applied in tests for measuring the permeation resistance of materials. Since most inorganic chemicals involve some ionic potential, water is used as the collection media with pH meters, ion-specific electrodes, atomic absorption, or ion chromatography used as the detectors. These systems are operated in a closed-loop mode.

Permeation tests involving volatile organic chemicals usually employ gas chromatograph detectors such as thermal conductivity, flame ionization, electron capture, or photoionization detectors. In most cases when gas chromatography is used, it is not necessary to use the column unless mixture permeation studies are being performed. Because these detectors are destructive in their analysis of the sample, they can only be used in an open-loop mode or closed loop if samples are withdrawn without replenishment of the collection medium. Perkins and Ridge [52] first described the use of infrared spectroscopy in permeation tests using a closed system test configuration. The advantage of this system is that it allows continuous recirculation of the collection medium through the detector and test cell. Similar systems have been used by Berardinelli et al. [35].

In tests involving nonvolatile, non-water-soluble chemicals, wipe samples of material interior surfaces, solid collection media, or liquid splash collec-

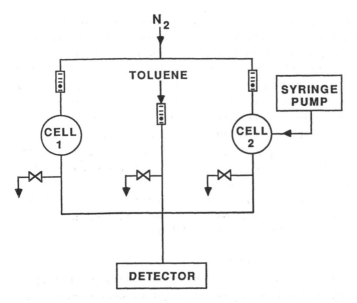

Figure 31   Configuration of permeation test system for measuring system detection limits.

tion aliquots are generally evaluated using either gas chromatography, high-performance liquid chromatography, or mass spectroscopy depending on the collection/extraction solvent used and the analytes being detected. Ehntholt et al. [46] used radiochemical labeling techniques for evaluating pesticide concentrations in isopropanol. Other techniques have been shown viable for difficult-to-evaluate chemicals by using small amount of collection medium [53] with ultraviolet (UV) spectroscopy.

It is important that the detector response remain linear within the range of chemical concentration to be evaluated. In some systems, rapid permeation at high rates can saturate the detector and provide meaningless data. Some laboratories have used sorbent tubes for collecting permeating chemical. This approach can also be used to determine the total or cumulative permeation when an open-loop test system is chosen, and allow for the separation and identification of specific components within challenge mixtures.

5.   Effects of Temperature

Spence [54] first showed significant changes in the permeation resistance of protective clothing materials with increasing temperature as evidenced by shorter breakthrough times and larger permeation rates. Changes in temperature may have an influence on permeation by several mechanisms. Increased temperatures may increase the concentration of the challenge chemical absorbed onto the material surface by increasing the solubility of the material–chemical matrix or by increasing the vapor pressure of the chemical [55] The rate-of-diffusion step in the permeation process may also increase with temperature following an Arrhenius equation type of relationship [54–58]. Temperature, therefore, exhibits its effect on breakthrough time and permeation rate through the diffusion coefficient (D) and solubility (S). The expected effect manifests itself in a logarithmic-like relationship between permeation rate and temperature. Figure 32 shows this relationship for several material–chemical pairs and temperatures. Zellers and Sulewski [59] modeled the temperature dependence of N-methylpyrrilidone through different gloves using Arrenhius corrections to both the diffusion coefficient and solubility originally proposed by Perkins and You [60]. Even small differences in temperature have been shown to significantly affect permeation breakthrough times as shown in Table 13. As a consequence, permeation testing must be performed under tightly controlled temperature conditions.

6. Effect of Multicomponent Challenges

When permeation tests involve multicomponent chemical challenges, test configurations must employ detection techniques that permit the identification of each chemical in the permeating mixture. A number of researchers have investigated the effects of multicomponent chemical mixture permeation through barrier materials. Stampher et al. [61] investigated the permeation of PCB/paraffin oil and 1,2,4-trichlorobenzene mixtures through protective clothing. They used a small amount of isooctane in the collection medium to capture permeating PCBs. Schwope et al. [62] performed extensive testing with pesticides using different active ingredients and carrier solvents. Their tests demonstrated different breakthrough times and proportions of permeating chemicals between pesticide and carrier solvent. Bentz and Man [55] identified a case involving an acetone/hexane mixture where the mixture permeated a dual-elastomer-coated material at shorter breakthrough times than either of the pure components. This testing illustrated the potential synergistic permeation of mixtures. Mickelsen et al. [63] evaluated elastomeric glove materials against three different binary mixtures and found similar permeation behavior where mixture permeation could not be predicted on the basis of the individual mixture components. Ridge and Perkins [64] attempted to model mixture permeation using solubility parameters and found the technique to be only partially successful. Goydan et al. [17] were able to predict mixture permeation using a series of empirical rules when applied to a particular fluoropolymer laminate material.

7. Evaluation of Adsorptive-Base Materials

Some materials may be designed to prevent the penetration or permeation of vapors by using adsorptive components within their structure. In general, these materials are intended for use in environments where only low levels of chemical

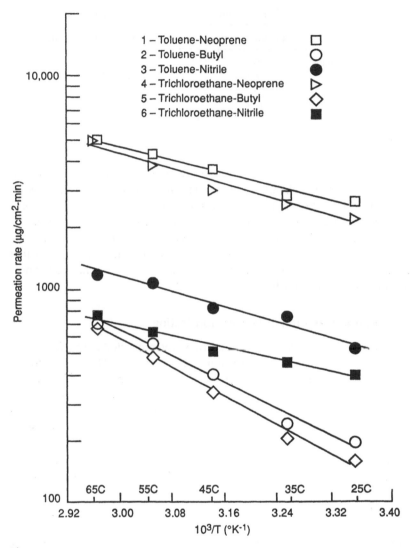

Figure 32 Plot showing effect of temperature on permeation rate for selected material-chemical combinations. (From Ref. 58.)

concentration are to be encountered. The evaluation of these materials may be conducted in a fashion similar to that described earlier for gases or chemical vapors. However, these fabrics are typically evaluated by passing an air stream through the material in a technique similar to that used for assessing the service life of sorbent-based respirator cartridges [65–67]. One such technique is described by

Table 13 Selected Indications of Temperature Effects on Breakthrough Time

Test material	Temperature (°C)	Acetone breakthrough time (min)
Viton/chlorobutyl laminate	20	95–98
	26.5	43–53
Chlorinated polyethylene	22	32–35
	24.5	27–31

Source: Ref. 55

Baars, et al. [68]. Their technique involves measurement of fabric sorption rate and capacity by passing the challenge gas at a constant flow rate and concentration through the material with continuous detection of the effluent chemical concentration. When the concentration of the effluent is normalized to the inlet chemical concentration, the area above the curve for normalized concentration versus time represents the cumulative chemical adsorbed, while the area under the curve is that of chemical that has penetrated the fabric. Fabric saturation is reached as the normalized concentration approaches 1.

D. Use and Interpretation of Permeation Testing

Of the chemical resistance data used in reporting protective clothing performance, the vast majority of test results are permeation resistance data. These data accompany most product performance data sheets and are provided in a number of data compilations [69–72]. However, a review of this data generally indicates that consistent comparison cannot be made and that suppositions for discerning material performance may not be properly based unless an understanding of the test conditions is realized. Much of the data is reported generically for material classes. Yet Michelsen and Hall [73] showed significant differences in chemical permeation through elastomers that were generically the same in composition and thickness. This illustrates that permeation data must be specific to the material–chemical combination being evaluated.

1. Reporting of Permeation Data

As originally indicated, breakthrough time and steady-state or maximum permeation rate are typically provided as permeation test data. ASTM F739 as well as CEN and other test methods also require reporting of key test parameters. In general, these include a complete description of the test material, test chemical, and test system configuration. Table 14 lists test parameters that should be reported with each test. ASTM F1194, Guide for Documenting the Results of Chemical Permeation Testing on Protective Clothing Materials, provides a more extensive list of testing reporting requirements.

Table 14 Permeation Test Report Parameters

Area	Reporting requirements
Test chemical	Components
	Concentration
	Source
Test material	Identification
	Source
	Condition at time of testing
	Thickness
	Unit area weight
Test system	Overall configuration (open or closed loop)
	Type of test cell
	Type of challenge (continuous or intermittent)
	Collection medium
	Collection medium flow rate
	Detector or analytical technique
Test results	Breakthrough time
	Normalized breakthrough time
	Test system sensitivity
	Steady-state or maximum permeation rate
	Cumulative permeation

Since sensitivity significantly affects breakthrough times, ASTM, CEN, and ISO have adopted reporting requirements that are intended to normalize the effect of test system parameters on this measurement. Currently, ASTM F739 and ASTM F1383 specify reporting of the "normalized" breakthrough time in addition to actual breakthrough time. Normalized breakthrough time is defined as the time when the permeation rate is equal to 0.10 $\mu g/cm^2$ min. International and European test methods specify reporting breakthrough times as the rate equals 1.0 $\mu g/cm^2$ min, a single order of magnitude difference from U.S. test methods. Therefore, it is important that permeation breakthrough time data be only compared when the respective sensitivities of the test laboratories are the same or if data is normalized on the same basis [74].

2. Interpretation of Permeation Data

Permeation resistance testing is the appropriate test when vapor protection is required. This does not mean that the test can only be applied for gas or vapor challenges, but rather that the test discriminates among chemical hazards at a molecular level owing to the sensitivity for detecting permeating chemical in its vapor form (as opposed to liquids or solids). As such, permeation testing represents the most rigorous of chemical resistance test approaches.

Within the protective clothing industry, many end users judge the acceptability of a material on the basis of how its breakthrough time relates to the expected period of exposure. Reporting of permeation rate offers a more consistent and reproducible means of representing material permeation. The inherent variability and test system dependence on breakthrough times make these data a less then satisfactory choice for characterizing material performance. Permeation rate data can be used to show subtle changes in material characteristics and determine cumulative (total) permeation when acceptable "dose" levels of the test chemical can be determined. On the other hand, some material-chemical systems take a long time to reach steady state or exceed the capacity of the detector. In addition, the lack of widespread data on acceptable dermal exposure levels for most chemicals leads many specifiers to rely on breakthrough times exclusively.

The flexibility of most permeation tests allows testing laboratories to choose those conditions that best represent the expected performance of the material. Usually, the primary decisions in specifying permeation test involve the following:

- The chemical and its concentration
- The state and periodicity for contacting the chemical with the material
- The material and its condition prior to exposure
- The environmental conditions of the exposure
- The length of the test
- Sensitivity of the test system

The majority of permeation tests in the protective clothing industry are conducted using neat chemical continuously contacting pristine material at room temperature for a period of 8 h. Test sensitivities are at $0.10 \ \mu g/cm^2$ min or better but may be higher for difficult-to-evaluate chemicals. Other barrier materials are generally evaluated against chemicals for longer period of times at slightly elevated temperatures for examining steady-state permeation rates and cumulative permeation. These test conditions are considered worst case, because constant contact of the material with the chemical is maintained, which may or may not be representative of actual use. When specific barrier product applications are identified, it is best to model the conditions of use through the selection of test parameters. If general performance is to be determined, using industry practices for test setup is preferred so that material performance may be compared against other available data.

V. RECOMMENDED TESTING APPROACHES

As products differ, so must the test strategies that are designed to evaluate barrier material characteristics of these products. Obviously, different tests are needed to evaluate different products based on how they are designed, what performance is intended, their required durability, and the expected application. The requirements for conducting barrier testing arise from several needs:

- Demonstrating product distinctive advantages (marketing)
- Meeting customer needs or demands
- Complying with appropriate standards
- Determining the viability of new products
- Documenting product quality control

While quality design and fabrication of product clothing items are paramount to offering a good product line, the majority of testing efforts are directed to evaluating the materials used in the construction of barrier products. This is partly because whole product performance may be difficult to assess in simulated "use" tests.

The use of standardized test methods provides advantages over in-house procedures. The latter procedures are often simple in design, done internally, and not always reproducible. Manufacturers tend to choose those methods that best represent their product's performance. Comparison of product performance on this basis is virtually impossible. Acceptance and use of ASTM and other recognized standards overcomes these problems and helps end users in their evaluation of product performance.

A.   Selection of Test Methods to Characterize Barrier Methods

The appropriateness of a specific chemical resistance test is dependent on the product's application and expected performance. The test selected should also consider the nature of the material to be evaluated. Some materials should not be evaluated in certain barrier tests, because the methods do not allow for discriminating their performance [75].

- Degradation testing may show how product materials deteriorate or are otherwise affected, but will not always demonstrate retention of barrier characteristics with respect to specific chemicals. Degradation testing is most useful when retention of specific physical properties (e.g., strength) is desired or as a screening technique for other chemical resistance (barrier) tests. This type of testing may be applied to all types of materials.
- Runoff-based penetration testing should only be used if the wetting or repellency characteristics of material surfaces are to be evaluated. Like degradation testing, runoff-based penetration testing does not offer an adequate assessment of material barrier performance. This testing approach can be used with all types of materials, but is best applied to textile materials or lightly coated fabrics for determining surface finish characteristics.
- Hydrostatic-based penetration tests are designed to evaluate water or chemical barrier performance of materials. Only ASTM F903 allows testing with chemicals other than water. As such, this test is appropriate for the evaluation of material performance against liquid chemicals and can be used to distinguish adequate chemical resistance for specific chemicals. This testing easily accommodates microporous and continuous film-based materials.

- Vapor transmission tests can be used for chemical vapor or gas challenges only for measuring gross vapor penetration or permeation over relatively long periods of time. Measurement of vapor transmission rates applies to any film-based materials, although some tests only apply to homogeneous films. Adsorbent-based textile fabrics or films can also be tested using this technique.
- Chemical permeation testing provides an assessment of a barrier material's total chemical resistance, permitting the measurement of relatively small amounts of permeating chemical. This test is best suited when only extremely small levels of chemical are permitted to pass through a material. Permeation testing should be employed for any type of continuous barrier material.

In the protective apparel industry, the decision to apply permeation or penetration data therefore requires a careful review of:

- The hazards associated with the chemical
- The intended duration of exposure
- The work environment

There are many cases where materials that resist liquid penetration are suitable for working environments, particularly when chemical exposure is unlikely and there is relative little hazard from wearer contact with chemical vapors. Some sample situations include working with dilute acids and bases. Having both penetration and permeation data for specific clothing products and chemicals in combination with a thorough knowledge of the chemical hazards and working environments allows the safety specialist to choose protective clothing that provides the needed level of performance. Table 15 provides sample data for three different materials including a microporous laminate. This testing illustrates how dissimilar materials can provide significantly different performance against a standard set of chemicals.

A strategy for testing a product will typically involve choosing the test method and establishing the specific conditions for the test. Many barrier tests are conducted under room temperature and humidity. Vapor transmission and permeation tests are very sensitive to these conditions, requiring a controlled environment for testing. In most cases, the duration of the test is set by the procedure, but degradation and permeation testing periods are not always specified. For these tests, test duration should be set to the maximum period of expected product use and chemical exposure. No studies have established conditions where accelerated testing provides good correlation to full-duration testing.

Test materials may also be subjected to various preconditioned prior to testing, especially for barrier testing. Preconditions may involve certain exposures intended to simulate product use such as abrasion, flexing, cleaning, or heat aging. Use of preconditioning may help to identify product barrier performance more consistent with use expectations.

Table 15 Comparison of Penetration and Permeation Resistance for Representative Liquid Splash-Protective Barrier Materials

Chemical	PVC/nylon			Saranex/Tyvek laminate			Microporous film/ nonwoven laminate		
	F903(C) Result	F739 B.T.	F739 P.R.	F903(C) Result	F739 B.T.	F739 P.R.	F903(C) Result	F739 B.T.	F739 P.R.
Acetone	Pass	8	>50	Pass	28	3.4	Pass	<4	>50
Acetonitrile	Pass	12	25	Pass	88	0.27	Pass	<4	>50
Carbon disulfide	Fail (6)	4	>50	Pass	4	>50	Pass	<4	>50
Dichloromethane	Fail (6)	4	>50	Pass	4	>50	Pass	<4	>50
Diethylamine	Fail (20)	8	>50	Pass	20	20	Pass	<4	>50
Dimethylformamide	Fail (40)	28	>50	Pass	72	1.8	Pass	<4	>50
Ethyl acetate	Pass	8	>50	Pass	20	1.5	Pass	<4	>50
Hexane	Fail (40)	20	8	Pass	None	N/A	Pass	<4	>50
Methanol	Fail (55)	16	13	Pass	None	N/A	Pass	<4	>50
Nitrobenzene	Pass	32	>50	Pass	120	6.0	Pass	<4	>50
Sodium hydroxide	Pass	None	N/A	Pass	None	N/A	Pass	<4	>50
Sulfuric acid	Pass	120	6	Pass	128	1.3	Pass	<4	>50
Tetrachlorethylene	Fail (30)	16	>50	Pass	4	>50	Pass	<4	>50
Tetrahydrofuran	Pass	8	>50	Pass			Pass	<4	>50
Toluene	Fail (25)	12	>50	Pass	24	40	Pass	<4	>50

Note: B.T., breakthrough time in minutes; P.R., permeation rate in $\mu g/cm^2$ min; N/A, not applicable. Penetration results provided as pass or fail with penetration in parentheses; permeation tests per ASTM F739 at ambient temperature for 3 h.
Source: Ref. 25.

B. Selection of Test Chemicals

Manufacturers may be faced with a myriad of chemicals to select for testing their products. If the products are intended for broad applications among several markets, these decisions become increasingly difficult. In the past, end users in some industries were unable to compare product performance, since all manufacturers tested their materials with different sets of chemicals. The ASTM F-23 Committee sought to overcome this problem in the protective clothing industry by devising a standard battery of chemicals that represented the following [76]:

- A wide range of chemical classes
- High volume usage in the chemical industry
- Varying levels of toxicity
- Aggressive interactions with most materials

The result was a 15-liquid-chemical battery established in ASTM Standard Guide F 1001 during 1986. This battery was expanded in 1989 to include six gases as well. The standard guide permits testing groups to test either the liquid battery, gas battery, or both, as applicable. These chemicals and relevant properties are listed in Table 16. Spence [77] developed procedures that permit simultaneous permeation testing with the 13 liquid organic chemicals in the battery as a means of quickly screening material performance.

VI. CONCLUSIONS

Various chemical barrier tests are available for establishing the performance of textile and related materials against specific chemicals. Of the three different chemical resistance approaches available, only penetration and permeation testing provide an assessment of the material's barrier performance in preventing the passage of chemical through the material. Degradation testing, while useful for examining chemical effects on materials, cannot provide information that assures barrier performance of materials. Penetration testing is usually qualitative, and the test conditions must be tailored to emulate the expected exposure conditions for discriminating material performance. While penetration testing is suitable for establishing the liquid barrier of materials, it takes consistency in operator interpretations and techniques to obtain precision in test results. It is also important to distinguish between tests that evaluate liquid repellency and those that measure liquid penetration as a result of applied pressure. Permeation testing provides the most rigorous of all chemical resistance test methods, and several techniques are available to provide flexibility in test conditions and applications. Permeation testing is suitable for establishing the vapor-barrier performance of materials. Taken collectively, the test methods described in this chapter provide a number of tools to characterize the barrier performance of materials. As with any testing, the

Table 16 ASTM F1001 Chemicals and Key Properties

Chemical	Class	Molecular weight	Vapor pressure (mm Hg)	Molar volume (cm³/mol)	Specific gravity	TLV (ppm)*
Acetone	Ketone	58	266	74.0	0.791	750
Acetonitrile	Nitrile	41	73	53.0	0.787	40
Ammonia	Inorganic gas	17	>760	—	N/A	25
1,3-Butadiene	Alkene	54	910	87.0	N/A	10, cancer
Carbon disulfide	Sulfur compound	76	300	62.0	1.260	10, skin
Chlorine	Inorganic gas	70	>760	—	N/A	0.5
Dichloromethane	Halogen compound	85	350	63.9	1.336	50, cancer
Dimethylformamide	Amide	73	2.7	77.0	0.949	10, skin
Ethyl acetate	Ester	88	76	99.0	0.920	400
Ethylene oxide	Heterocyclic	44	>760	—	N/A	1, cancer
Hexane	Aliphatic	86	124	131.6	0.659	50
Hydrogen chloride	Inorganic gas	37	>760	—	N/A	5
Methanol	Alcohol	32	97	41.0	1.329	200, skin
Methyl chloride	Halogen compound	51	>760	—	N/A	50, skin
Nitrobenzene	Nitro compound	123	<<1	102.7	1.203	1, skin
Sodium hydroxide	Inorganic base	40	≈0	N/A	2.130	2 mg/m³
Sulfuric acid	Inorganic acid	98	<0.001	N/A	1.841	1 mg/m³
Tetrachloroethylene	Halogen compound	166	14	101.1	1.631	25, cancer
Tetrahydrofuran	Ether	72	145	81.7	0.888	200
Toluene	Aromatic	92	22	106.8	0.866	50, skin

*Threshold Limit Value (TLV): maximum concentration of chemical permitted in air for 8 hour exposure without toxic health effects.

careful selection of the test and its parameters depends mostly on the understanding of the test product (or material) and its application.

APPENDIX A: REFERENCED EDITIONS OF STANDARD TEST METHODS

American Association of Textile Chemist and Colorist (AATCC) Test Methods

AATCC 42, Water Resistance: Impact Penetration Test, 1989
AATCC 118, Oil Repellency: Hydrocarbon Resistance Test, 1989
AATCC 127, Water Resistance: Hydrostatic Pressure Test, 1989

American Society for Testing and Materials (ASTM) Test Methods

ASTM D471, Test Method for Rubber Property—Effect of Liquids, 1991
ASTM D543, Test Method for Resistance of Plastics to Chemical Reagents, 1987
ASTM D751, Methods for Testing Coated Fabrics, 1989
ASTM D814, Test Method for Rubber Property—Vapor Transmission of Volatile Liquids, 1991
ASTM D1003, Test Method for Haze and Luminous Transmission of Transparent Plastics, 1992
ASTM D1434, Test Method for Determining Gas Permeability Characteristics of Plastic Film and Sheeting, 1992
ASTM D1683, Test Method for Failure of Sewn Seams of Woven Fabrics, 1990
ASTM D1975, Practice for Environmental Stress Crack Resistance of Plastic Injection-Molded Open-Head Pails, 1991
ASTM D2062, Test Methods for Operability of Zippers, 1992
ASTM D2240, Test Method for Rubber Property—Durometer Hardness, 1991
ASTM D3132, Test Method for Solubility Range of Resins and Polymers, 1990
ASTM D3393, Specification for Coated Fabrics—Waterproofness, 1991
ASTM D3985, Test Method for Oxygen Gas Transmission Rate Through Plastic Film and Sheeting Using a Coulometric Sensor, 1988
ASTM E96, Test Methods for Water Transmission of Materials, 1994
ASTM F88, Test Methods for Seal Strength of Flexible Barrier Materials, 1985
ASTM F739, Test Method for Resistance of Protective Clothing Materials to Permeation by Liquids and Gases Under Conditions of Continuous Contact, 1991
ASTM F903, Resistance of Protective Clothing Materials to Penetration by Liquids, 1995
ASTM F1001, Guide for Selection of Liquid and Gaseous Chemicals to Evaluate Protective Clothing Materials, 1993
ASTM F1194, Guide for Documenting the Results of Chemical Permeation Testing on Protective Clothing Materials, 1994

ASTM F1383, Test Method for Resistance of Protective Clothing Materials to Permeation by Liquids and Gases Under Conditions of Intermittent Contact, 1992

ASTM F1407, Test Method for Resistance of Chemical Protective Materials to Liquid Permeation—Permeation Cup Method, 1992

European Community (CEN) Test Methods

EN 368, Protective clothing—Protection against liquid chemicals—Test Method: Resistance of materials to penetration by liquids, 1992

EN 369, Protective clothing—Protection against liquid chemicals—Test method: Resistance of materials to permeation by liquids, 1993

EN 374-3, Protective gloves against chemicals and microorganisms—Part 3: Determination of resistance to permeation by chemicals, 1994

International Standards Organization (ISO) Test Methods

ISO 2556, Plastics—Determination of the gas transmission rate of films and thin sheets under atmospheric pressure—Manometric method, 1974

ISO 6529, Protective clothing—Protection against liquid chemicals—Determination of resistance of air-impermeable materials to permeation of liquids, 1990

ISO 6530, Protective clothing—Protection against liquid chemicals—Determination of resistance of materials to penetration by liquids, 1990

ISO 8096, Rubber- or plastics coated fabrics water-resistant clothing—Specification—Part 1: PVC-coated fabrics; Part 2: Polyurethane- and silicone elastomer-coated fabrics, 1989

U.S. Federal Government Test Methods (FTMS)

FTMS 191A,5512—Water Resistance of Coated Cloth; High Range, Hydrostatic Pressure Method, 1990

FTMS 191A,5514—Water Resistance of Coated Cloth; Low Range, Hydrostatic Pressure Method, 1990

FTMS 191A,5516—Water Resistance of Cloth; Water Permeability. Hydrostatic Pressure Method, 1990

FTMS 191A,5520—Water Resistance of Cloth; Drop Penetration Method, 1990

FTMS 191A,5522—Water Resistance of Cloth: Water Impact Penetration Method, 1990

FTMS 191A,5524—Water Resistance of Cloth; Rain Penetration Method, 1990

REFERENCES

1. J. O. Stull, M. G. Steckel, and R. E. Jamke, Evaluating a new material for use in totally-encapsulating chemical protective suits, *Performance of Protective Clothing: Second Symposium, ASTM STP 989*(S. A. Mansdorf, R. Sager, and A. P. Nielsen, eds.), American Society for Testing and Materials, Philadelphia, 1988, pp. 847–861.

2. *Introducing Gore-Gard Fabric*, Promotional literature, W. L. Gore & Associates, Elkton, MD, 1991.

3. D. W. Jones and P. Watts, Kinetics of vapor sorption by latex-bonded carbon particles, *Performance of Protective Clothing: Second Symposium, ASTM STP 989* (S. Z. Mansdorf, R. Sager, and A. P. Nielsen, eds.), American Society for Testing and Materials, Philadelphia, 1988, pp. 832–846.

4. J. O. Stull, Performance standards and testing of chemical protective clothing, Proceedings for Clemson University's Sixth Annual Conference on Protective Clothing, Greenville, SC, 1992.

5. NASA Test Procedure for Hypergol Compatibility Testing of Materials, MTB-175-88, NASA Kennedy Space Center, FL, 1988.

6. G. C. Coletta, S. Z Mansdorf, and S. P. Berardinelli, Chemical protective clothing standard test method development: Part II. Degradation test method. *Am. Ind. Hyg. Assoc. J. 49*:26–33 (1988).

7. A. D. Schwope, T. R. Carroll, R. Haung, and M. D. Royer, Test kit for evaluation of the chemical resistance of protective clothing, *Performance of Protective Clothing: Second Symposium, ASTM STP 989* (S. Z. Mansdorf, R. Sager, and A. P. Nielsen, eds.), American Society for Testing and Materials, Philadelphia, 1988, p. 314.

8. J. O. Stull, J. H. Veghte, and S. B. Storment, An evaluation of a permeation field test kit to aid selection of chemical protective clothing, *Performance of Protective Clothing: Fourth Volume, ASTM STP 1133* (J. P. McBriarity and N. W. Henry, eds.), American Society for Testing and Materials, Philadelphia, 1992, p. 382.

9. J. O. Stull, Early development of a hazardous chemical protective ensemble, *Final Report CG-D-24-86*, AD A174,885, Washington, DC, 1986.

10. C. J. Bryan and M. D. Hampton, A method to determine propellant handlers ensemble fabric degradation, *Chemical Protective Clothing Performance in Chemical Emergency Response, ASTM STP 1037* (J. L. Perkins and J. O. Stull, eds.), American Society for Testing and Materials, Philadelphia, 1989, pp. 185–194.

11. H. R. Henriksen, Selection of materials for protective gloves, polymer membranes to protect against epoxy products, *Danish National Labor Inspection Service*, Lynby, Copenhagen, 1982.

12. C. M. Hansen, Diffusion in polymers, *Polym. Eng. Sci. 20*:252–258 (1980).

13. C. M. Hansen, The three dimensional solubility parameter—Key to paint component affinities: I. Solvents, plasticizers, polymers, and resins, *J.Paint Tech. 39*:104–117 (1976).

14. A. Holcombe, Use of solubility parameters to predict glove polymer permeation by industrial chemicals, Master's thesis, University of Alabama at Birmingham, 1983.

15. A. P. Bentz and C. L. Billing, Jr., Determination of solubility parameters of new suit materials, *Performance of Protective Clothing, Second Symposium, ASTM STP 989* (S. Z. Mansdorf, R. Sager, and A. P. Nielsen, eds), ASTM, Philadelphia, 1988, p. 209.

16. J. L. Perkins and A. Tippitt, Use of three-dimensional solubility parameter to predict glove permeation, *Am. Ind. Hyg. Assoc. 46*:455 (1985).

17. R. Goydan, J. R. Powell, A. P. Bentz, and C. B. Billing, Jr., A computer system to predict chemical permeation through fluoropolymer-based protective clothing, *Performance of Protective Clothing: Fourth Volume, ASTM STP 1133* (J. P. McBriarity and N. W. Henry, eds.), American Society for Testing and Materials, Philadelphia, 1992, pp. 969–971.

18. J. O. Stull, Limitations of chemical permeation modeling as applied to protective clothing performance evaluations, Paper presented at 1990 American Industrial Hygiene Conference, Orlando, FL, 1990.

19. J. O. Stull, The direction of protective clothing testing in the 1990's, *Proceedings of IFIA High Performance Fabrics Conference*, Industrial Fabrics International Association, St. Paul, MN, 1990.

20. G. J. van Amerongen, Diffusion in elastomers, *Rubber Chem. Technol. 37*:1065 (1964).

21. C. R. Dodgen, TRI/Environmental, Inc. Internal Test Report, August 1994.

22. S. Z. Mansdorf and S. P. Berardinelli, Chemical protective clothing standard test method development, Part 1. Penetration method, *Am. Ind. Hyg. Assoc. J. 49*:21–25 (1988).

23. S. P. Berardinelli and L. Cottingham, Evaluation of chemical protective garment seams and closures for resistance to liquid penetration, *Performance of Protective Clothing, ASTM STP 900* (R. L. Barker and G. C. Coletta, eds.), American Society for Testing and Materials, Philadelphia, 1986, pp. 263–275.

24. J. O. Stull and D. F. White, Selecting chemical protective clothing on the basis of overall integrity and material performance tests, *Am. Ind. Hyg. Assoc. J. 53*:455–462 (1992).

25. J. O. Stull, D. F. White, and T. C. Greimel, A comparison of the liquid penetration test with other chemical resistance tests and its application in determining the performance of protective clothing, *Performance of Protective Clothing Fourth Volume, ASTM STP 1133* (J. P. McBriarity and N. W. Henry, eds.), American Society for Testing and Materials, Philadelphia, 1992, pp. 123–138.

26. J. O. Stull, Performance standards for improving chemical protective suits, *Chemical Protective Clothing Performance in Chemical Emergency Response, ASTM STP 1037* (J. L. Perkins and J. O. Stull, eds.), American Society for Testing and Materials, Philadelphia, 1989, pp. 245–264.

27. J. L. Perkins, Solvent–polymer interactions, *Chemical Protective Clothing*, Vol. 1 (J. S. Johnson and K. J. Anderson, eds.), American Industrial Hygienists Association, Akron, OH, 1990, p. 40.

28. C. M. Hansen, The universality of the solubility parameter, *IEC Prod. Res. Dev. 8*:2 (1968).

29. J. O Stull, *Development of a Model to Predict Permeation Properties of Chemical Warfare Protective Dive Suite Materials*, Final Report, Contract N60921-86-D-A142, Del. Order No. 15, Naval Coastal Systems Center (1989).

30. J. Crank and G. S. Park, Methods of measurement, *Diffusion in Polymers* (J. Crank and G. S. Park, eds.), Academic Press, New York, 1968, pp. 132–170.

31. J. Crank, *The Mathematics of Diffusion*, 2nd ed., Clarendon Press, Oxford (1975).

32. N. W. Henry and C. N. Schlatter, The development of a standard method for evaluating chemical protective clothing to permeation by liquids, *Am. Ind. Hyg. Assoc. J. 42*:202–207 (1981).

33. C. B. Billing, Jr., and A. P. Bentz, Effect of temperature, material thickness, and experimental apparatus on permeation measurement, *Performance of Protective Clothing: Second Symposium, ASTM STP 989* (S. Z. Mansdorf, R. Sager, and A. P. Nielsen, eds.), American Society for Testing and Materials, Philadelphia, 1988, p. 226.

34. N. W. Henry, Comparative evaluation of a small version of the ASTM permeation test cell versus the standard cell, *Performance of Protective Clothing: Second Symposium*,

ASTM STP 989 (S. Z. Mansdorf, R. Sager, and A. P. Nielsen, eds.), American Society for Testing and Materials, Philadelphia, 1988, p. 236.

35. S. P. Berardinelli, R. L. Mickelsen, and M. M. Roder, Chemical protective clothing; A comparison of chemical permeation tests cells and direct-reading instruments, *Am. Ind. Hyg. Assoc. J. 44*:886 (1993).

36. G. L. Patton, M. Conoley, and L. H. Keith, Problems in determining permeation cell equivalency, *Performance of Protective Clothing: Second Symposium, ASTM STP 989* (S. Z. Mansdorf, R. Sager, and A. P. Nielsen, eds.), American Society for Testing and Materials, Philadelphia, 1988, p. 243.

37. J. O. Stull, A Comparison of U.S. and CEN Performance Standards for Chemical Protective Clothing, Proceedings of the Third Scandinavian Symposium on Protective Clothing Against Chemicals and Other Health Risks (NOKOBETEF IV), Finnish Institute of Occupational Health, 1992, pp. 28–33.

38. A. D. Schwope, R. Goydan, R. C. Reid, and S. Krishnamurthy, State-of-the-art review of permeation testing and the interpretation of its results, *Am. Ind. Hyg. Assoc. J. 45*:634 (1984).

39. J. O. Stull and M. F. S. Pinette, Special considerations for conducting permeation testing of protective clothing materials with gases and chemical vapors, *Am. Ind. Hyg. Assoc. J. 51*:378 (1990).

40. J. O. Stull, M. F. S. Pinette, and R. E. Green, Permeation of ethylene oxide through protective clothing materials for chemical emergency response, *Appl. Ind. Hyg. 5*:448–462 (1990).

41. J. Lara and D. Drolet, Testing the resistance of protective clothing materials to nitroglycerin and ethylene glycol dinitrate, *Performance of Protective Clothing: Fourth Volume, ASTM STP 1133* (J. P. McBriarity and N. W. Henry, eds.), American Society for Testing and Materials, Philadelphia, 1992, pp. 153–161.

42. A. D. Schwope, R. Goydan, T. Carroll, and M. D. Royer, Test methods development for assessing the barrier effectiveness of protective clothing materials, Proceedings of the Third Scandinavian Symposium on Protective Clothing Against Chemicals and Other Health Risks, Norwegian Defence Research Establishment, Gausdal, Norway, September 1989.

43. V. L. Man, V. Bastecki, G. Vandal, and A. Bentz, Permeation of protective clothing materials: Comparison of liquid contact, liquid splashes, and vapors on breakthrough times, *Am. Ind. Hyg. Assoc. J. 48*:551 (1987).

44. M. F. S. Pinette, J. O. Stull, C. R. Dodgen, and M. G. Morley, A new permeation testing collection method for non-volatile, non-water soluble chemical challenges of protective clothing, *Performance of Protective Clothing: Fourth Volume, ASTM STP 1133* (J. P. McBriarity and N. W. Henry, eds.), American Society for Testing and Materials, Philadelphia, 1992, pp. 339–349.

45. C. Fricker and J. K. Hardy, The effect of an alternative environment as a collection medium on the permeation characteristics of solid organics through protective glove materials, *Am. Ind. Hyg. Assoc. J. 55*:738–742 (1994).

46. D. J. Ehntholt, D. L. Cerundolo, I. Dodek, A. Schwope, M. D. Royer, and A. P. Nielsen, A test method for the evaluation of protective glove materials used in agricultural pesticide operations, *Am. Ind. Hyg. Assoc. J. 51*:462–468 (1990).

47. P. M. Swearengen, J. S. Johnson, and S. J. Priante, A modified method for fabric permeation testing, Paper presented at 1991 American Industrial Conference, Salt Lake City, UT, 1991.

48. M. W. Spence, An analytical technique for permeation testing of compounds with low volatility and water solubility, *Performance of Protective Clothing: Second Symposium, ASTM STP 989* (S. Z. Mansdorf, R. Sager, and A. P. Nielsen, eds.), American Society for Testing and Materials, Philadelphia, 1988, p. 277–285.

49. R. L. Mickelsen, R. C. Hall, R. T. Chern, and J. R. Myers, Evaluation of a simple weight-loss method for determining the permeation of organic liquids through rubber films, *Am. Ind. Hyg. Assoc. J. 52*:445–447 (1991).

50. H. S. A. Al-Hussaini, W. J. Koros, M. Howard, and H. B. Hopfenberg, A simple apparatus for measurement of liquid permeabilities through polymeric films, *Ind. Eng. Chem. Proc. Res. Dev. 23*:317–320 (1984).

51. K. L. Verschoor, L. N. Britton, and E. D. Golla, Innovative method for determining minimum detectable limits in permeation testing, *Performance of Protective Clothing: Second Symposium, ASTM STP 989* (S Z. Mansdorf, R. Sager, and A. P. Nielsen, eds), ASTM, Philadelphia, 1988, pp. 252–256.

52. J. L. Perkins and M. C. Ridge, Use of infrared spectroscopy in permeation tests, *Performance of Protective Clothing, ASTM STP 900* (R. L. Barker and G. C. Coletta, eds.), American Society for Testing and Materials, Philadelphia, 1986, pp. 22–31.

53. R. P. Moody, Analysis of 2,4-D glove permeation under controlled environmental conditions, *Performance of Protective Clothing: Fourth Volume, ASTM STP 1133* (J. P. McBriarity and N. W. Henry, eds.), American Society for Testing and Materials, Philadelphia, 1992, pp. 189–197.

54. M. W. Spence, Chemical permeation through protective clothing materials: An evaluation of several critical variables, Presented at Am. Ind. Hyg. Conference, Portland, OR, 1981.

55. A. P. Bentz and V. L. Man, Critical variables regarding permeability of materials for totally encapsulating suits, Proceedings of the 1st Scandinavian Symposium on Protective Clothing Against Chemicals, Copenhagen, Denmark, 1984.

56. G. O. Nelson, B. Y. Lum, and G. J. Carlson, Glove permeation by organic solvents, *Am. Ind. Hyg. Assoc. J. 42*:217–224 (1981).

57. R. Y. Huang and V. J. C. Lin, *J. Appl. Polymer Sci. 12*:2615 (1968).

58. N. Vahdat and M. Bush, Influence of temperature on the permeation properties on protective clothing materials, *Chemical Protective Clothing Performance in Chemical Emergency Response, ASTM STP 1037*, (J. L. Perkins and J. O. Stull, eds.), American Society for Testing and Materials, Philadelphia, 1989, pp. 132–145.

59. E. T. Zellers and R. Sulewski, Modeling the temperature dependence of *N*-methyl-pyrrolidone permeation through butyl and natural rubber gloves, *Am. Ind. Hyg. Assoc. J. 54*:465–479 (1993).

60. J. L. Perkins and M. Yu, Predicting temperature effects on chemical protective clothing permeation, *Am. Ind. Hyg. Assoc. J. 53*:77–83 (1992).

61. J. F. Stampher, M. J. McLeod, M. R Bettis, A. M. Martinez, and S. P. Berardinelli, Permeation of polychlorinated biphenyls and solutions of these substances through selected protective clothing materials, *Am. Ind. Hyg. Assoc. J. 45*:634–641 (1984).

62. A. D. Schwope, R. Goydan, D. J. Ehntholt, U. Frank, and A. P. Nielsen, Permeation resistance of glove materials to agricultural pesticides, *Performance of Protective Clothing: Fourth Volume, ASTM STP 1133* (J. P. McBriarity and N. W. Henry, eds.), American Society for Testing and Materials, Philadelphia, 1992, pp. 198–209.

63. R. L. Mickelsen, M. M. Roder, and S. P. Berardinelli, Permeation of chemical protective clothing by three binary solvent mixtures, *Am. Ind. Hyg. Assoc. J. 47*:236–240 (1986).

64. M. C. Ridge and J. L. Perkins, Permeation of solvent mixtures through protective clothing elastomers, *Chemical Protective Clothing Performance in Chemical Emergency Response, ASTM STP 1037* (J. L. Perkins and J. O. Stull, eds.), American Society for Testing and Materials, Philadelphia, 1989, pp. 113–131.

65. G. O. Nelson and A. N. Correia, Respirator cartridge efficiency studies: VIII. Summary and conclusions, *Am. Ind. Hyg. Assoc. J. 37*:514–525 (1976).

66. Y. H. Yoon and J. H. Nelson, Application of gas adsorption kinetics I. A theoretical model for respirator cartridge service life, *Am. Ind. Hyg. Assoc. J. 45*:509–516 (1984).

67. G. O. Wood, Estimating service lives of organic vapor cartridges, *Am. Ind. Hyg. Assoc. J. 55*:11–15 (1994).

68. D. M. Baars, D. B. Eagles, and J. A. Emond, Test method for evaluating adsorptive fabrics, *Performance of Protective Clothing, ASTM STP 900* (R. L. Barker and G. C. Coletta, eds.), American Society for Testing and Materials, Philadelphia, 1986, pp. 39–50.

69. A. D. Schwope, P. P. Costas, J. O. Jackson, J. O. Stull, and D. J. Weitzman, *Guidelines for the Selection of Chemical Protective Clothing*, 3rd ed., Volumes I and II, American Conference of Government Industrial Hygienists, Akron, OH, 1988.

70. K. Forsberg and L. H. Keith, *Chemical Protective Clothing Performance Index*, Wiley-Interscience, New York, 1989.

71. K. Forsberg and S. Z. Mansdorf, *Quick Selection to Chemical Protective Clothing*, 2nd ed., Van Nostrand Reinhold, New York, 1993.

72. *CPC Base*, Arthur D. Little, Cambridge, MA, 1990.

73. R. L. Mickelsen and R. C. Hall, A breakthrough time comparison of nitrile and neoprene glove materials produced by different glove manufacturers, *Am. Ind. Hyg. Assoc. J. 48*:941–947 (1987).

74. R. A. Jamke, Understanding and using chemical permeation data in the selection of chemical protective clothing, *Chemical Protective Clothing Performance in Chemical Emergency Response, ASTM STP 1037* (J. L. Perkins and J. O. Stull, eds.), American Society for Testing and Materials, Philadelphia, 1989, pp. 11–22.

75. J. O. Stull, Developing test programs to demonstrate protective clothing material performance, *Proceedings of INDA IMPACT '90 Conference*, Association of the Nonwoven Fabrics Industry, Cary, NC, February 1990.

76. M. W. Spence, A proposed basis for characterizing and comparing the permeation resistance of chemical protective clothing materials, *Performance of Protective Clothing, ASTM STP 900* (R. L. Barker and G. C. Coletta, eds.), American Society for Testing and Materials, Philadelphia, 1986, pp. 32–38.

77. M. W. Spence, Streamlined methodology for permeation testing using the ASTM F-1001 standard test chemical battery, Paper presented at 1987 American Industrial Hygiene Conference, Montreal, Quebec, May 1987.

13

Assessment of the Protective Properties of Textiles Against Microorganisms

PETER L. BROWN   W. L. Gore & Associates, Inc., Elkton, Maryland

I. SCOPE

A. Types of Textiles

There are many different types of textiles used today in an attempt to either limit or prevent the transmission of hazardous microorganisms. These materials range from solitary layers of nonwoven single-use products to composites of woven and knitted multiple-use products that include film reinforcements. There is a multitude of various different fiber types, ways in which fiber assemblies and textile and film structures can be bonded together, and different chemical finishing applications and additives that can be used in textile constructions to impart varying degrees of protection against microorganisms. Often, there are entirely different performance objectives for textiles, depending on the nature of their intended application, which can dictate which types of textile structures are appropriate for use as microbial barriers.

Some structures must allow for the transfer of fluids, such as air or liquid, while limiting or preventing the transfer of potentially pathogenic microbes being transported within them. (The fluids transporting microorganisms are often referred to as vehicles.) These materials are generally characterized as being porous, where the pores would allow the transfer of the vehicles and the surrounding structures would act to impede the penetration of the microorganisms to varying degrees. The resistance of the materials to the penetration of both the fluids and the microbes will depend on a whole host of very important factors, many of which are discussed later in this chapter.

Other structures are made in an attempt to limit or prevent the transfer of the vehicles and thereby indirectly prevent the transfer of the microbes. Materials

469

protecting against airborne biohazards in this way would generally be characterized as being nonporous. Textiles used for this purpose can be augmented by film reinforcements, which will not allow the bulk flow of air through the material and thereby prevent the transfer of microorganisms. In the case of protecting against liquid-borne biohazards in this manner, both porous and nonporous structures can be utilized; however, the objective for both structures would be to prevent the bulk flow of liquid through the materials and thereby prevent the transfer of microorganisms. Depending on the nature of the liquid-borne biohazard, various textile or textile and film structures can be used.

These two basic performance objectives, allowing fluid flow and not allowing fluid flow, are fundamentally different and require utilizing different experimental approaches to the analysis of the barrier properties of the respective materials to microorganisms.

B.   Types of Applications

Applications requiring fluid flow while limiting or excluding microbiological penetration would include such things as high-efficiency particulate air (HEPA) filtration for sensitive environments (operating and clean rooms), microfiltration of heat-labile pharmaceuticals (disk and cartridge filter media), sterile packaging for the steam sterilization of medical devices (vents for pouches and container systems, wrappers for surgical instrument trays and linen packs), and respiratory protection for health-care workers (surgical masks and respirators). Alternatively, applications that do not require fluid flow while limiting or excluding microbiological penetration include such things as providing certain types of personal protection (garments, headwear, gloves, and footwear), providing or maintaining an aseptic environment for infection control purposes (medical device and pharmaceutical clean-room apparel and surgical gowns, patient drapes, table covers, and equipment covers), and caring for wounds (occlusive wound dressings).

This chapter focuses mainly on strategies for the laboratory analysis and decision logic related to textile structures used in clothing systems for personal protection and textile structures used to provide or maintain an aseptic environment for infection control purposes. This chapter does not cover all of the potential applications for textiles as barriers to microorganisms or the associated strategies for the laboratory analysis and decision logic related to each one. However, similar strategies for laboratory analyses may be applied to textile structures used in many other applications where microbial barrier properties are important.

C.   Types of Hazards

There are many different types of microorganisms that are important when discussing textile structures used in clothing systems for personal protection and providing and maintaining aseptic environments for infection control. These microbes fall into three major categories; viruses, bacteria, and fungi. Obviously, whole intact organisms in each of these categories can cause problems; however, there are also some subcellular components and by-products of metabolism that can cause problems as well. The strategy of experimentation used to evaluate the barrier properties of textile structures against these agents has typically been specific for each agent, although in some circumstances it may be possible to develop a model that has broad application and predictive capabilities.

There are many important factors in determining the nature of the biohazards: the type of vehicle by which the microbe is being transported, the physical and chemical characteristics of any carriers associated with the microbe (particles, cells, tissue fragments, hair, etc.), the concentration of the microbe in the vehicle, the state of the vehicle (dynamic or static), the forces or pressures associated with the vehicle, the physical and chemical characteristics of the microbe, the stability of the microbe (the viability of the microorganism under various environmental conditions), the virulence of the microbe and the dose necessary to cause infection, the susceptibility of the potential host, and the resistance of the microbe to disinfection, sterilization, and antimicrobial therapy. Modes of transport for microbes, other than common vehicles, must also be considered. It is possible for microbes to also be transported via direct contact with contaminated animate (bites, scratches, etc.) and inanimate objects (needle sticks, cuts, etc.).

The route of transmission plays a significant roll in the biohazard exposure assessment. Relative to the prevention of infection, each pathway—oral, respiratory, mucous membrane, percutaneous (surgery, needle sticks, cuts, nonintact skin, etc.)—that would allow a susceptible host to become infected should be recognized and dealt with in the assessment of each respective textile barrier. In many cases, textile barriers are not the first line of defense against biohazards and must be used in conjunction with other means of mitigating biohazard risks, such as engineering controls (laminar-flow biohazard exhaust hoods, clean or steam in-place bioreactor designs, self-sheathing needles, etc.), work-practice controls (double gloving, use of eye or face shields, use of sharps containers, removal of overtly contaminated protective clothing, etc.), and immunization (against diphtheria/tetanus, rubella, rubeola, varicella, mumps, measles, polio, influenza, and hepatitis B).

D. Significance of Laboratory Test Data

Textile structures that are used to limit or prevent the transmission of microorganisms play an ever-increasing role of importance in our society today. The strategies that are employed in the laboratory analysis of these materials and the resulting understanding of their performance expectations are of paramount importance when deciding which materials are fit for what application. These strategies should be relative to the perceived risk associated with the transmission of each microbe or class of related microbes. Even though it may be virtually impossible to duplicate the myriad of physical, chemical, and thermal stresses placed on textile materials in the real world, the goal of laboratory testing should be to provide information that would allow a realistic estimation of the performance of barrier textiles during actual use. Risk reduction decisions are likely to be made based on the laboratory data, and the conclusions that are drawn from the laboratory data should in fact provide a reduction in risk during actual use.

II.   GENERAL CHARACTERIZATION OF TEXTILES

A.   End-Use Requirements

The overall performance requirements of textiles in each personal protection and infection control end-use application can be quite different. Understanding the performance requirements for each end-use application is the key to developing a successful strategy of experimentation and defining a risk reduction decision logic. In each end-use application all of the technical attributes that are required for the textile to perform adequately should be identified. Two different end uses may require the same protection against the penetration of microorganisms; however, there may be completely different requirements for physical, chemical, and thermal properties. In some cases, one attribute may have to be sacrificed in order to obtain the goal for another attribute. As an example, strength objectives might dictate that the weight of the textile has to be increased; however, increasing the weight can negatively impact the hand (stiffness) of the textile. When considering the performance objectives for textiles in just a few end uses, such as firefighter turnout clothing, emergency medical response clothing, and surgical apparel, it becomes obvious that the biobarrier demands for the textiles used in these applications will have to be integrated and balanced with a multitude of other technical attributes. Other significant influences, which are not technical performance attributes but can have a direct impact on deciding which textile to use, are cost and environmental impact.

It is beyond the scope of this chapter to address all of the end-use requirements for textile structures used in personal protective clothing and infection control applications. Outlined next is a discussion of a few physical, chemical, and thermal properties that should be considered when evaluating textile materials as barriers to microorganisms.

B.   Physical Properties

Four of the easiest and most objective analyses for textiles are weight, thickness, bulk density, and microscopy. Together, these tools can serve to benchmark different textiles for comparison and help to predict other important microbial barrier and physical attributes.

1. Weight

The weight of textiles is often determined in order to be used with thickness measurements to obtain a bulk density. Weight measurements for similar textile structures can also be related to other physical attributes, such as strength, abrasion resistance, and stiffness. The international units of measurement for weight are grams per square meter (Fig. 1).

2. Thickness

As stated earlier, thickness determinations can be used with weight measurements to calculate bulk density, which can be used for a variety of purposes. Thickness measurements for similar textile structures can also be related to other physical attributes, such as strength, abrasion resistance, and stiffness. The compressibility

Figure 1 The weight of textiles can be determined using ASTM D 3776-85 [Standard Test Methods for Mass Per Unit Area (Weight) of Woven Fabric: Option C—Small Swatch of Fabric]. (Photo courtesy of the Institute for Environmental Research, Kansas State University, Manhattan, KS.)

Figure 2 The thickness of textiles can be determined using ASTM D 1777-64 (Re-approved 1975) (Standard Test Method for Measuring Thickness of Textile Materials). (Photo courtesy of the Institute for Environmental Research, Kansas State University, Manhattan, KS.)

of various textile structures may be different and need to be taken into account when making this measurement for comparative purposes. The international units of measurement for thickness are millimeters (Fig. 2).

3.   Bulk Density

Bulk density calculations for textiles can be related to the insulative properties and may also be useful in helping to understand the liquid, air, moisture vapor, and microbial penetration resistance characteristics. Generally speaking, for textiles with similar physical and chemical structures, as the bulk density increases the penetration resistance increases. The international units of measurement for density are kilograms per cubic meter [1].

$$\text{Bulk density (kilograms/cubic meter)} = \frac{\text{mass (grams)/area (square meters)}}{\text{thickness (millimeters)}}$$

(1)

When textiles include a film reinforcement, the characteristics of the film will be the overriding factor in determining the penetration resistance properties. Film reinforcements may have little impact on the overall bulk density calculations while significantly changing the penetration resistance properties.

4. Microscopy

Visually observing the magnified images of textiles is perhaps the most interesting and informative analysis that can be made in order to better understand the physical structures. Virtually all of the physical elements can be investigated, including the type of structure (type of weave, knit, or nonwoven), additional features of the structure (calendered, texturized, entangled, point bonded) the complexity of the structure (multiple layers, fiber blends, film reinforcements), the porosity of the structure (define pathways, approximate the size of yarn and fiber interstices, approximate the pore size of some films), yarn characteristics (filaments/yarn), fiber characteristics (diameter, cross-sectional shape, classification of some types), film bonding techniques (adhesive laminated, direct coated/extruded, point bonded), and film types (monolithic, bicomponent, microporous). Many of these features can be determined using either a stereomicroscope or a mono-objective compound microscope; however, scanning electron microscopy (SEM) has the flexibility to determine all of them and more.

In order to illustrate the diversity of some of the structures, products that represent the most common types of textile structures currently being used in personal protective clothing and infection control applications were chosen and SEM was employed to view their outside surface and cross-section. Refer to Figures 3–10. (Please note that different magnifications would be necessary to make some of the determinations listed in the preceding paragraph.) (Figures 34 and 35 denote the physical and microscopic characterization of two of the most common types of film-reinforced textile structures.)

C. Antimicrobial Properties

Incorporating antimicrobial compounds into textile fibers and finishes has been practiced for many years. Desirable features for an antimicrobial textile include durability of activity (including laundering and sterilization or dry cleaning if necessary), selective activity against undesirable microorganisms, acceptable moisture transport properties (important for agents that rely on a controlled release mechanism), compatibility with other finishing agents, absence of any toxic effects to the wearer or user, and commercial availability [4].

Textiles that include antimicrobial additives require special consideration when assessing the microbiological barrier properties. These textiles may confound normal microbial challenge testing by eliminating the challenging organisms. Most of the standard microbial challenge tests evaluate the ability of a textile

Figure 3 Scanning electron micrograph: surface at 100× magnification, Tyvek, supplied by E. I. du Pont de Nemours & Co., described as spun-bonded plexifilamentary linear high-density polyethylene.

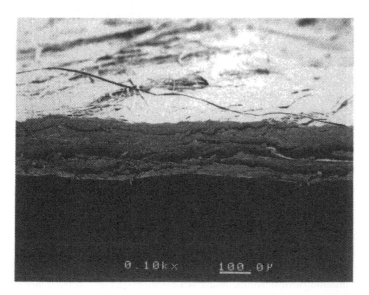

Figure 4 Scanning electron micrograph: cross section at 100×magnification, Tyvek, supplied by E. I. du Pont de Nemours & Co., with weight 41.32 g/m^2, thickness 0.24 mm, and bulk density 172.17 g/m^3.

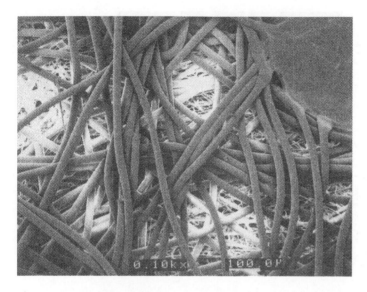

Figure 5 Scanning electron micrograph: surface at 100× magnification, Evolution, supplied by Kimberly-Clark Corp., described as spunbond/meltblown/spunbond (SMS) polypropylene (point bonded).

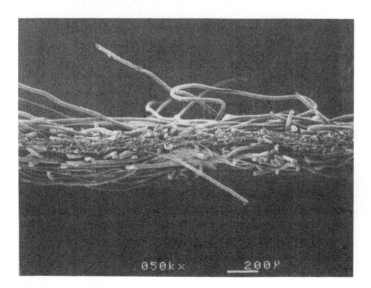

Figure 6 Scanning electron micrograph: cross section at 50× magnification, Evolution, supplied by Kimberly-Clark Corp., with weight 92.95 g/m², thickness 0.67 mm, bulk density 138.73 g/m³.

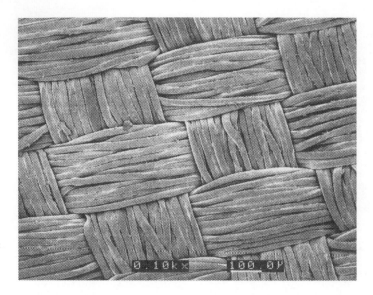

Figure 7   Scanning electron micrograph: surface at 100× magnification, Compel, supplied by Standard Textile Co., described as woven continuous filament polyester microfiber (calendered).

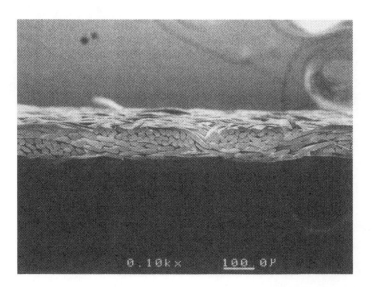

Figure 8   Scanning electron micrograph: cross section at 100× magnification, Compel, supplied by Standard Textile Co., with weight 117.90 g/m^2, thickness 0.20 mm, bulk density 589.50 g/m^3.

Figure 9 Scanning electron micrograph: surface at 100× magnification, Optima, supplied by Baxter Healthcare Corp., described as spunlace woodpulp/polyester nonwoven (entangled).

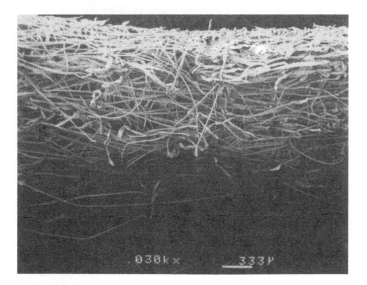

Figure 10 Scanning electron micrograph: cross section at 30× magnification. Optima, supplied by Baxter Healthcare Corp., with weight 69.85 g/m², thickness 0.34 mm, bulk density 205.44 g/m³.

to prevent penetration and assume that the challenging organism remains viable throughout the process. If the mechanism by which the textile acts to protect against the penetration of microorganisms is to inactivate the microorganisms, then this fact should be fully understood and appropriate means to evaluate the protective properties of that textile should be utilized in the laboratory. Textiles incorporating antimicrobial agents might not prevent the penetration of or direct contact with viable pathogenic microbes during use, as the effectiveness of antimicrobial agents requires direct contact with the microbe for a specified period of time under defined conditions. The antimicrobial agent can be ineffective if penetration of the microbe through the textile occurs quickly or if the activity of the antimicrobial agent is diluted or inactivated by the vehicles carrying the microbe (such as with blood or body fluids).

Certain specific textile end uses may not require microbial penetration resistance, but may rely on antimicrobial treatments to reduce the overall bioburden in the work environment or to reduce the likelihood of cross contamination. This chapter does not address test methods for assessing the antimicrobial activity of textiles. It is assumed, and required by some microbial challenge tests, that the textile materials will be compatible with the challenging microorganisms.

D. Chemical Resistance Properties

In virtually every end-use application for textiles being used as microbiological barriers there exists the potential for exposure to various types of chemical agents. A complete list of all chemical agents that are likely to come into contact with textiles in each personal protection and infection control end-use application should be made. Some of these agents will be hazardous and require that the textile also act as a barrier to them. This list could include many different chemical types, such as disinfectants, acids, bases, solvents, fuels, and lubricants. If dual protection to both microorganisms and chemicals is required, then both types of laboratory analyses for textiles in these applications will need to be performed. Depending on the type of chemical that the textile is exposed to and the chemical resistance properties of the textile, the chemical may seriously compromise the ability of the textile to effectively limit or prevent microbial penetration. Whether the chemical is hazardous or nonhazardous, some types of textiles and film reinforcements, particularly those that are porous in nature, can be made to allow microbial penetration to occur after contact with chemical prewetting and contaminating agents. If dual exposure situations can be clearly defined, then preconditioning textile samples with the chemical agents of concern prior to microbial challenge testing would be appropriate. For more information on assessing the chemical barrier properties of textiles, refer to Chapter 12.

E. Thermal Properties

Most of the personal protection and infection control end-use applications have standards and regulations governing flammability requirements for textile-based products. These requirements should be recognized and appropriate flammability testing should be performed. In each end-use application, situations involving possible exposure to heat and ignition sources should be recognized and appropriate strategies should be developed in an effort to avoid overexposure. The thermal stability of textiles, including flame resistance, dimensional stability, insulative properties, etc., should be well characterized for those end-use applications requiring thermal protection. The microbiological barrier properties of some textile materials may degrade with repeated or prolonged exposure to hot environments. For those applications requiring both thermal protection and microbiological barrier properties, preconditioning textile samples with thermal exposures prior to assessing the microbial barrier properties should be considered. For further references see the Appendix.

III. THEORETICAL BASIS FOR PREVENTING PENETRATION OF MICROBES

A. Modelling the Real World

Perhaps one of the most important concepts in determining how to evaluate the barrier properties of textiles against microorganisms, in the laboratory, is that the evaluation should relate to the real-world application of the product in such a way as to allow a meaningful judgment to be made regarding the risk of transmission. This assessment requires a very thorough understanding of the end use of each product, including all of the various factors that could stress the barrier integrity and negatively impact performance. Certainly, the more risk associated with the transmission of any given microbe, the more rigorous the laboratory analysis required.

There are two fundamentally different approaches that can be taken with regard to evaluating the microbial barrier properties of textiles in the laboratory. The first approach would be to define test conditions whereby the barrier properties are evaluated on a continuous scale of measurement to allow a relative comparison to be made between the breakthrough points for all products. This approach is often employed when the use conditions for products are not well defined or not controlled enough to determine predictive performance limits and when absolute barrier properties cannot be achieved because of the need to balance them against other technical requirements (such as low air penetration resistance). The second approach would be to define laboratory test conditions based on a thorough examination of the application for each product, whereby the barrier properties are evaluated and determined to be adequate (passing the test) or inadequate (failing the test). Depending on the limitations of the test device and procedure used, the failure point may or may not be identified in the laboratory with either approach. However, acceptance and rejection judgments should be made based on sound logic related to an understanding of the actual use conditions for each textile application.

Recognizing the complexities of the different end-use environments for microbial barrier textiles and the various stresses that can be imposed on their barrier integrity is the first step in developing a logic related to product evaluation in the laboratory. Most likely there will be no perfect strategy; however, the means with which to characterize the physical, chemical, and thermal properties of textiles appear to be very abundant. Therefore, it would seem feasible that a hierarchy or decision tree could be built based on combinations of various tests, some of which may need to be used as preconditioning steps prior to barrier testing, with the ultimate goal of reducing the risk of product failure during actual use.

The degree of hazard associated with exposure to the microbe(s) will dictate how carefully the end use application for the textile will need to be studied, how conservative the modeling and experimental approach should be in the laboratory, and the definitions for adequate versus inadequate microbial barrier performance. Many of the key variables that should be identified in situations where engineering controls, work practices, immunization, and antimicrobial therapy cannot reduce the risk associated with the transmission of hazardous microorganisms to an acceptable level are discussed in Section III. J.

B. Behavior of Liquids

In considering the various performance requirements for textiles intended to be used in personal protective clothing products and products used to provide or maintain an aseptic environment, it is important to understand the behavior of liquids as potential vehicles for microbial transport. Depending on the application, there could be a variety of different liquids that could challenge the integrity of textile barriers. Liquid challenge sources can vary from single insults with contaminated pure liquids to multiple insults with contaminated mixtures. Textile barriers can be confronted by liquids in many forms: splashing, spraying, pooling, and soaking. There may be cases where the sequence of liquid challenges can allow wetting of otherwise nonwetting liquids to occur, and there may be circumstances where the exact composition of the challenging liquids is not known or cannot be predicted. The severity of liquid challenges can also be greatly influenced by the pressure and time of the exposures. Once the general behavior characteristics of liquids are understood and possible exposure scenarios have been clearly thought through, worst-case modeling in the laboratory is appropriate. In most situations, due to the fact that the microbes of concern are so small in comparison to the interstices between the yarns/fibers of the textile or the pores in porous film reinforcements, the goal will be to prevent the wetting and subsequent penetration of the microbiohazardous liquids through the materials.

There are a number of important factors that can impact the resistance of textiles to liquid wetting and penetration. The literature is replete with references to demonstrate the importance of variables such as the surface tension, viscosity, and density of the challenging liquids, the contact angle of the liquids against the textiles, the porosity of the textiles, and the pressure and time of the liquid exposures.

1. Surface Tension of Liquids

Surface tension of liquids is a property that results from unbalanced intermolecular cohesive forces, such as electromagnetic interactions (γ_{LW}), whether due to oscillating temporary dipoles (London), or permanent dipoles (Keesom), or induced dipoles (Debye), and acid–base interaction, including hydrogen bonding (γ_{AB}), at or near the surface, that causes the surface to contract. This theory of surface tension was pioneered by Fowkes and is expressed by Good [5] as:

$$\text{Surface tension } (\gamma) = \gamma_{LW} + \gamma_{AB} \quad \text{(newtons/meter)} \tag{2}$$

One means of measurement of the surface tension of liquids is the du Nouy ring method. This test is illustrated in Figure 11.

The surface tensions for a variety of liquids, including human body liquids and liquids that could commonly be found in the laboratory and clinical settings, are listed in Table 1. This list is far from complete, but serves to illustrate the range of some of the liquid surface tension values that can be found in the environments requiring personal protective clothing and aseptic barriers. As mentioned previously, it would not be difficult to imagine situations where textile barriers could be exposed to a complex array of liquids such as these, in diverse forms with different pressure and time factors. Some of the lower surface tension liquids, such as 70% isopropyl alcohol, can prewet certain textiles and create a pathway for other higher surface tension liquids to follow. Other liquids may leave behind contaminating residues and surface active agents that can compromise the liquid resistance properties of certain textiles at a later time.

2. Contact Angle: Hydrophilicity vs. Hydrophobicity of Textiles

The behavior of liquids on solid surfaces can be illustrated by Young's [12] stated equation:

$$\gamma_{SL} = \gamma_{SV} - \gamma_{LV} \cos \theta \tag{3}$$

This equation contains components for deriving surface tension. These components include the surface free energy of the solid–liquid interface (γ_{SL}), of the solid–vapor interface (γ_{SV}), of the liquid–vapor interface (γ_{LV}), and the projection of the vector for γ_{LV} ($\cos \theta$) on the plane of the surface [13]. Young's equation assumes an ideal solid—chemically homogeneous, rigid, and flat—on an atomic

Figure 11 The surface tension of challenging liquids can be determined using ASTM D 1331-89 (Standard Test Methods for Surface and Interfacial Tension of Solutions of Surface-Active Agents). Here the test liquid has been placed in a glass petri dish on an elevator platform. The strain required to pull an immersed 6.0-cm platinum iridium ring, which is suspended from a balance beam that is connected to a pressure transducer, out of the liquid is recorded. The direct reading (apparent surface tension) can be converted to obtain the absolute (corrected) surface tension of the liquid. The temperature of test liquids must be controlled; 25°C is prefered. The correction factor must be adjusted to compensate for temperature changes, as lower temperatures will raise the apparent surface tension and higher temperatures will lower the apparent surface tension of the test liquids.

scale. Contact angle hysteresis, as evident by comparing the difference between advancing and receding contact angle measurements of liquid drops on tilted textile surfaces, theoretically occurs because textile surfaces are not ideal solids and because some liquids and solids may chemically interact.

In order to determine the relative hydrophilicity or hydrophobicity of textiles, contact-angle measurements can be made for sessile drops of pure water placed on

Table 1 Surface Tension Measurements for a Variety of Liquids

Liquids	Surface tension; γ (N/m)
Human body liquids	
Sweat (37–38°C)	0.069–0.070
Urine (temperature not specified)	0.064–0.069
Cerebrospinal fluid (20°C)	0.060–0.063
Semen (15°C)	0.052–0.060
Tears (30°C)	0.040–0.050
Whole blood, fasting (20°C)	0.056
Blood serum (25°C)	0.047
Bile, hepatic and gallbladder (37°C)	0.040–0.044
Saliva (temperature not specified)	0.015–0.026
Laboratory-grade reagents and media	
Saline, 0.6–2.8% (20°C)	0.073–0.074
Water (20°C)	0.073
Trypticase soy broth (temperature not specified)	0.059
AOAC letheen broth (temperature not specified)	0.045
Phi-X174 nutrient broth, with 0.1% Polysorbate 80 (23°C)	0.042
Mineral oil (23°C)	0.031
Isopropyl alcohol, 70% (23°C)	0.024
Isopropyl alcohol, 100% (23°C)	0.021
Clinical liquids	
5% Dextrose + 0.45% NaCl inj., USP (23°C)	0.046
Sterile water for irrigation, USP (23°C)	0.046
5% Hypochloride (23°C)	0.044
Lactated Ringer's inj., USP (23°C)	0.044
5% Dextrose inj., USP (23°C)	0.043
5% Dextrose in lactated Ringer's inj. (23°C)	0.042
0.9% NaCl inj., USP (23°C)	0.041
Amphyl (23°C)	0.035
Metaquat (23°C)	0.033
2% Gluteraldehyde (23°C)	0.033
0.75% Iodine scrub (23°C)	0.032
1% Topical iodine paint (23°C)	0.031
3% Lysol (23°C)	0.031
Ultradex, PCMX (23°C)	0.030
Iodofore (23°C)	0.029
4% Chlorohexidine gluconate (23°C)	0.028

Source: Refs. 6–11, 23.

Figure 12 The contact angle of a sessile drop of liquid on the surface of a textile can be determined using TAPPI T 458 om-89 [Surface wettability of paper (angle of contact method), also known as the Goniometer contact angle test]. Here a droplet of water has been placed on the surface of a repellent textile. The Goniometer (scope) is used to magnify the droplet and determine the contact angle at the liquid/solid textile interface.

the surface of textile materials. Two examples are given in Figure 13 to illustrate textiles exhibiting hydrophilic behavior, where γ_{SL} is less than 90 degrees, and hydrophobic behavior, where γ_{SL} is greater than 90 degrees. Contact angle measurements for other potential challenging liquids can be made in this fashion and used to help determine their relative resistance to wetting. The higher the contact angle measurement between a liquid and a textile, the more resistant the textile will be to wetting and penetration of the liquid.

Figure 13 Contact angle measurements of water ($\gamma \approx 0.072$ N/m): left, water placed on a hydrophilic surface, contact angle less than 90 degrees; right, water placed on a hydrophobic surface, contact angle greater than 90 degrees.

In order to further illustrate how different textiles and different liquids can interact Figures 14–17 were prepared. Four liquids, which span a broad range of surface tensions, were selected and sessile drops were placed on the surfaces of textiles. (The textiles used are illustrated in Figures 3–10.) The liquids are water ($\gamma \approx 0.072$ N/m), synthetic blood ($\gamma \approx 0.042$ N/m), mineral oil ($\gamma \approx 0.031$ N/m), and 70% isopropyl alcohol ($\gamma \approx 0.024$ N/m).

Some of the conclusions that can be drawn about the interactions between each textile and each liquid from this simple demonstration are:

1. Lower surface tension liquids developed lower contact angles (the sessile drops were more flattened on the surface of the textiles) than higher surface tension liquids on all four textiles.
2. Water behaved similarly on the surface of all four textiles, developing a high contact angle (the sessile drops stood higher on the surface of the textiles).
3. The behavior of water on a textile may not predict the behavior of other liquids. Some textiles exhibiting hydrophobic behavior allowed lower surface tension liquids to spontaneously wet and penetrate through while other textiles did not allow spontaneous wetting and penetration of those same liquids. The most obvious examples of the same liquids exhibiting different contact angles on different textiles are mineral oil and isopropyl alcohol.

Figure 14 The wettability behavior of four liquids placed on the surface of spunbonded plexifilamentary linear high-density polyethylene nonwoven. From left to right, the liquids are 70% isopropyl alcohol ($\gamma \approx 0.024$ N/m), mineral oil ($\gamma \approx 0.031$ N/m), synthetic blood ($\gamma \approx 0.042$ N/m), and water ($\gamma \approx 0.072$ N/m).

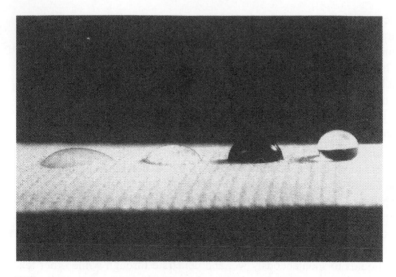

Figure 15 The wettability behavior of four liquids placed on the surface of spunbond/ meltblown/spunbond (SMS) polypropylene nonwoven. From left to right, the liquids are 70% isopropyl alcohol ($\gamma \approx 0.024$ N/m), mineral oil ($\gamma \approx 0.031$ N/m), synthetic blood ($\gamma \approx 0.042$ N/m), and water ($\gamma \approx 0.072$ N/m).

Figure 16 The wettability behavior of four liquids placed on the surface of woven continuous filament polyester microfiber. From left to right, the liquids are 70% isopropyl alcohol ($\gamma \approx 0.024$ N/m), mineral oil ($\gamma \approx 0.031$ N/m), synthetic blood ($\gamma \approx 0.042$ N/m), and water ($\gamma \approx 0.072$ N/m).

Figure 17 The wettability behavior of four liquids placed on the surface of spunlace woodpulp/polyester nonwoven. From left to right, the liquids are 70% isopropyl alcohol ($\gamma \approx 0.024$ N/m), mineral oil ($\gamma \approx 0.031$ N/m), synthetic blood ($\gamma \approx 0.042$ N/m), and water ($\gamma \approx 0.072$ N/m).

3. Breakthrough Pressure

Breakthrough pressure refers to the pressure required to force liquid to penetrate through a textile. A very succinct treatment of the important variables was put together by Olderman [6]. Olderman expressed these variables in a word equation as follows:

> The resistance of a textile to liquid penetration varies as
>
> $$\frac{\text{surface tension, viscosity, contact angle, and pore length}}{\text{hydrostatic pressure, time, pore radius, and number of pores}} \qquad (4)$$

It is generally accepted that those liquids that develop low contact angles (less than 90 degrees) on the surface of a textile can wet and penetrate through the textile via capillary pressure and wicking forces more easily than those liquids that develop high contact angles (greater than 90 degrees). The capillary forces are described by the Laplace theory of capillarity, which defines the pressure that is necessary to push or draw a liquid through a uniform channel or pore [14,15].

$$P = \text{capillary pressure} = \frac{2\gamma \cos \theta}{r} \qquad (5)$$

where

γ = surface tension of the liquid
θ = contact angle of the liquid on the surface of the textile
r = pore radius

Further modifications can be made to the Laplace theory of capillarity to account for the relationship between capillary pressure (P), the height of the liquid column (h), the gravitational acceleration constant (g), and the density of the liquid (ρ): $P = hg\rho$ [14]. This produces the following equation:

$$h = \frac{2\gamma \cos \theta}{rg\rho} \tag{6}$$

From this equation it is apparent that as the challenging liquid surface tension and contact angle values get lower, the resistance of the textile to wetting and penetration is also reduced. Challenging liquid surface tension can be plotted against breakthrough pressure to graphically illustrate this relationship. Figure 18 graphically depicts the theoretical impact of decreasing the surface tension of the challenging liquid on the break-through pressure for porous textiles. Reductions in the resistance of the textile to wetting and penetration can also result from increasing the pore radius.

When a liquid either spontaneously wets or is forced to wet the textile by hydrostatic pressure, the rate of liquid penetration can be described by the law of Poiseuille and the Washburn equation [16]. The Washburn equation introduces two

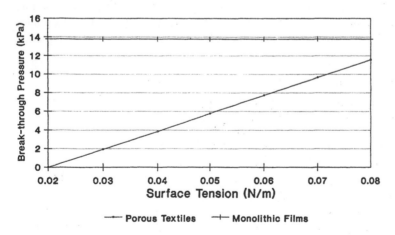

Figure 18 The relationship between the surface tension of various challenging liquids and the penetration resistance for any given textile can be plotted as illustrated. This graph shows the contrast between the theoretical behavior of porous textiles and the theoretical behavior of textiles that are reinforced with monolithic (nonporous) films.

new variables: the length of the pore and the viscosity of the liquid. As the values for pore length (influenced by the thickness of the textile and tortuosity of the path) and viscosity increase, the rate of entry for the liquid decreases. This equation is stated as follows:

$$V = \frac{r\gamma \cos \theta}{4\ell n} + \frac{P_A r^2}{8\ell n} \qquad (7)$$

where

V = rate of entry of a liquid in a capillary
r = pore radius
γ = surface tension of the liquid
θ = contact angle of the liquid on the surface of the textile
P_A = hydrostatic pressure
ℓ = length or depth of the pore
n = viscosity of the liquid or resistance of the liquid to flow

C. Air Penetration Resistance

In most cases involving known biohazardous aerosol generation, rigorous engineering controls, such as negative-air-pressure rooms and biosafety cabinets, are employed in order to isolate and/or eliminate the risk. However, if textiles are intended to limit or prevent the penetration of hazardous airborne microorganisms, understanding whether those textiles allow air to penetrate through is the first analytical step. If the textiles or the films used to reinforce the textiles are porous, the pores will allow air to penetrate through the structure. Textiles that are reinforced with monolithic (nonporous) films that are free from defects will not allow air to penetrate. Two of the standard test methods used to characterize the air penetration resistance of textiles are outlined next.

1. Low Air-Penetration Resistance

The air-penetration resistance of low resistance textiles can be determined using ASTM D 737-75, Standard Test Method for Air Permeability of Textile Fabrics (Fig. 19). Some textile structures, such as the plexifilamentary linear high-density polyethylene nonwoven (Figs. 3 and 4) and the woven continuous-filament polyester microfiber (Figs. 7 and 8), can yield false negative results using this test procedure. High-density/low-porosity textile structures like these, which yield results of less than 0.5 cm^3/cm^2·s (cubic centimeters per square centimeter per second), should be evaluated with the high-resistance test method.

2. High Air-Penetration Resistance

The air-penetration resistance of high-resistance textiles can be determined using Federal Test Method Standard 191A, Method 5452 (Permeability to Air; Cloth;

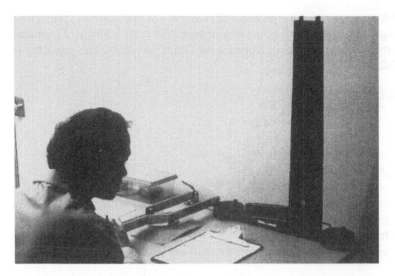

Figure 19   The air penetration characteristics of textiles with low resistance can be determined using ASTM D 737-75 (Standard Test Method for Air Permeability of Textile Fabrics, also known as the calibrated orifice method). This method determines the volume of air that can penetrate through a textile with a pressure differential of 0.12 kPa. The higher the volume of air the more permeable the textile. (Photo courtesy of the Institute for Environmental Research, Kansas State University, Manhattan, KS.)

Falling Cylinder Method) (Fig. 20). Typically results are recorded as the time (in seconds) necessary to pass 300 cc (cubic centimeters) of air through the textile. The test method can be modified to accommodate textiles or textiles with various film reinforcements with very high resistance to air penetration by reducing the air volume penetration end point (<300 cc) and by extending the observation time period (hours). In order to obtain accurate measurements for some of the higher resistance textiles with film reinforcements, special sealing techniques may need to be employed in order to ensure that the air is flowing through the film and not escaping through the textile interface. Microscopy may also be used to confirm that the textile structure contains a monolithic (nonporous) element.

Other methods of determining air penetration resistance of textiles are being developed in order to document flow resistance changes with varying pressure drops [17]. The filtration of biohazardous aerosols through textiles is discussed later in this chapter.

D. Moisture Vapor Permeability and Thermal Insulative Properties

Depending on the end-use application for the textile, it may be necessary to characterize the moisture vapor permeability and thermal insulative properties. Bal-

Figure 20   The air penetration characteristics of textiles with high resistance can be determined using Federal Test Method Standard 191A, Method 5452 (Permeability to Air; Cloth; Falling Cylinder Method). This method determines the time that is necessary to allow a certain volume of air to penetrate through the textile. The shorter the time interval, the more permeable is the textile.

ancing the thermal comfort properties against the microbial barrier properties can be very important for textiles intended to be used in personal protection and infection control clothing. Allowing the human body to maintain thermal equilibrium can lower heat stress and result in better job performance (physically and mentally), improve productivity (work longer with fewer breaks), reduce the risk of noncompliance (not wearing protective clothing because it is too hot), and increase job satisfaction (make work more enjoyable). Different work environments, physical labor, and clothing styles (gown vs. coverall, zoned vs. complete reinforcement) can place different demands on textiles in order to achieve the thermal comfort balance. The comfort versus protection paradigm has presented a long-standing problem in personal protection and infection control clothing textile applications; however, this paradigm can be broken with breathable film reinforced textiles.

Institutes developing excellence in this area are utilizing two main research tools: the sweating guarded hot plate and the thermal (heated) manikin [1,2,3,18]. When textiles are required to minimize or prevent the penetration of hazardous microorganisms and allow the human body to maintain thermal equilibrium, the resistance to evaporative heat transfer should be determined using the sweating guarded hot plate. The sweating guarded hot plate is preferred over other simpler methods for making comparative analyses of the moisture vapor permeability of textiles because it simulates the heat and mass transfer characteristics of the human body fairly accurately. Textiles exhibiting lower resistance values R(et) against the transfer of moisture vapor with this text can be utilized to construct more comfortable garments.

Generally speaking, most textiles that do not include film reinforcements exhibit low resistance values R(et) against the transfer of moisture vapor. However, new developments in the field of monolithic film reinforcements can also provide low resistance values R(et) against the transfer of moisture vapor. As an example, one of these film-reinforced textiles that exhibits low R(et) can be found in Figure 36.

The thermal manikin is used to assess the impact of the end garment as influenced by the geometry of the human body (clothing fit, design, and layering) on heat and mass transfer. Human subjects are also used to evaluate clothing systems to more directly measure the physiologic factors important to thermal comfort and to qualitatively assess comfort variables that are difficult to determine in the laboratory.

E. Penetration Versus Permeation

The terms penetration and permeation are often used interchangeably when describing the transfer of air, liquids, and microorganisms from one side of a textile barrier to the other side. There is a fundamental difference between penetration and permeation that should be understood when evaluating the barrier properties of textiles in the laboratory. Penetration is defined as the bulk flow of gases, vapors, or liquids through porous materials and is driven by a pressure gradient across the barrier. Permeation can be defined as the diffusion of gases or vapors through porous materials and dissolved gases, vapors, or liquids through nonporous materials on a molecular level, and is driven by a concentration gradient across the barrier.

Permeation testing is usually employed on textiles that are intended to protect against the diffusion of hazardous gases and vapors. The basic equation for diffusion can be derived from Fick's law, stated as follows [18]:

$$\frac{m}{A} = \frac{\Delta c}{R} \tag{8}$$

where

m = mass flow
A = area
Δc = concentration difference
R = resistance to diffusion

If penetration can occur, through the interstices between the yarns/fibers in a textile or through the pores in a porous film reinforcement, then permeation can also occur. However, the rate of transfer will most likely be more dependent on bulk flow rather than diffusion. Currently, microorganisms are thought to penetrate and not permeate through materials, mainly due to their large size in comparison to gas and vapor molecules. Even the smallest known human pathogenic viruses, as depicted in Figure 21, are almost two orders of magnitude larger in diameter than the molecular diameters of CO_2 and H_2O. (The molecular diameters for gas molecules, such as carbon dioxide and water vapor, have been calculated and determined to be in the range of 0.0004 μm [19].) This size difference is one of the reasons why monolithic (nonporous) films can act as such effective

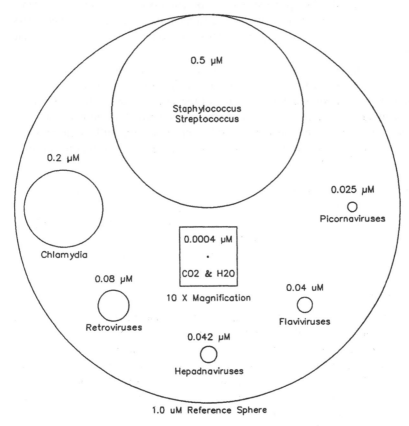

1.0 uM Reference Sphere

Figure 21 Examples of the sizes for various selected microorganisms are depicted to illustrate the difference between viruses, bacteria, and fungi. Even the smallest fungal spores would be larger than the outer 1.0-μm reference sphere. The molecular diameter for carbon dioxide and water has been added in the box at the center of the diagram for comparison.

microbial barriers. This is also one of the reasons why it is possible to create "breathable" monolithic film reinforcements, as the small size of water vapor allows it to diffuse through on a molecular level, while microbes are excluded due to their larger size.

Microbial challenge testing has fundamentally different requirements than permeation testing. Some of the important differences are as follows.

1. For many types of microorganisms, given the correct inoculation route, very few infectious units are necessary to cause infection [20,21]. Under these circumstances, where the acceptable number of viable infectious microbes that can penetrate through a textile is extremely small (i.e., 1–100), determining the penetration rate is of little or no real value.
2. Infection is an absolute process. Either infection occurs or it does not occur. Beyond a certain number, related to the number of infectious units necessary to cause infection to occur, quantification and determination of steady state are moot points.
3. Molecular diffusion is currently not recognized as a mode of transfer for microorganisms through textile barriers. The major mechanism of transport for microbes is to move with the bulk flow of the air and liquid vehicles. Pressure gradients across the textiles can cause the penetration of the air and liquid vehicles, whereas permeation testing can be performed with little or no pressure differential across the barrier as permeation is the result of a concentration gradient.

Test cells that are designed to measure the chemical permeation resistance of textiles, such as ASTM F739-91 (Standard Test Method for Resistance of Protective Clothing Materials to Permeation by Liquids or Gases Under Conditions of Continuous Contact), unless modified to apply pressure, are not appropriate for use in assessing the microbial barrier characteristics of textiles.

F.   Time of Test

Time is an important parameter to understand when conducting challenge tests of various types on textiles in the laboratory. Some penetration mechanisms are time dependent and others are not time dependent. When considering air-based challenge testing, depending on the filtration mechanisms acting on the aerosols, the time to travel through the textile can influence the filtration efficiency. The time necessary to travel through will be proportional to the pressure differential, which influences the flow rate of the air.

When considering liquid-based challenge testing, the time of the tests is secondary in importance as compared to pressure. If the hydrostatic pressure exerted on the liquid does not cause the liquid to overcome the resistance of the textile, then the time variable could be infinite, as wetting and penetration may never occur. It has been suggested that if the test pressure exceeds the average applied pressure that would normally be exerted on the textile during use, then the test time may be shorter than the time of use [22]. Historically, the pressures used in liquid based microbial challenge tests have been quite low. As an example, in one study, when using a constant, very low hydrostatic test pressure ($\approx$0.06 kPa), time periods greater than 60 min did not increase bacterial penetration through barrier materials [23]. However, this conclusion is only valid if the hydrostatic pressures challenging the textiles during use do not exceed 0.06 kPa. Depending on the end-use application, hydrostatic pressures far in excess of 0.06 kPa and probably as high as 13.8 kPa can be exerted on textile barriers during use [24,25,26].

Assuming that the hydrostatic pressure exerted on the liquid causes the liquid to wet and penetrate into the textile barrier, the Washburn equation identifies two variables that can influence the flow rate of liquids through the textile: the pore length (influenced by thickness of the textile and tortuosity of the path) and the viscosity of the liquid. As the hydrostatic pressure on the penetrating liquid is increased, the flow rate of the liquid through the textile barrier will increase and the corresponding time to penetrate through the textile will decrease.

Each personal protective clothing and infection control end-use application may have different requirements for the period of time that the textile barriers are expected to meet the barrier performance expectations. The true goal of laboratory testing is to discriminate among the barrier properties in a meaningful way. This does not mean that the challenge time in the laboratory has to equal the challenge time in each end-use application, but that the challenge time in the laboratory is controlled in a way that provides data that can be used to make reasonable predictions of barrier performance in each end-use application.

If the reason for considering increasing the time interval of the test in the laboratory is to determine how long the textile will act as an effective microbial barrier in actual use, then other types of use factors that can significantly impact barrier integrity and induce other modes of failure should be considered. Factors influencing microbial penetration in use could include a whole list of physical, chemical, and thermal stresses. These stresses may need to modeled in the laboratory and used as preconditioning steps prior to microbial barrier integrity testing. As an example, flexing or abrading a textile sample for 1 min may cause an immediate failure at very low pressure during barrier integrity testing. However, without flexing or abrading, that same textile may not demonstrate failure over long periods of time at high pressure.

G. Liquid Challenge Testing

Liquid challenge testing has been used over the years as a means to predict the liquid-borne microbial barrier properties of textiles [6,24]. A lot of the work done in this area focused on the need for infection control in the surgical end-use applications. However, when health-care workers became aware of the personal risks associated with direct contact with blood and body fluids, textile and garment manufacturers, academia, regulatory agencies, and the medical community became much more critical of barrier integrity testing. Recent work has demonstrated that there are significant limitations in the industry standard liquid challenge test methods [24,27]. This work has led to the conclusion that liquid challenge tests can be useful prescreening tools in determining which protective fabrics warrant further investigation with microbiological challenge tests, but should never be used alone to infer absolute microbial barrier properties. The most common liquid challenge test methods are briefly reviewed and some of the more significant limitations are discussed next.

1.   Review of Common Standard Liquid Challenge Tests

The INDA Standard Test IST 80.5-92 (Saline Repellency of Nonwovens) is illustrated in Figure 22. This test is designed to measure the amount of time required for saline to penetrate through textile barriers under defined conditions. Specimens of the textile are cut and fit into the lid of a Mason jar, which is inverted and placed on a glass surface over a mirror. Saline is added through a hole in the bottom of the jar and adjusted to a height of 115 mm (1.13 kPa hydrostatic pressure). The test is terminated when visible penetration of the saline through the textile is observed and the time (in minutes) is recorded. The longer the time interval the more repellent the textile. (Normally this test is terminated if no visible penetration occurs in 60 min.) Here the saline has been substituted with synthetic blood as part of a study to determine the effect of liquid type on the outcome of the results.

AATCC Test Method 42-1989 (Water Resistance: Impact Penetration Test) is illustrated in Figure 23. This test is designed to measure the amount of water that can penetrate through a textile under defined conditions. A preweighed piece of blotter paper is placed under a specimen of the textile barrier that is oriented at 45 degrees to a funnel situated 61 cm above. After 500 ml of distilled water is poured through the funnel, impacting the top surface of the textile, the blotter is reweighed. Any weight gain in the blotter is attributed to water penetrating through the textile barrier. The lower the weight gain in the blotter, the more water impact resistant the textile is. Here the water has been substituted with synthetic blood as part of a study to determine the effect of liquid type on the outcome of the results.

AATCC Test Method 127-1989 (Water Resistance: Hydrostatic Pressure Test) is illustrated in Figure 24. This test is designed to measure the hydrostatic pressure necessary to force water to penetrate through a textile under defined conditions. A specimen of the textile barrier is clamped over the end of a water column. The height of the water in the column is raised at the rate of 1.0 cm/sec until penetration of the water is visible through the textile. The standard column height is 100 cm (maximum hydrostatic pressure = 9.8 kPa). The higher the column height achieved before water penetration, the greater the water resistance of the textile is.

Figure 22   The saline repellency of textiles can be determined using INDA Standard Test IST 80.5-92 (Saline Repellency of Nonwovens, also known as the Mason jar test). (Photo courtesy of the Institute for Environmental Research, Kansas State University, Manhattan, KS.)

2.   Limitations of Standard Liquid Challenge Tests

Detecting liquid penetration through the use of the naked eye or by weight gain in a paper blotter is significantly less sensitive than a microbiological assay. A significant number of microorganisms can be carried in a very minute volume of liquid, which may not be visible to the naked eye or measured by weight gain in a blotter (refer to Fig. 25).

The liquids normally used in these liquid challenge tests, water and saline, have high surface tensions, exhibit high contact angles (>90 degrees) on most textile barriers, and consequently do not wet or penetrate through textile barriers as easily as some of the liquids that are potentially contaminated with hazardous microorganisms.

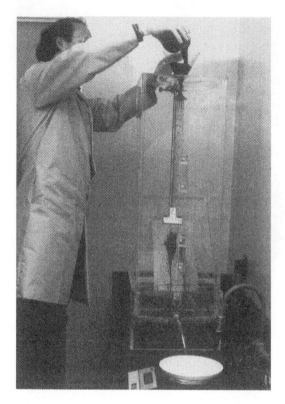

Figure 23 The water impact resistance of textiles can be determined using AATCC Test Method 42-1989 (Water Resistance: Impact Penetration Test). (Photo courtesy of the Institute for Environmental Research, Kansas State University, Manhattan, KS.)

These test devices have limitations on the amount of pressure that is applied to the liquid during the challenge procedure and may not be indicative of the pressures that can be exerted on liquids in contact with textile barriers during use.

These liquid challenge tests are normally conducted for shorter periods of time than the anticipated time of liquid challenge in some end-use applications. A short time in combination with low liquid challenge pressure may render misleading test results.

3. Development of New Tests

Recognizing the inadequacies of the industry standard liquid challenge tests prompted two manufacturers to respond and attempt to develop new liquid challenge tests. Each one of these new test methods is briefly described next.

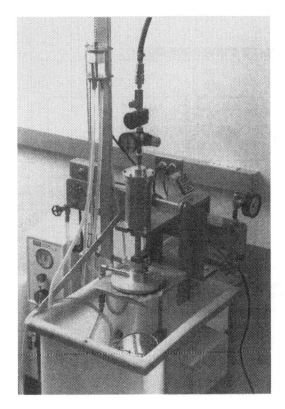

Figure 24 The penetration resistance of textiles to water at low pressure can be determined using AATCC Test Method 127-1989 (Water Resistance: Hydrostatic Pressure Test).

The Kimberly-Clark blood strikethrough test is illustrated in Figure 26. This test was designed to measure the amount of heparinized bovine blood that could penetrate through textile barriers under defined conditions. This procedure is performed by placing a small amount (1.4 g) of bovine blood on the surface of the textile. The jack stand is then raised, compressing the textile between the water bottle and the top plate of the test apparatus. The pressure on the water bottle is increased to 6.9 kPa. After a specified period of time the pressure is released and a paper blotter is removed from under the textile and weighed. The lower the weight gain in the blotter, the more resistant the textile is to the penetration of bovine blood.

This test procedure was considered for standardization through the American Society for Testing and Materials (ASTM); however, it was found that certain types of absorbent nonbarrier fabrics, such as surgical gauze, would pass this test by preventing the penetration of bovine blood. Therefore, ASTM discontinued its effort with this method [28].

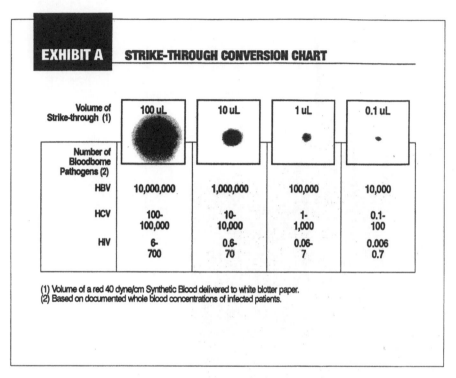

EXHIBIT A **STRIKE-THROUGH CONVERSION CHART**

Volume of Strike-through (1)	100 uL	10 uL	1 uL	0.1 uL
Number of Bloodborne Pathogens (2)				
HBV	10,000,000	1,000,000	100,000	10,000
HCV	100-100,000	10-10,000	1-1,000	0.1-100
HIV	6-700	0.6-70	0.06-7	0.006 0.7

(1) Volume of a red 40 dyne/cm Synthetic Blood delivered to white blotter paper.
(2) Based on documented whole blood concentrations of infected patients.

Figure 25 Strike-through conversion chart.

The GORE elbow lean test is illustrated in Figure 27. This test was designed to determine if a body fluid model (synthetic blood) could visibly penetrate through textile barriers during simulated pressing and leaning activities. The body fluid model was selected because the whole blood of humans or other animals may not be predictive of the wetting and penetration characteristics of the entire range of potentially infectious human body fluids. Excluding saliva, the surface-tension range of human blood and body fluids is 0.042–0.060 N/m (refer to Table 1). A more appropriate body fluid model would have a surface tension approximating the lower end of the blood and body fluid range and would be more predictive of the penetration characteristics of body fluids and other liquids with higher surface tensions. Researching the literature led to a synthetic blood formulation [29]. The synthetic blood contains 10 g direct red 081 dye (CI 28160, colorant with surfactant), 25 g Acrysol G-110 (thickening agent), and 1.0 L normalized distilled water. The surface tension of the synthetic blood is 0.042 N/m. The synthetic blood has been evaluated in comparison to other test liquids, including heparinized whole bovine blood, for penetration through a variety of commercially available textile barrier fabrics and has been found to be the preferred test liquid [27].

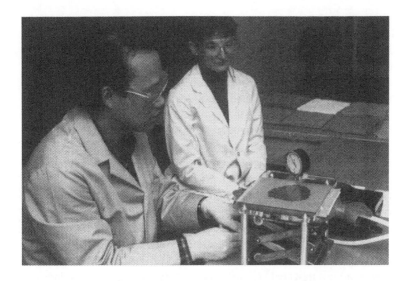

Figure 26 Kimberly-Clark blood strike-through test. (Photo courtesy of the Institute for Environmental Research, Kansas State University, Manhattan, KS.)

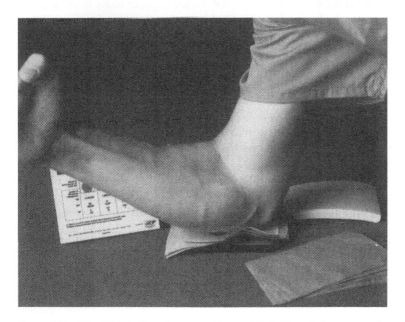

Figure 27 The GORE elbow lean test. Contact W. L. Gore & Associates, Inc., for details on how to order test kits [11].

The GORE elbow lean test is performed by saturating a foam pad with the synthetic blood, laying the textile barrier over the soaked pad (face side down), placing a white paper blotted over the textile (the paper is backed by a polyethylene film to prevent penetration of the synthetic blood if the textile fails), and then pressing and leaning on the textile. Pressing and leaning can be done in various ways to simulate the types of stresses and pressures that the textile barrier would be exposed to during use. Leaning with the elbow can easily apply direct mechanical pressures on the textile in excess of 345 kPa. Visible penetration of synthetic blood through the textile to the paper blotter denotes failure. If desired, the volume of penetrating blood and the respective number of infectious microorganisms can be approximated by comparing the blotter to the strike-through conversion chart (refer to Fig. 25). Textiles that prevent the visible penetration of synthetic blood during pressing and leaning are considered to be more protective against the penetration of blood and body fluids.

Although the GORE elbow lean test has not been standardized, the results of this simple pressing and leaning experiment have served as a benchmark to compare with other more complicated scientific laboratory tests and to develop new standardized liquid and microbial challenge procedures (ASTM ES21-92 and ASTM ES22-92). This test serves as a classic example of how to apply knowledge of the real-world end-use application for textile barriers to the development of challenge test methods in the laboratory that would be more likely to reduce risk. The GORE elbow lean test can also be used by end users in the field, simulating many different types of personal protection and infection control end-use applications, to help determine the visible liquid barrier properties of textiles.

Typically, textile barriers that are not film-reinforced will fail and textile barriers that are film-reinforced will pass the GORE elbow lean test [2,3,24,27]. (Those textiles found in Figs. 3–10 consistently demonstrate failing results, and those textiles found in Figs. 35 and 35 consistently demonstrate passing results.) Illustrations of failing and passing results are given in Figures 28 and 29.

The resistance of textiles to the visible penetration of synthetic blood can also be determined by using ASTM ES 21-92 (Emergency Standard Test for Resistance of Protective Clothing Materials to Synthetic Blood). The development of this method is based on correlation studies that were done, comparing the results obtained when testing the synthetic blood resistance of a variety of commercially available textile barriers, using the GORE elbow lean test and the ASTM ES 21-92 test [24,27] (refer to Table 2).

The test was developed as a prescreening test for ASTM ES22-92 (discussed later in this chapter). ASTM ES21-92 is designed to determine if textiles can prevent the visible penetration of synthetic blood under defined test conditions. Textile specimens are clamped into the penetration cell with the face side oriented toward the cell cavity. Then the penetration cell in attached to the apparatus and the cell cavity is filled with 60 ml of synthetic blood. An air line is connected to

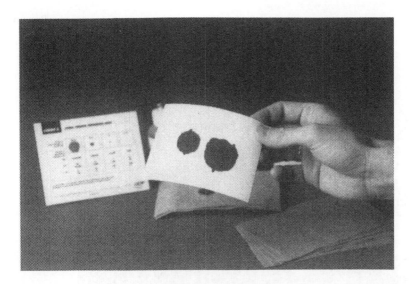

Figure 28 A failing result on the GORE elbow lean test. The specimen tested was cut from a fabric-reinforced surgical gown. The textile type is spunlace woodpulp/polyester nonwoven (refer to Figs. 9 and 10). The penetrating synthetic blood on the left of the paper blotter is the result of one finger press, and on the right is the result of one elbow lean.

Figure 29 A passing result on the GORE elbow lean test. The specimen tested was cut from a film-reinforced surgical gown. The textile type is spunlace woodpulp/polyester nonwoven with film reinforcement (refer to Fig. 35). No visible penetration of the synthetic blood is noted on the paper blotter as a result of the finger press or the elbow lean. Multiple leans can be performed and yield the same results.

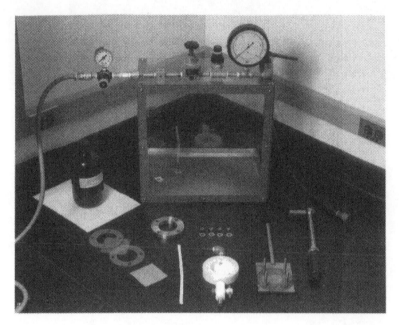

Figure 30 The resistance of textiles to the visible penetration of synthetic blood can be determined by ASTM ES21-92 (Emergency Standard Test for Resistance of Protective Clothing Materials to Synthetic Blood).

the cell and a specific time and hydrostatic pressure protocol is followed: 5 min at ambient pressure, 1 min at 13.8 kPa, and 54 min at ambient pressure. The back side of the textile specimen is observed through the viewing port, and any visible sign of synthetic blood penetration through the textile denotes a failure. Examples of failing and passing results are illustrated in Figures 31 and 32. Textiles that demonstrate passing results should then be tested with ASTM ES22-92 to confirm the microbial barrier integrity.

The results of the ASTM ES21-92 test (Figs. 31 and 32) can be directly compared to the results for the GORE elbow lean test (Figs. 28 and 29). Once again, textile barriers that are not film-reinforced typically fail and textile barriers that are film-reinforced typically pass the ASTM ES21-92 test [2,3,24,27]. (Those textiles found in Figs. 3–10 consistently demonstrate failing results and those textiles found in Figs. 35 and 36 consistently demonstrate passing results.)

A comparison of the six liquid challenge tests already listed was performed using synthetic blood and nine different single-use and multiple-use barrier textiles. (This necessitated replacement of the standard liquid challenge reagents with synthetic blood for some of the tests.) The results of this analysis can be found in Table 2. The highest correlation found between any two tests was between the

Figure 31   Failing test result for ASTM ES21-92. The specimen tested was cut from a fabric reinforced surgical gown. The textile type is spunlace woodpulp/polyester nonwoven (refer to Figs. 9 and 10).

Figure 32   Passing test result for ASTM ES21-92. The specimen tested was cut from a film-reinforced surgical gown. The textile type is spunlace woodpulp/polyester nonwoven with film reinforcement (refer to Fig. 35).

Table 2 Spearman Correlation Coefficients for Liquid Barrier Tests Using Synthetic Blood

Comparative liquid barrier tests	KC blood strike-through test	AATCC 42-1989 impact penetration test	IST 80.5-92 Mason jar test	ASTM ES21-92 synthetic blood resistance test at 6.9 kPa	ASTM ES21-92 synthetic blood resistance test at 13.8 kPa	AATCC 127-1989 hydrostatic resistance test
KC blood strike-through test	N/A					
AATCC 42-1989 impact penetration test	0.92[b]					
IST 80.5-92 Mason jar test	0.69[a]	0.75[a]				
ASTM ES21-92 synthetic blood resistance test at 6.9 kPa	0.38	0.25	0.54[b]			
ASTM ES21-92 Synthetic blood resistance test at 13.8 kPa	0.60	0.49	0.37	0.59		
AATCC 127-1989 hydrostatic resistance test	0.92[b]	0.75[a]	0.56	0.63	0.74[a]	
GORE elbow lean test	0.61	0.50	0.38	0.60	0.98[b]	0.75[a]

[a]Significant to the .05 level.
[b]Significant to the .01 level.
Source: Ref. 27.

GORE elbow lean test (dubbed the "human factors test") and the ASTM ES21-92 test. This high correlation (.98 correlation coefficient, significant at the .01 level) demonstrates that these two tests were able to discriminate among the nine different barrier textiles in the same way. The ASTM ES21-92 test, a laboratory benchtop test, discriminated among the nine different barrier textiles in the same way as the GORE elbow lean test, a simulated use or "human factors" validation test. The other five tests were not able to discriminate among the nine different barrier textiles in the same way as the GORE elbow lean test. This finding is very important for the laboratory evaluation and determination of which types of textiles might be capable of resisting visible liquid penetration in end-use applications involving soaking with blood and body fluids and pressing and leaning pressures.

H. Pressure

As mentioned previously, the pressure placed directly on the challenging air and liquids can significantly influence the flow rate of those fluids through porous textile barriers. However, quantifying the pressures actually applied on these fluids in the real world during exposure situations is difficult. Each end-use application will bring a different set of dynamic uncontrolled circumstances that can change the factors that are important in calculating the pressures applied to the air and liquids challenging the barrier textiles. Pressure is defined [9] as:

Pressure (P) = force per unit area

SI unit: Pascal (Pa)   (9)

1 Pa = 1 N/m^2 (newton per square meter)

Most of the work done to date has focused on quantifying the mechanical pressures applied to the textiles as they are compressed between the human body and the work space. For example, the pressures exerted on surgical gowns during pressing and leaning activities in surgery can range from less than 6.9 kPa to more than 414 kPa [31]. The hydrostatic pressures exerted on the liquids in contact with the surgical gowns during pressing and leaning have not yet been quantified, but are thought to easily exceed 6.9 kPa [24,25].

As discussed earlier, when comparing the resistance of many different barrier textiles against synthetic blood, there is a strong correlation between the 13.8 kPa hydrostatic pressure used in the ASTM ES21-92 test and the much higher direct mechanical pressure ($\approx$345 kPa) used in the GORE elbow lean test. This correlation was confirmed again in a more recent study involving an even higher direct mechanical pressure (427 kPa) on the textile [26]. Under most end-use conditions the hydrostatic pressure exerted on the challenging liquid will be lower than the direct mechanical pressure applied to the textile as the liquid is not contained but free-flowing and able to escape from the forces being applied to it. In an attempt to illustrate this point, Figures 33 and 34 were prepared. Figure 33 is a

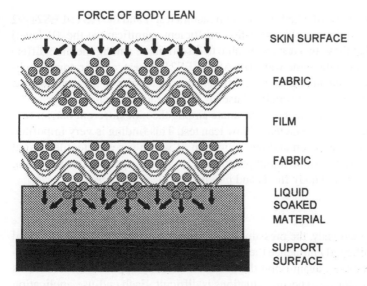

FORCE OF BODY LEAN

SKIN SURFACE

FABRIC

FILM

FABRIC

LIQUID
SOAKED
MATERIAL

SUPPORT
SURFACE

Figure 33 Orientation of a textile (three-layer film-reinforced laminate) between the human body and a wet contaminated work environment. Here the liquid is adsorbed into a porous compressible material, such as a foam pad. As the force of the body increases, the textile is pressed into the liquid-soaked pad. While a direct mechanical pressure is exerted on the textile and the foam, a corresponding hydrostatic pressure is exerted on the liquid. Since the liquid is free flowing, pressure will only remain on the liquid until the structural elements of the textile align themselves to provide a solid support for the force of the body against the work surface. The range of direct mechanical pressures exerted on the textile can be calculated and controlled; however, the range of hydrostatic pressures exerted on the liquid has only been approximated and cannot be controlled.

representation of the GORE elbow lean test and Figure 34 is a representation of the ASTM ES21-92 test.

I. Textile Composites

Film-reinforced textiles, as a group, can offer a much higher level of resistance to airborne and liquid-borne microbial penetration. When textiles include film reinforcements, the films become the overriding factor in determining the air, liquid, and microbial barrier properties. The textiles mainly serve to support and protect the films. The types of textiles that are used for this purpose and the ways in which the textiles and films are combined can have a direct influence on the initial barrier properties and the in-use durability of the those barrier properties. As an example, two-layer composites (film combined with one textile) where the film is oriented out to the work environment would subject the film structure directly to potentially damaging physical stresses, such as abrasion, scoring, etc.

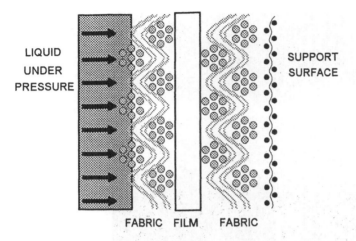

LIQUID UNDER PRESSURE

SUPPORT SURFACE

FABRIC FILM FABRIC

Figure 34 This diagram illustrates the orientation of a textile (three-layer film-reinforced laminate) in a hydrostatic test device such as ASTM ES21-92 and ASTM ES22-92. Here the liquid represents the contaminated work environment and is placed in direct contact with the textile. The hydrostatic pressure placed on the liquid is known and can be very precisely controlled. As the hydrostatic pressure on the liquid is increased, the liquid is compressed against the textile. In some cases, if the textile is weak or subject to distortion-related failures, a retaining screen can be used to add support. Liquid penetration through the textile occurs if the hydrostatic pressure exceeds the ability of the textile to resist wetting.

Concerning the evaluation of the protective properties of films, there are two major film reinforcement categories, monolithic and microporous, which are defined by fundamental structural differences that can influence the airborne and liquid-borne microbial barrier properties. Other subcategories exist that are defined more by the chemical nature of the films and how the films interact with liquids, such as hydrophobic, hydrophilic, oleophobic, etc., which may further influence the airborne and liquid-borne microbial barrier properties.

The inherent differences between the two major categories of textile film reinforcements are that (1) porous film reinforcements allow the bulk flow of air through the pores, relying on filtration mechanisms against air-borne biohazards, and resist the penetration of challenging liquids to varying degrees based on the laws and equations discussed earlier [Eq. (2)–(7)] and (2) monolithic (nonporous) textile film reinforcements do not allow the bulk flow of challenging air or challenging liquids.

The major types of film reinforcements being used today for microbial barrier textiles rely either entirely or partially on a monolithic structure to impart the air, liquid, and microbial barrier properties. Representative structures are depicted in Figures 35 and 36. Three major studies have been conducted, using the most

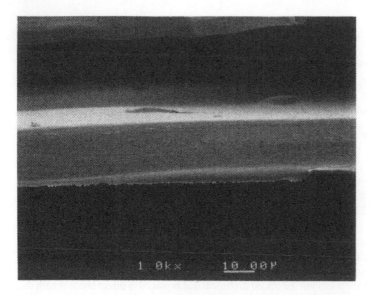

Figure 35 Scanning electron micrograph: cross section at 1000× magnification, Optima (polyreinforced), supplied by Baxter Healthcare Corp., described as spunface woodpulp/polyester nonwoven with film reinforcement; weight 115.44 g/m², thickness 0.65 mm, bulk density 177.60 g/m³. Description of film: monolithic (nonporous) polyethylene.

stringent liquid and liquid-borne microbial challenge procedures available today, which have demonstrated that these two film-reinforced textiles provide excellent barriers to liquids and liquid-borne microorganisms [2,3,26].

Unlike porous film reinforcements, the liquid penetration resistance of monolithic film reinforcements is only influenced by one variable in Eq. (4), hydrostatic pressure. In reference to Figure 18, the line that has been plotted for monolithic films simply represents the strength of the film. The breakthrough pressure of monolithic film reinforcements, across the whole range of liquid surface tensions, will be the result of the film bursting. Porous films, on the other hand, will behave similarly to the line represented for porous textiles as the breakthrough pressure will be proportional to the wetting characteristics of the challenging liquid and subject to all of the variables in Eq. (4).

1. Film Strength

One method of distinguishing between monolithic film reinforcements is to evaluate the bursting strength. Bursting strength may also be determined after various different types of preconditioning steps, such as flexing, abrading, etc., to simulate the impact of actual use. However, the bursting strength test is not intended as a measurement of the integrity of monolithic film reinforcements against airborne

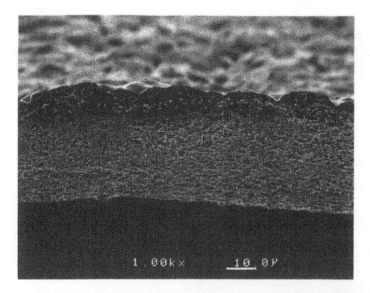

Figure 36 Scanning electron micrograph: cross section at 1000× magnification, GORE surgical barrier fabric, supplied by W. L. Gore & Associates, Inc. described as woven polyester/film/knitted polyester laminate; 184.88 g/m², thickness 0.47 mm, bulk density 393.36 g/m³. Description of film: bicomponent; expanded polytetrafluoroethylene membrane partially impregnated with a monolithic (nonporous) hydrophilic/oleophobic polymer.

or liquid-borne microbial penetration. With the exception of pressure, this test suffers from all the same limitations of many of the other liquid challenge tests discussed previously. Also, the results obtained on liquid challenge tests of this type have been found to be sensitive to the ramping speed [32].

The burst strength of film-reinforced textiles can be determined by using ASTM D751–89 (Standard Test Methods for Coated Fabrics; Procedure A). The test device is depicted in Figure 37. Figure 38 shows a specimen being observed for leakage.

2. Film Integrity Testing

Characterizing the maximum pore size in porous film reinforcements or the size of defects in either monolithic (nonporous) or porous film reinforcements can be done by using ASTM F316–86 (Standard Test Methods for Pore Size Characteristics of Membrane Filters by Bubble Point and Mean Flow Pore Test). This procedure is illustrated in Figures 39 and 40. The maximum pore or defect size is calculated using the following equation:

$$d = C\gamma/p \tag{10}$$

Figure 37 The penetration resistance of textiles to water at high pressure can be determined using ASTM D751-89 (Standard Test Methods for Coated Fabrics; Procedure A). The hydrostatic pressure of the water on the textile is steadily increased until visible penetration occurs or until the 6984 kPa pressure limit is reached. The higher the pressure achieved before water penetration, the greater is the water resistance of the textile. (This test is also known as Mullen's burst test.)

where

      d = limiting diameter
      C = constant (2860 when p is in pascals)
      γ = surface tension, mN/m
      p = pressure, Pa

Results of this test may not be a direct indication of the particle retention characteristics of a film but may be used as a quality control tool or for comparative

Figure 38   ASTM D751–89, procedure A. Test specimen is observed until leakage or bursting is determined. Note how the textile specimen is distorted (ballooning up) by the hydrostatic pressure of the water.

analysis. The microbial retention characteristics of porous film reinforcements should always be validated using filtration test methods.

J.   Microbial Challenge Testing

1.   A Brief Historical Review

Historically, even with the urging of such noted authorities as William C. Beck, M.D., F.A.C.S., dating back to 1952, the medical fabric industry was unable to reach a consensus regarding microbiological barrier performance standards [33]. The plethora of different test methods being used to assess the "barrier" properties of materials, including both industry standard and corporate-sponsored methods, resulted in a significant state of confusion among the members of the health-care community concerning product performance. The literature was replete with comparative analyses of scientific laboratory bench-top liquid and microbiological barrier evaluations [6,22]. However, the objective was always to compare and rank product performance and not to identify which products might actually be capable of preventing microbial penetration. In the past, laboratory testing was mainly intended to identify those materials that might reduce postoperative wound infection rates but not to identify those materials that could prevent the transmission of infectious microorganisms in use. Comparative in-use analyses of barrier products followed a similar theme [34].

Figure 39 The maximum pore size of nonfibrous porous film reinforcements used in textiles can be approximated using ASTM F316-86 (Standard Test Methods for Pore Size Characteristics of Membrane Filters by Bubble Point and Mean Flow Pore Test).

Recently, the focus for preventing transmission of infectious microorganisms through barrier materials has grown to include both infection control and personal protection. One major reason for this growth is the significant risk associated with occupational exposure to bloodborne pathogens perceived by the health care community. The Occupational Safety and Health Administration (OSHA), under the U.S. Department of Labor, published its Final Rule on protecting health care workers from occupational exposure to bloodborne pathogens. The OSHA Final Rule [35] states:

When there is occupational exposure, the employer shall provide at no cost to the employee, appropriate personal protective clothing, such as, but not

Figure 40 Test result for ASTM F316-86. This is an example of the first stream of continuous air bubbles breaking through a microporous membrane. Calculations would be made as to the maximum pore size of this film and compared to filtration test results using particles and/or microorganisms of various types and sizes.

limited to, gloves, gowns, laboratory coats, face shields or masks, and eye protection, and mouthpieces, resuscitation bags, pocket masks, or other ventilation devices. Personal protective equipment will be considered "appropriate" only if it does not permit blood or other potentially infectious materials to pass through to or reach the employees' work clothes, street clothes, undergarments, skin, eyes, mouth, or other mucous membranes under normal conditions of use and for the duration of time in which the protective equipment will be used.

In the last decade the medical literature has exploded with research studies reporting on bloodborne pathogen exposure rates based on various occupational risk factors, on risk reduction strategies, and on compliance issues [36–43]. There have also been a number of significant new standards, books, recommended practices, and technical reports published. Many of these references are listed in the Appendix.

Beyond bloodborne pathogens looms the threat of other potentially hazardous microorganisms: prions (Creutzfeldt-Jakob agent), *Escherichia coli* 0157:H7, Muerto Canyon virus (hantavirus), and multiple-drug-resistant forms of *Mycobacterium tuberculosis*, staphylococci, and enterococci, to name a few.

Workers can also face other significant microbiological hazards than bloodborne pathogens, such as biotechnology workers dealing with recombinant DNA (rDNA), laboratory workers handling concentrated cultures of human pathogens (other than bloodborne), and veterinary or agriculture workers dealing with zoonotic agents. Each work environment and potential microbiological hazard may require a different strategy of experimentation and risk reduction decision logic. Similar strategies and risk reduction decisions may be made for microbes falling into similar classifications.

2. Classification Schemes for Biohazards

Many of the key variables important to assessing the degree of hazard associated with exposure to biohazards have already been discussed. Specifically, with respect to determining the most appropriate type of microbial challenge, the following factors are important:

1. The type of microbe(s): size, shape, concentration, environmental viability/ stability, resistance to inactivation, compatibility with the textile material (no antimicrobial effects), binding mechanisms, motility, and limit of detection.
2. The susceptibility of the host: host immunity, virulence of the microbe(s), the dose necessary to cause infection, and the risk to the lab technician.
3. The nature of the exposure(s): transport modes (direct or vehicle dependent), vehicle type (air or liquid), associated carriers, forces and pressures applied directly to the vehicles.
4. The state of the textile when exposure occurs: environmental conditions, and physical, chemical, and thermal stresses.

   With respect to handling the actual human pathogenic microbial agents there are a number of good references that should be consulted prior to beginning any experimentation (refer to the Appendix).

3. Transfer of Vehicles Versus Transfer of Microbes

Internal research at W. L. Gore & Associates, Inc., and research published by others regarding the viral barrier properties of protective clothing products have demonstrated that viral penetration can occur in the absence of any perceivable liquid penetration [2,11,44]. Similar results have also been found with latex surgical gloves [45,46]. Therefore, testing the penetration characteristics of textiles to air and liquids does not rule out the possibility of the transmission of infectious microorganisms. Certainly, those materials that appear to be highly resistant are probably much better barriers to microorganisms. However, since the real hazard is infectious microorganisms, the goal should be to demonstrate that the textiles are effective microbiological barriers. The only truly definitive test is a microbiological challenge.

The one significant limitation of air and liquid penetration testing of textile barrier products is the limit of detection for fluid transfer. With respect to liquid penetration, this can be graphically illustrated by referring to Figure 25; the strike-through conversion chart. This chart converts the amount of strike-through to the amount of bloodborne pathogen contamination. The four spots at the top of the chart were formed from premeasured droplets of synthetic blood and are marked in microliters, ranging from 100 µl to 0.1 µl. Listed on the left are the three primary bloodborne pathogens: hepatitis B (HBV), hepatitis C (HCV), and human immunodeficiency (HIV) viruses. The approximate number of infectious units that could be present in each spot was calculated from the known blood serum concentration of infected patients and is shown for each type of virus. For example, the number of infectious units of hepatitis B virus in a 0.1 µl droplet of infected blood serum can be 10,000 or higher; this is one of the reasons why hepatitis B is so highly infectious and easily transmitted.

When considering the high concentrations of hepatitis B virus (HBV) [20], hepatitis C virus (HCV) [47], and the human immunodeficiency virus (HIV) [48] in the whole blood and blood serum of infected patients, it is clear that very minute amounts of liquid, which may not be visible to the naked eye, can carry significant quantities of infectious disease. (Other microorganisms with concentrations of 1×10^8 infectious units/ml would be comparable to the HBV results on this chart.) When thinking in these terms, even very small amounts of liquid strike-through may represent an unacceptable risk.

4.   Characteristics of Pathogenic Microbes

It has been estimated that there are at least 193 important biological agents that show infectious, allergenic, toxic, or carcinogenic activities in the working population [43]. The following diagrams and charts are not represented to be complete, but are based on the references cited. They are provided in an attempt to illustrate the magnitude of the problem associated with hazardous microorganisms and to illustrate the diversity of shapes and sizes.

Figure 41 illustrates the morphology of animal virus families including human pathogens, and Table 3 gives other details of pathogenic viruses. Figure 42 shows the morphology of bacterial ultrastructure, with Table 4 giving for other details of pathogenic bacteria. In Figure 43 are shown different types of spores in fungi, accompanied by Table 5 for other details of pathogenic fungi.

5.   Possible Models for Pathogenic Microbes

With the impending OSHA Final Rule focusing on reducing the risk of occupational exposure to hepatitis B and C and the human immunodeficiency viruses, the medical fabric industry expended significant efforts to develop a new barrier integrity test. One of the primary factors in the design of this new test was the

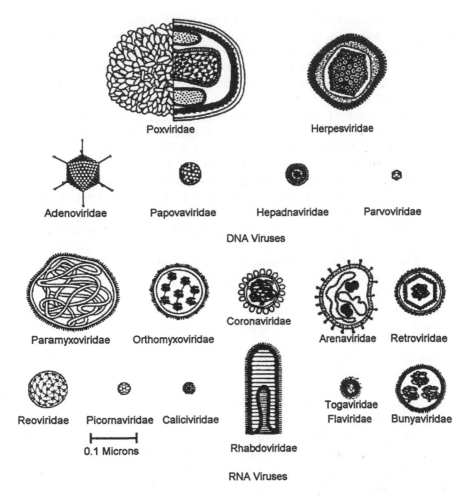

Poxviridae   Herpesviridae

Adenoviridae   Papovaviridae   Hepadnaviridae   Parvoviridae

DNA Viruses

Paramyxoviridae   Orthomyxoviridae   Coronaviridae   Arenaviridae   Retroviridae

Reoviridae   Picornaviridae   Caliciviridae

0.1 Microns   Rhabdoviridae

Togaviridae
Flaviridae   Bunyaviridae

RNA Viruses

Figure 41 Morphology of animal virus families including human pathogens. (From Ref. 49.) Refer to Table 3 for other details of pathogenic viruses.

selection of an appropriate model that was capable of determining the "viral" barrier properties of protective clothing materials. Table 6 provides a brief summary of some of the various types of options that were considered.

The phi-X174 bacteriophage was selected as the most appropriate bloodborne pathogen model because it has no envelope (similar to HCV), is 27 nm in size (similar to HCV, the smallest pathogen), has icosahedral or nearly spherical morphology (similar to all three pathogens), can be cultivated to reach very high titers [$>10^8$ placque-forming units (PFU)/ ml, similar to HBV, the most concentrated

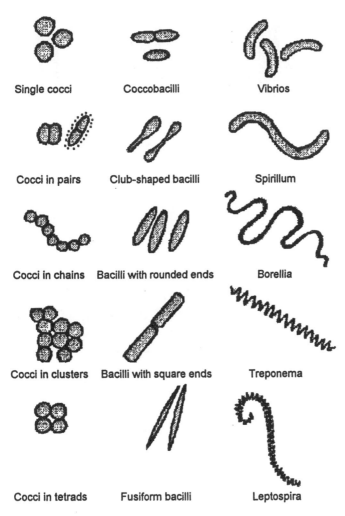

Single cocci   Coccobacilli   Vibrios

Cocci in pairs   Club-shaped bacilli   Spirillum

Cocci in chains   Bacilli with rounded ends   Borellia

Cocci in clusters   Bacilli with square ends   Treponema

Cocci in tetrads   Fusiform bacilli   Leptospira

Figure 42 Morphology of bacterial ultrastructure. (From Ref. 69.) Refer to Table 4 for other details of pathogenic bacteria.

pathogen], has excellent environmental stability (similar to HBV), is noninfectious to humans (low risk), has an extremely low limit of detection (approaching 1 viral particle/ml), and grows very rapidly (assay results can be read in as few as 6–18 hrs). Further research has also demonstrated that the phi-X174 bacteriophage has fewer compatibility problems than alternative surrogate viruses, is less subject to binding, and the filtration behavior of phi-X174 is as expected (based on the size determined by scanning electron microscopy, SEM) [56,57].

Table 3 Characteristics of Pathogenic Viruses

Viral organisms	Associated disease(s)	Morphology	Smallest diameter (μm)	Shortest length (μm)	Other important factors
Adenoviruses	Pharyngitis, respiratory disease, keratoconjunctivitis, hemorrhagic cystitis, gastroenteritis, cervicitis, urethritis	Icosahedral	0.07	N/A	No envelope, DNA genome
Arenaviruses	Lassa fever, hemorrhagic fever, lymphocytic choriomeningitis	Spherical (pleomorphic)	0.05	N/A	Envelope, RNA genome
Bunyaviruses	Hemorrhagic fever (Hataan), encephalitis	Spherical	0.09	N/A	Envelope, RNA genome
Caliciviruses	Hepatitis (E), Norwalk gastroenteritis	Icosahedral	0.035	N/A	No envelope, RNA genome
Coronaviruses	Respiratory, colds	Spherical (pleomorphic)	0.06	N/A	Envelope RNA genome
Delta virus	Hepatitis (D) (superinfection of HBV carriers)	Spherical	0.036	N/A	No envelope, RNA genome
Filoviruses	Marburg and Ebola hemorrhagic fevers	Filamentous	0.08	0.8	Envelope, RNA genome
Flaviviruses	Hepatitis (C), yellow fever, hemorrhagic fever, encephalitis	Spherical	0.04	N/A	Envelope, RNA genome
Hepadnaviruses	Hepatitis (B)	Spherical	0.042	N/A	Envelope, DNA genome
Herpesviruses	Herpes, chicken pox, cytomegalovirus infection, mononucleosis	Spherical	0.12	N/A	Envelope, DNA genome
Orthomyoviruses	Influenza	Spherical or filamentous	0.08	N/A	Envelope, RNA genome

Papovaviruses	Warts, cancer	Icosahedral	0.045	N/A	No envelope, DNA genome
Paramyxoviruses	Measles, mumps, parainfluenza, respiratory syncytial virus infection,	Spherical (pleomorphic)	0.15	N/A	Envelope, RNA genome
Parvoviruses	Erythema infectiosum, rheumatoid arthritis	Icosahedral	0.018	N/A	No envelope, DNA genome
Picornaviruses	Hepatitis (A), polio, colds, meningitis, paralysis, myocarditis, leurodynia, herpangina, acute hemorrhagic conjunctivitis, hand, foot, and mouth disease, pancreatitis, gastroenteritis	Icosahedral	0.025	N/A	No envelope, RNA genome
Poxviruses	Smallpox, *molluscum contagiosum*, monkeypox, cowpox and milkers nodes, Orf	Brick	0.1×0.24	0.3	Envelope, DNA genome
Reoviruses	Infantile gastroenteritis, diarrhea, Colorado tick fever	Icosahedral	0.06	N/A	No envelope, RNA genome
Retroviruses	Cancer, acquired immune deficiency syndrome	Spherical	0.08	N/A	Envelope, RNA genome
Rhabdoviruses	Rabies, hemorrhagic fever, vesicular stomatitis	Bullet	0.075	0.18	Envelope, RNA genome
Togaviruses	Rubella, fever, hemorrhagic fever, arthritis, eastern and Venezuelan equine encephalitis	Spherical	0.06	N/A	Envelope, RNA genome

Source: Refs. 49–51.

Table 4 Characteristics of Pathogenic Bacteria

Bacterial organisms	Associated disease(s)	Morphology	Smallest diameter (μm)	Shortest length (μm)	Other important factors
Acinetobacter calcoaceticus	Pneumonia, sepsis, endocarditis	Rod, coccoid rod	0.9	1.5	Gram negative, nonmotile, endotoxin
Aeromonas hydrophila	Wound infection, gastroenteritis, septicemia	Rod	0.3	1.0	Gram negative, motile, endotoxin
Alcaligenes faecalis	Urogenital tract, wound infections, abscesses, pleuritis	Rod, coccoid rod, or cocci	0.5	0.5	Gram negative, motile, endotoxin
Bacillus spp.	Anthrax, eye infections, urinary tract infections, septicemia, meningitis, otitis	Rod	0.5	1.2	Gram positive, sporulating, motile
Bordetella pertussis	Pertussis (whooping cough)	Coccoid rod	0.2	0.5	Gram negative, motile and nonmotile
Brucella spp.	Brucellosis, Mediterranean fever, Bang's disease	Coccoid or rodlike	0.5	0.5	Gram negative, nonmotile
Campylobacter spp.	Enteritis, septicemia, meningitis	S-shaped rod	0.2	1.5	Gram negative, motile
Chlamydia psittaci	Psittacosis	Coccoid or pear	0.2	N/A	Gram negative, nonmotile
Citrobacter spp.	Urinary tract and wound infection, septicemia, nosocomial infections	Rod	1.0	2.0	Gram negative, motile

Organism	Disease	Shape			Characteristics
Corynebacterium diptheriae	Diptheria	Rod	0.3	1.0	Gram positive, nonmotile
Clostridium spp.	Tetanus, botulism	Rod	0.5	1.3	Gram positive, sporulating, motile
Enterobacter spp.	Opportunistic, septicemia, nosocomial infections, meningitis	Rod	0.6	1.2	Gram negative, motile, endotoxin
Enterococcus spp.	Endocarditis, septicemia, urinary tract infection, nosocomial infections	Cocci, ovoid	0.5	N/A	Gram positive, some are motile
Escherichia coli	Diarrhea, dysentery, urinary tract infection, meningitis, nosocomial infections	Rod	1.1	2.0	Gram negative, some are motile
Francisella tularensis	Tularemia	Rod	0.2	0.2	Gram negative, nonmotile
Haemophilus spp.	Respiratory infections	Pleomorphic rod	0.3	0.5	Gram negative, nonmotile
Klebsiella pneumoniae	Pneumonia, urinary tract infection, meningitis, nosocomial infections	Rod	0.3	0.6	Gram negative, nonmotile
Legionella pneumophila	Legionnaires disease, pneumonia	Rod shaped or filamentous	0.3	2.0	Gram negative, motile
Leptospira interrogans	Leptospirosis (anicteric and icteric)	Helical rod	0.1	6.0	Gram negative motile

Table 4 Continued

Bacterial organisms	Associated disease(s)	Morphology	Smallest diameter (μm)	Shortest length (μm)	Other important factors
Listeria monocytogenes	Listeriosis, meningitis, meningo-encephalitis, septicemia, endocarditis	Rod	0.4	0.5	Gram positive, motile
Mycobacterium spp.	Tuberculosis, skin ulcers, leprosy, soft tissue infections	Rod	0.2	1.0	Gram positive nonmotile
Mycoplasma spp.	Pneumonia, arthritis, pelvic inflammatory disease, postpartum fever, wound infections	Pleomorphic	0.3	N/A	Gram negative, nonmotile
Neisseria spp.	Gonorrhea, meningoencephalitis	Cocci	0.6	N/A	Gram negative, nonmotile
Proteus spp.	Urinary tract infections, septicemia, meningitis, nosocomial infections	Rod	0.4	1.0	Gram negative, motile
Pseudomonas spp.	Glanders, endocarditis, respiratory and urogenital tract infection, septicemia, nosocomial infections	Straight or curved rod	0.5	1.5	Gram negative, motile
Rickettsia spp.	Q fever, typhus, Rocky Mountain spotted fever	Pleomorphic coccoid or rod	0.3	1.5	Gram negative nonmotile

Salmonella spp.	Food poisoning, typhoid fever	Rod	0.7	2.0	Gram negative, motile, endotoxin
Serratia spp.	Wound, urinary tract, and pulmonary infections, septicemia, meningitis, nosocomial infections	Rod	0.5	0.9	Gram negative, motile
Shigella spp.	Dysentery	Rod	1.0	2.0	Gram negative, nonmotile
Staphylococcus spp.	Inflammation, suppuration, nosocomial infections	Cocci	0.5	N/A	Gram positive, nonmotile
Streptobacillus monili-formis	Rat bite fever	Pleomorphic rod	0.1	1.0	Gram negative, nonmotile
Streptococcus spp.	Inflammation, scarlet fever, pneumonia, toxic shock, nosocomial infections, endocarditis	Cocci	0.5	N/A	Gram positive, nonmotile
Treponema pallidum	Syphillus	Spirochete	0.1	6.0	Giensa stain, motile
Vibrio spp.	Cholera, gastroenteritis	Curved rod	0.5	1.4	Gram negative, motile
Yersinia pestis	Bubonic plague	Pleomorphic rod	0.5	1.0	Gram negative, nonmotile

Source: Refs. 52–55.

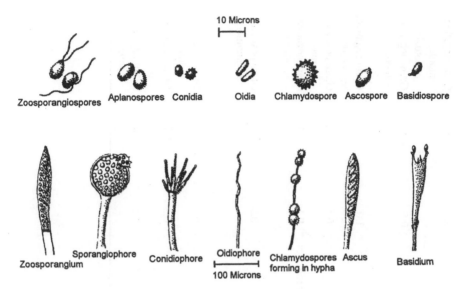

Figure 43 Different types of spores in fungi. (From Ref. 70.) Refer to Table 5 for other details of pathogenic fungi.

The phi-X174 bacteriophage could potentially be used as a conservative model for a variety of other viruses (other than HBV, HCV, and HIV) and larger pathogenic microorganisms.

K. Liquid-Borne Microbial Challenge Testing

Historically, the hydrostatic pressures used to evaluate the liquid-borne microbial barrier properties of textiles have been very low. Most of the work in this area was done in an effort to try to reduce the likelihood of postoperative wound infection rates associated with bacteria. Therefore, the laboratory devices were designed to simulate the pooling of fluids on surgical drapes. Figure 44 illustrates the GORE diffusion technique of bacterial challenge testing [65]. Techniques such as this were proposed for standardization; however, as previously mentioned, no consensus could be reached [33]. This test was used to assess the bacterial barrier properties of textiles in both surgical and clean-room applications, becoming recognized in the industry and adopted for use by such prestigious organizations as the Shirley Institute and the University of Massachusetts [66,67]. However, like other bacterial challenge test devices of that period of time, the GORE diffusion technique was designed to apply only minimal hydrostatic pressure.

Recently, a new and highly stringent test method has been developed and approved as an emergency standard test method by the American Society for Testing and Materials (ASTM). This effort was undertaken by ASTM in response to questions regarding the performance of personal protective clothing posed in the Announced Notice of Public Rulemaking by the U.S. Department of Labor, Occupational Safety and Health Administration (OSHA), concerning occupational exposure to bloodborne pathogens. The objective was to develop a test method that took the most important variables, relevant to bloodborne pathogen exposure, into consideration and provided a higher level of assurance in the barrier qualities of protective clothing for health-care workers [24].

The following ASTM emergency standard test method, ASTM ES22-92 (Emergency Standard Test Method for Resistance of Protective Clothing Materials to Penetration by Bloodborne Pathogens Using Viral Penetration as a Test System) uses a specific time and pressure protocol that is identical to the ASTM ES21-92 test (which has been highly correlated to the GORE elbow lean test). ASTM ES22-92 also uses appropriate body liquid and microbial models to overcome some of the most significant weaknesses that were outlined in the industry standard liquid challenge test methods. The ASTM ES22-92 procedure can be used alone or as a confirmatory test for those textile barriers which pass ASTM ES21-92. ASTM ES21-92 can be used to prescreen textiles, eliminating those textiles that are easily penetrated.

The ASTM ES22-92 is illustrated in Figure 45. This test is designed to determine if textiles can prevent the penetration of liquid-borne viruses under defined test conditions. This method has been specifically designed for modeling the viral penetration of hepatitis (B and C) and the human immunodeficiency viruses. The utility of the method for other purposes (modeling the penetration characteristics of other microbes) must be assessed on a case-by-case basis. The microorganism that is utilized is the phi-X174 bacteriophage at a minimal challenge concentration of 1×10^8 PFU/ml. The bacteriophage is suspended in a liquid media with a surface tension that simulates the lower end of the surface tension range for blood and body fluids, 0.042 N/m.

To perform the test, textile specimens are clamped into the penetration cell with the face side oriented toward the cell cavity. Then the penetration cell is attached to the apparatus and the cell cavity is filled with 60 ml of phi-X174 bacteriophage challenge suspension. An air line is connected to the cell and a specific time and hydrostatic pressure protocol is followed: 5 min at ambient pressure, 1 min at 13.8 kPa, and 54 min at ambient pressure. The back side of the textile specimen is observed through the viewing port. The test is terminated if there are any visible signs of liquid penetration through the textile or at 60 min. At the end of the test a very sensitive microbial assay is performed to determine if viruses penetrated through, even in the absence of visible liquid penetration. Any evidence of viral penetration constitutes a failure. Examples of failing and passing results are illustrated in Figures 46 and 47.

Textiles that pass the ASTM ES22-92 test are considered to be highly protective against liquid and microbial penetration. Textile barriers that are not

Table 5   Characteristics of Pathogenic Fungi

Fungal Organisms	Associated disease(s)	Sporophore or spore type(s)	Smallest spore diameter (μm)	Shortest spore length (μm)	Other important factors
Absidia spp.	Zygomycosis	Sporangiosphore	N/A	N/A	
Alternaria spp.	Skin nodules, lesions	Conidiophore	N/A	N/A	
Aspergillus spp.	Aspergillosis; pneumomycosis, bronchmycosis; pulmonary infiltration	Conidiophore	N/A	N/A	Mycotoxin, carcinogenic
Blastomyces dermatitidis	North American blastomycosis, Gilchrist's disease	Conidia, budding cells	3.0 5.0	N/A N/A	
Candida spp.	Candidiasis, endocarditis (involving prosthetic devices)	Blastospore, chlamydospore, budding cells	4.0 N/A 4.0	6.0 N/A N/A	
Chaetomium spp.	Onchomycosis	Ascus	5.0	8.5	
Cladosporium spp.	Subcutaneous phaeohyphomycosis and possible systemic infection, chromoblastomycosis	Conidiophore	8.0	N/A	
Coccidioides immitis	Coccidioidomycosis, San Joaquin Valley fever	Arthrospore, spherules with endospores	2.0	N/A	

Organism	Disease	Morphology			
Cryptococcus neoformans	European blastomycosis, cryptococcosis, cryptococcal meningitis	Budding cells with capsules	2.5	N/A	
Fusarium spp.	Localized infections of skin, implicated in esophageal cancer	Conidiophore	N/A	N/A	Mycotoxin
Histoplasma capsulatum	Histoplasmosis, Darling's disease	Microconidia, macroconidia, budding cells	2.0 7.0 1.5	N/A N/A 3.0	
Mucor spp.	Zygomycosis	Sporangiophore	N/A	N/A	
Penicillium spp.	Generalized reticulosis, subcutaneous mycosis	Conidiophore	N/A	N/A	Mycotoxin
Sporothrix schenckii	Sporotrichosis	Conidia, cigar-shaped budding cells	3.0	6.0	
Trichophyton spp.	Skin infections, athlete's foot	Chlamydospore, microconidia, macroconidia	2.0 5.0 N/A N/A	6.0 N/A N/A N/A	

Source: Refs. 51, 52, and 55.

Table 6 Comparison of Possible Models for Human Pathogenic Viruses

Class of model	Type of model	Advantage(s)	Disadvantage(s)	Other factors to consider
Microbial	Actual pathogen	True measure	Health hazard, long analysis time, expense	Viability, compatibility
	Bacteriophage (phi-X174)	Microbe (virus), spherical shape, 0.027 μm in size, high titer possible, robust/stable, no health hazard, short analysis time, LOD (limit of detection) ≈ 1.0 PFU/ml		Viability, compatibility, nonmotile, no envelope
Biological; nonmicrobial	Antibodies (ELISA)	Short analysis time	Not a microbe, development cost, high LOD	Small size; permeability behavior
	Blood (hemoglobin)	Short analysis time, refined method, LOD ≈ 5–20 cell/ml	Not a microbe, large size of red blood cell	
Particles	Latex spheres (laser diffraction)	Short analysis time, variety of sizes; 0.014 μm and larger (ultraclean)	Not a microbe, development cost, high LOD	Penetration behavior
Chemicals	Indicators and dyes	Short analysis time	Not a microbe, high LOD	Small size; permeability behavior
Gases	Helium Nitrogen	Short analysis time	Not a microbe, high LOD	Small size; permeability behavior

Source: Refs. 11, 56–64.

Figure 44 The GORE diffusion technique of bacterial challenge testing. This test device serves as an example of the state of the art in bacterial challenge testing in the 1980s. The textile sample acts as a partition between the two halves of the cell, with the left side containing the bacterial challenge (normally *Escherichia coli* at 10^8 CFU/ml) and the right side containing sterile detector media. A low hydrostatic pressure (~ 1.0 kPa) was placed on the bacterial challenge media in left side of the cell, and at the end of a predetermined time interval the detector media in the right side of the cell was assayed for the presence of the challenge bacteria.

film-reinforced typically fail and textile barriers that are film reinforced (with high quality continuous films) typically pass the ASTM ES22-92 test [2,3,24,26]. (Those textiles found in Figs. 3–10 consistently demonstrate failing results and those textiles found in Figs. 35 and 36 consistently demonstrate passing results.)

Two ASTM emergency standard test methods (ASTM ES21-92 and ASTM ES22-92) were developed and published to meet a demand for more rapid issuance. The Executive Committee of ASTM Committee F23 on Protective Clothing recommended this publication, and the ASTM Committee on Standards concurred in the recommendation. These emergency standards differ from full

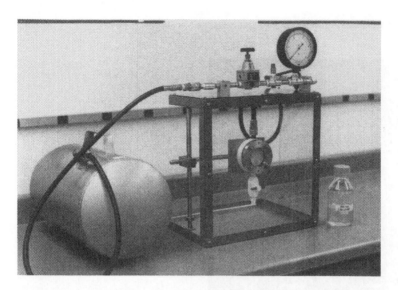

Figure 45 The resistance of textiles to viral penetration can be determined by ASTM ES22-92 (Emergency Standard Test Method for the Resistance of Protective Clothing Materials to Penetration by Blood-borne Pathogens Using Viral Penetration as a Test System). (Photo courtesy of Nelson Laboratories, Inc., Salt Lake City, UT.)

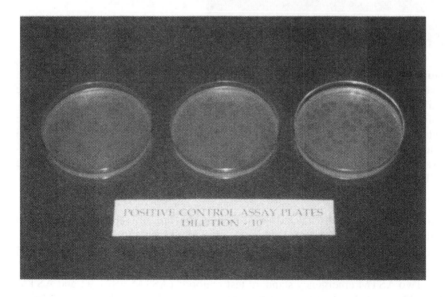

Figure 46 Failing test result for ASTM ES22-92. This is an example of the type of result typically exhibited by the positive control (0.04 μm) hydrophilic nylon microfiltration membrane). The clear spots, known as plaque-forming units, that are apparent in the surface of the lawn of *Escherichia coli* C (host bacterium) in the petri dishes were made by penetrating bacteriophage. (Photo courtesy of Nelson Laboratories, Inc., Salt Lake City, UT.)

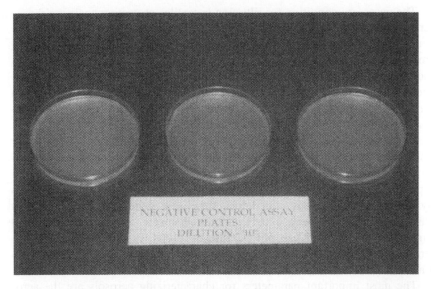

Figure 47 Passing test result for ASTM ES22-92. This is an example of the type of result typically exhibited by the negative control (medical packaging grade of monolithic non-porous polyester film). Note the absence of any plaque-forming units in the surface of the lawn of *Escherichia coli* C in the petri dishes. (Photo courtesy of Nelson Laboratories, Inc., Salt Lake City, UT.)

consensus standards in that they are only balloted through the subcommittee level as per the regulations governing ASTM technical committees. These documents were published by ASTM through August 1994. At the time of this writing, ASTM was in the process of developing full-consensus standard test methods for the evaluation of the resistance of protective clothing materials to synthetic blood and to bloodborne pathogens. As with any standard, these test methods may be revised to some extent as a result of this process.

L. Aerosol-Borne and Liquid-Borne Microbial Filtration Testing

1. The Mechanisms of Filtration

The process of filtration is complicated, and although the general principles are well known there is a substantial gap between theory and experiment. A common misconception is that aerosol filters work like microscopic sieves in which only particles smaller than the holes can get through. This view may be appropriate for the liquid filtration of solid particles, but it is not how aerosol filtration typically works [19]. An aerosol is defined as a suspension of solid or liquid particles in a gas that range in size from 0.001 to 100 μm in diameter. Microbes that are suspended in air are, by definition, an aerosol.

The necessary condition for filtration in textiles protecting against airborne microbiological agents is a pressure drop across the material causing air to flow through. If there is no air flow (as with monolithic nonporous films), the transport of microbes through the textile will not occur. (This assumes that molecular diffusion does not play a roll in transport.) Porous textiles and porous film reinforcements, with pores larger than the size of the aerosol particles, will depend on the mechanisms of filtration for determining the degree of separation of the biohazardous particles from the air passing through.

In considering airborne transmission of microbes, although it is certainly possible that individual microorganisms can be transported alone, it is generally agreed that in most cases the organisms are associated with larger aerosol particulates and droplets. However, the size of these particles and droplets may depend on the means with which they were generated. Recent studies have shown that the median aerodynamic diameter of particles generated during laser surgery is 0.3 μm, with a range of 0.1–0.8 μm [68]. This finding illustrates how important it is to investigate the nature of each microbiological hazard.

The most important parameters for characterizing aerosols are the aerodynamic diameter of the particles and the velocity of the particles in the air. Velocity is presumably very low in most cases of protection against biological agents, as the pressure drop on garments in most situations is thought to be very small. When the aerodynamic diameter of the aerosol is smaller than the pores through the textile, there are five basic mechanisms by which they can be stopped [19]:

1. Interception: Occurs when a particle follows an air streamline that happens to come within one particle radius of the surface of the fiber. The particle impacts the fiber and is captured.
2. Inertial impaction: Occurs when a particle, because of its inertia, is unable to follow abrupt changes in the flow of the air streamlines traveling around the fiber, crosses the streamline, and impacts the fiber. This mechanism is important for particles with a diameter >0.5 μm.
3. Diffusion: Occurs when the Brownian motion of the small particles (<0.5 μm in diameter) is sufficient to greatly enhance the probability of their impacting a fiber while traveling past in a nonintercepting streamline.
4. Gravitational settling: Occurs when the sedimentation velocity of a particle causes the trajectory of the particle to deviate from the air streamline and impact a fiber. This mechanism is important to particles with a diameter >3.0 μm.
5. Electrostatic attraction: Occurs when a charged particle is attracted to an oppositely charged fiber by coulombic attraction or when a charged particle induces an equal and opposite charge in an uncharged fiber surface

by image forces. This mechanism is important for particles with a diameter <3.0 μm.

Particles that impact a fiber surface must stick to the surface in order to be captured. Particles will be retained if the forces directing the particle away from the surface are smaller than the forces attracting the particle to the surface. Forces attracting the particle to the surface of the fiber are known as van der Waals intermolecular forces. Forces directing the particle away from the surface of the fiber are elastic energy stored when the particle impacts the fiber, macroscopic external forces (shaking, abrading, etc.), and the inherent mobility of the microorganisms. For particles with aerodynamic diameters <1.0 μm, the attractive forces are expected to overcome the dispersive forces.

In considering the filtration of liquid-borne microorganisms it is important to recognize that when microbes are suspended in a liquid medium they are surrounded by a boundary layer of liquid. This is one of the major differences that exists between airborne and liquid-borne microbial filtration testing. This boundary layer of liquid can nullify many of the capture mechanisms relied upon in the filtration of airborne particles. The predominant mechanism relied upon in liquid-borne microbial filtration is size exclusion (sieving). This means that the difference between the hydrodynamic diameter of the hazardous liquid-borne microbes and the pore size of the textiles or film reinforcements will be the major determining factor as to whether the microbes will be effectively retained. Therefore, porous textiles and porous film reinforcements that are intended to filter hazardous microorganisms out of a penetrating liquid stream should have the pore size well characterized. Pore size measurements should also be validated against performance in a liquid-borne microbial filtration test to demonstrate effective retention properties against the microbes of concern.

2. Filtration Testing

The aerosol-borne bacterial filtration efficiency of textiles can be determined by using the MIL-M-36954C (Military Specification; Mask, Surgical, Disposable) test illustrated in Figure 48. Although this test was originally designed for determining the protective qualities of masks against bacterial aerosols, it is currently being used to assess both the bacterial and viral (phi-X174 bacteriophage surrogate) filtration efficiencies of textiles for a variety of end-use applications. This test is designed to determine how many challenging microbes can penetrate through under defined test conditions. Textile samples are challenged with a microbial aerosol of a controlled concentration and a fixed air flow rate (28.3 L/min). After passing through the textile, the aerosol is separated according to aerodynamic size and is deposited onto agar surfaces in an Anderson particle fractionating viable sampler for enumeration. Results are recorded in terms of logarithmic reduction values (LRV). The higher the LRV, the higher is the filtration efficiency.

Figure 48   MIL-M-36954C (Military Specification; Mask, Surgical, Disposable) test device. This test is also known as the Anderson sampler test. (Photo courtesy of Nelson Laboratories, Inc., Salt Lake City, UT.)

The liquid-borne bacterial filtration efficiency of textiles can be determined by using ASTM F838-83 (Reapproved 1993) (Standard Test for Determining Bacterial Retention of Membrane Filters Utilized for Liquid Filtration). This test is normally employed to test the bacterial filtration characteristics of sterilizing liquid filtration membranes. ASTM F838-83 is designed to determine how many challenging organisms (*Pseudomonas diminuta*) can penetrate through under defined test conditions. Film samples are challenged with a known concentration of bacteria (to deliver 10^7 CFU/cm^2 of film surface area) and a maximum pressure differential (206 kPa) to deliver a controlled flow rate. The filtrate is collected for

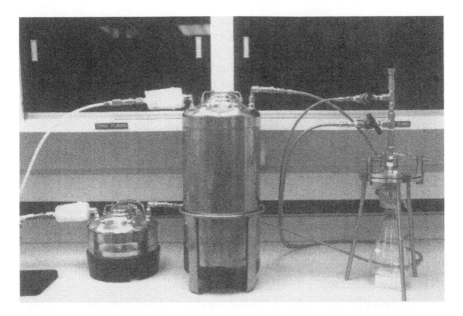

Figure 49 ASTM F838-83 (Reapproved 1993); Standard Test for Determining Bacterial Retention of Membrane Filters Utilized for Liquid Filtration. This test is also known as the HIMA challenge. (Photo courtesy of W. L. Gore & Associates, Inc., Flagstaff, AZ.)

enumeration, and the results are recorded in terms of logarithmic reduction values (LRV). The higher the LRV, the higher is the filtration efficiency.

IV. LIMITATIONS OF LABORATORY TEST METHODS

A. End-Use Conditions

Each intended end-use application for textiles in personal protection and infection control applications may impose different microbial, physical, chemical, and thermal stresses. Selecting a single laboratory test that simulates all of the possible combinations of these variables is practically impossible. Most of the laboratory tests that have been reviewed make a determination about the air, liquid, and microbial barrier properties of textiles under defined and controlled laboratory conditions. It is important to recognize the inherent limitations in each test, as well as the relationship between the laboratory test and the real world.

Most laboratory tests evaluate specimens (not whole end products) "in the condition in which they were received." Each end-use application will impose a distinct set of requirements, as there will very likely be exposures to different microbial, physical, chemical, and thermal stresses in each real-world environment. There are many examples of actual end-use stresses that should be considered when testing the barrier properties of textiles in the laboratory. Many of these stresses may be required as preconditioning options for barrier testing in order to make a more accurate prediction of actual performance in use. Some of the important variables that should be considered are:

1. Physical stresses: flexing, abrasion, puncture, tear, and tensile.
2. Chemical stresses: chemical resistance, chemical degradation, and contamination resistance; acids, bases, solvents, lubricants, disinfectants, surfactants, etc.
3. Thermal stresses: flame resistance, melt point, thermal insulation.
4. The impact of laundering (and sterilizing if required) should be assessed for multiple-use products. Process parameters that could accelerate the deterioration of the products should be identified (chlorine bleach, high temperature, high pH, long time, etc.)
5. The impact of sterilization (if required) should also be assessed for single-use products.
6. The impact of transportation, storage conditions (time, temperature, and relative humidity), and shelf life should be assessed. There may be geographical and seasonal fluctuations that will need to be considered.

B. Finished Products

Laboratory tests typically evaluate two-dimensional specimens and may not be indicative of the performance of finished products. Finished textile products are normally manufactured from roll goods by cutting and sewing (or gluing, fusing, etc.). The continuity of the two-dimensional structure of the textile is compromised as it is cut. When textile pieces are put together in a finished product, seams are created. If the seams are intended to possess the same barrier characteristics as the original textile, then the same laboratory tests that were performed on the textile should be performed on the seams.

Most end products are not used in a two-dimensional form. As an example, garments must conform to the human body, which requires that the design of closures (such as zippers) and interfaces (such as cuffs) also provide effective barriers. Sometimes end products are manufactured from different types and/or multiple layers of textiles. Each area of the end product, representing a different textile or layering, that is required to be a barrier will need to be evaluated.

C. Single-Use Versus Multiple-Use Products

Due to the demand for cost controls (or reductions) and the concern over the potential negative environmental impact of using single-use products (incineration and sterilized biohazardous solid waste), multiple-use products should be given serious consideration. Whether the multiple-use products are provided via contract services or processed for reuse in-house, there must be a reliable system of quality control in order to maintain barrier integrity. Appropriate systems would include dependable reprocessing controls, monitoring the number of uses (via check grids, bar codes, etc.), inspecting before reuse, barrier testing, repairing damaged items, and replacing worn out items.

D. Quality Assurance/Quality Control

Manufacturers of roll goods and manufacturers of finished products should recognize that proper statistical design and analysis of larger data sets than those specified in many of the laboratory test methods may be required in order to establish reasonable confidence limits concerning barrier properties. Many of the textile products intended for use in personal protection and infection control applications will be classified as medical devices and subject to the Medical Device Amendments of 1976 and the Safe Medical Devices Act of 1990. These products may also be subject to the FDA [510(k)] premarket notification and medical device reporting (MDR) regulations. In addition, good manufacturing practices (GMPs) may be necessary in the manufacture and reprocessing of these devices, and they may need to be labeled in accordance with FDA requirements.

V. CONCLUSION

Our society is becoming increasingly concerned about being protected from exposure to human pathogenic microorganisms. Many people are faced with having to make a decision about how to reduce the risk of exposure to potentially hazardous airborne and liquid-borne microbes every day. Textiles are often employed in an effort to reduce the risk of exposure by helping to avoid direct contact with the biohazards. Considering the variety of different types of textiles, types of end-use applications, and types of biohazards, the decision logic required to help reduce the risk may not be obvious.

Determining which textiles will perform adequately in each application against the biohazards of concern often involves the use of laboratory testing. Understanding the performance requirements of each personal protective clothing and infection control end use application is important in making this assessment. The list of technical attributes that results from the end-use survey will lead to the development of a strategy of experimentation in the laboratory. This attributes list often includes physical, chemical, and thermal performance requirements, as well as microbial barrier properties. Each personal protection and infection control application may require different strategies of experimentation to determine if the textiles will perform adequately against the anticipated

Table 7 Summary of Standardized Air, Liquid, and Microbial Challenge Tests

Class of test	Challenge type	Test method(s)	Important test variables
Air penetration resistance	Ambient air	ASTM D737-75 FTMS 191A method 5450 IST 70.1-92	Calibrated orifice (0.12 kPa ΔP)
	Ambient air	FTMS 191A method 5452	Falling cylinder
Porosity	Solvent/air	ASTM F316-86	Bubble point; maximum pore size determination
Contact angle	Various liquids	TAPPI T458 om-89	Sessile drops
Liquid repellency	Saline	IST 80.5-92	Direct contact 1.13 kPa hydrostatic)
	Water	AATCC 22-1989 FTMS 191A method 5526 IST 80.1-92	Spray/impact (15.2 cm height)
	Various Oils	AATCC 118-1992 IST 80.7-92	Sessile drops
	Alcohol	IST 80.6-92	Sessile drops
Liquid penetration resistance	Water	AATCC 127-1989 ASTM D751 EDANA 120.1 FTMS 191A method 5514 IST 80.4-92	Direct contact (0–9.8 kPa hydrostatic)
	Water	ASTM D751-89 FTMS 191A method 5512	Direct contact (0–6984 kPa hydrostatic)
	Water	AATCC 42-1989 FTMS 191A method 5522 IST 80.3-92	Spray/impact (61 cm height)
	Synthetic blood	ASTM ES21-92	Direct contact (0–13.8 kPa hydrostatic)
Bacterial penetration resistance	Dry particulate	EDANA 190.0-89	*Bacillus subtilis* (10^7 CFU, vibration)
	Liquid	EDANA 200.0-89	*Streptococcus faecalis* (rotating finger)
Viral penetration resistance	Liquid	ASTM ES22-92	Phi-X174 bacteriophage (10^8 PFU/ml, 0–13.8 kPa hydrostatic)
Bacterial filtration	Aerosol[a]	MIL-M-36954C 6/12/75 EDANA 180.0-89	*Staphylococcus aureus,* *S. aureus/epidermidis* (2200 CFU, 28.3 L/min)
	Liquid[a]	ASTM F838-83	*Pseudomonas diminuta* (10^7 CFU/cm^2, 206 kPa hydrostatic maximum)

Note: Acronyms are ASTM (American Society for Testing and Materials); AATCC (American Association of Textile Chemists and Colorists); EDANA (European Disposables and Nonwovens Association); IST (Association of the Nonwoven Fabrics Industry Standard Test); FTMS (Federal Test Method Standard); MIL (Military Specification); and TAPPI (Technical Association of the Pulp & Paper Industry).

[a]The protocols for these two tests can be modified in order to accommodate other types of bacteria. Viruses such as the phi-X174 bacteriophage can also be used in these tests.

microbiological hazards. Challenge testing in the laboratory should produce data that can be interpreted and used to help reduce the risk of microbial exposure or contamination during actual use.

Simulating each set of actual end-use conditions in the laboratory may not be possible. Nevertheless, risk reduction strategies may be developed based on simulated use testing and reasonable worst-case exposure modeling. Determining how the textile is constructed and why the textile should act as a microbial barrier may help in deciding what type of testing will be necessary. Depending on the structure of the textile (porous vs. nonporous), the behavior against airborne biohazards and liquid-borne biohazards may be different. Also, the failure mechanisms for these two types of constructions may be different.

Conclusions should never be drawn about the absolute microbial barrier properties of textiles based on air and liquid challenge testing. Microbial challenge testing is the only truly definitive measurement of performance. Significant progress has been made in the area of liquid-borne microbial challenge testing with the development of ASTM ES22-92 (Emergency Standard Test Method for the Resistance of Protective Clothing Materials to Penetration by Blood-borne Pathogens Using Viral Penetration as a Test System). The utility of this extremely sensitive method for determining which textiles are capable of resisting viral penetration has been rapidly recognized by a number of important agencies and incorporated into several influential standards, guidelines, and reports (NFPA 1999, NFPA 1973, FDA Draft Guidance on [510(k)] Submissions, AAMI TIR No. 11-1994, and CSA Z314.10; references for these documents can be found in the Appendix).

Other very important factors to consider in the risk reduction analysis are the performance of finished products (as opposed to roll goods), any special requirements for the maintenance of multiple-use products, and performing properly designed, statistically valid microbial barrier integrity testing. A concise summary of the various air, liquid, and microbial challenge tests is provided in Table 7.

VI. APPENDIX

A. Publications from Regulatory Agencies and Associations

Association for the Advancement of Medical Instrumentation (AAMI), AAMI TIR No. 11-1994—Technical Information Report, Selection of Surgical Gowns and Drapes in Health Care Facilities, AAMI, Arlington, VA.

Association for the Advancement of Medical Instrumentation (AAMI), AAMI Standards and Recommended Practices, Vol. 4, Biological Evaluation of Medical Devices, 1994, AAMI, Arlington, VA.

American Academy of Orthopaedic Surgeons (AAOS), Recommendations for the Prevention of Human Immunodeficiency Virus (HIV) Transmission in the Practice of Orthopaedic Surgery, 1989, AAOS, Park Ridge, IL.

American College of Surgeons (ACS), Statement on the Surgeon and HIV Infection, *Bulletin of the American College of Surgeons*, December 1991, 76(12):28–31, ACS, Chicago, IL.

American Dental Association (ADA), Statement of the American Dental Association on the OSHA Proposed Rule Regarding Occupational Exposure to Bloodborne Pathogens, September 21, 1989, ADA, Washington, DC.

American Industrial Hygiene Association (AIHA), AIHA Biosafety Manual, Chapter V, Personal Protective Equipment, 1994, AIHA, Fairfax, VA.

Association of Operating Room Nurses, Inc. (AORN), Recommended Practice for Protective Barrier Materials for Surgical Gowns and Drapes, October 17–19, 1993, AORN, Denver, CO.

Association for Practitioners in Infection Control, Inc. (APIC), Universal Blood and Body Substance Precautions, Specific Guidelines, Revised April, 1992, Barbara Hendrickson, Dublin, Augusta, KS, APIC, Washington, DC.

Association for Practitioners in Infection Control, Inc. (APIC), Resource List For Standards & Guidelines, 1993, APIC Guidelines Committee, APIC, Washington, DC.

American Public Health Association, *Control of Communicable Diseases in Man*, 15th Ed. (A. S. Beneson, ed.), 1990, APHA, Washington, DC.

American Society for Microbiology (ASM), *Laboratory Safety; Principals and Practices*, 2nd ed., 1995, ASM Press, Washington, DC.

American Society for Testing and Materials (ASTM), ES21-92: Emergency Standard Test Method for Resistance of Protective Clothing Materials to Synthetic Blood, 1992, ASTM, Philadelphia, PA.

American Society for Testing and Materials (ASTM), ES22-92, Emergency Standard Test Method for Resistance of Protective Clothing Materials to Penetration by Blood-Borne Pathogens Using Viral Penetration as a Test System, 1992, ASTM, Philadelphia, PA.

Canadian Standards Association (CSA), Final Draft of Standard Z314.10, Subcommittee on Reusable Textiles, May 4, 1993, CSA, Ontario, Canada.

Centers for Disease Control (CDC), Recommendations for Prevention of Transmission of Human Immunodeficiency Virus and Hepatitis B Virus to Patients During Exposure-Prone Invasive Procedures, 1991, CDC, Atlanta, GA.

Centers for Disease Control (CDC), National Institutes of Health (NIH), *Biosafety in Microbiological and Biomedical Laboratories*, 3rd ed., May 1993, U.S. Government Printing Office, Washington, DC.

International Society for Clinical Laboratory Technology (ISCLT), A Guideline to OSHA Requirements for Hospital, Independent, and Physician Office Laboratories, 1990, ISCLT, St. Louis, MO.

National Committee for Clinical Laboratory Standards (NCCLS), Document M29-P, Vol. 7, No. 9, November 1987, Protection of Laboratory Workers from Infectious Disease Transmitted by Blood and Tissue, NCCLS, Villanova, PA.

National Fire Protection Association (NFPA), NFPA 1999 Standard; Protective Clothing for Emergency Medical Operations, 1992 ed., NFPA, Quincy, MA.

National Fire Protection Association (NFPA), NFPA 1973 Standard; Gloves for Structural Fire Fighting; 1993 ed., NFPA, Quincy, MA.

National Research Council, *Biosafety in the Laboratory; Prudent Practices for the Handling and Disposal of Infectious Materials*, 1989, National Academy Press, Washington, DC.

NSF International (NSF), Class II (Laminar Flow) Biohazard Cabinetry, (NSF) 49, 1992, NSF, Ann Arbor, MI.

Textile Rental Services Association of America (TRSA), Guidelines for OSHA's Bloodborne Pathogens Standard Employers Manual, 1992, TRSA, Hallendale, FL.

U.S. Department of Health and Human Services (HHS), *Federal Register—* II Classification of Oncogenic Viruses on the Basis of Potential Hazard, 46(233), 1981, U.S. Government Printing Office, Washington, DC.

U.S. Department of Health and Human Services (HHS), CDC Guidelines for Isolation Precautions in Hospitals and CDC Guidelines for Infection Control in Hospital Personnel, part of the manual entitled *Guidelines for the Prevention and Control of Nosocomial Infections*, July 1983, National Technical Information Service (NTIS), Springfield, VA.

U.S. Department of Health and Human Services (HHS), *Morbidity and Mortality Weekly Report* (MMWR), Summary: Recommendations for Preventing Transmission of Infection with Human T-Lymphotropic Virus Type III/ Lymphadenopathy-Associated Virus in the Workplace, 34(45), 1985, CDC, Atlanta, GA.

U.S. Department of Health and Human Services (HHS), *Federal Register*—NIH Guidelines for Research Involving Recombinant DNA Molecules, 51(88), 1986, U.S. Government Printing Office, Washington, DC.

U.S. Department of Health and Human Services (HHS), *Morbidity and Mortality Weekly Report* (MMWR), Recommendations for Prevention of HIV Transmission in Health-Care Settings, 36(2S), 1987, CDC, Atlanta, GA.

U.S. Department of Health and Human Services (HHS), Guidelines for Protecting the Safety and Health of Health Care Workers, September 1988, NIOSH, Washington, DC.

U.S. Department of Health and Human Services (HHS), *Morbidity and Mortality Weekly Report* (MMWR), Guidelines for Prevention of Transmission of Human Immunodeficiency Virus and Hepatitis B Virus to Health-Care and Public-Safety Workers, 38(S-6), 1989, CDC, Atlanta, GA.

U. S. Department of Health and Human Services (HHS), *Federal Register*—Respiratory Protective Devices; Proposed Rule, 42 CFR Part 84, 1994, U.S. Government Printing Office, Washington, DC.

U. S. Department of Labor, OSHA, Memo, Subject: Enforcement Policy and Procedures for Occupational Exposure to Tuberculosis—Guidelines for Preventing the Transmission of Tuberculosis in Health-Care Settings with Special Focus on HIV-Related Issues, 39(RR-17), December 1990, U.S. Department of Labor, Washington, DC.

U.S. Department of Labor, OSHA, *Federal Register*—Occupational Exposure to Bloodborne Pathogens Final Rule, Part II, 29 CFR Part 1910.1030, December 6, 1991, U.S. Government Printing Office, Washington, DC.

U.S. Department of Labor, OSHA, *Federal Register*—Personal Protective Equipment for General Industry, 29 CFR Part 1910, 59(66), April 6, 1994, U.S. Government Printing Office, Washington, DC.

U.S. Food and Drug Administration (FDA), Draft Guidance on Premarket Notification [510(k)] Submissions for Surgical Gowns and Surgical Gowns and Surgical Drapes, August 1993, Center for Devices and Radiological Health, Infection Control Devices Branch, Division of General and Restorative Devices, FDA, Rockville, MD.

World Health Organization (WHO), *Laboratory Biosafety Manual*, 2nd ed., 1993, World Health Organization, Geneva, Switzerland.

B. Books

Chemical Safety Associates, *OSHA Bloodborne Pathogens Exposure Control Plan*, Chemical Safety Associates, San Diego, CA (Lewis Publishers), 1992.

C. H. Collins, *Laboratory-Acquired Infections*, 3rd ed., Butterworth-Heinemann Ltd., Oxford, 1993.

L. G. Donowitz, M.D., *Infection Control for the Health Care Worker*, Williams & Wilkins, Baltimore, MD, 1994.

Doan J. Hansen, ed., *The Work Environment; Healthcare Laboratories and Biosafety*, Vol. 2, Lewis Publishers, Boca Raton, FL, 1993.

Infection Control for Prehospital Care Providers, 2nd edition, Mercy Ambulance, Grand Rapids, MI, 1993.

Safety Equipment Institute, *Certified Product List; Personal Protective Equipment*, Safety Equipment Institute, Arlington, VA, 1994.

Universal Precautions Policies, Procedures and Resources, American Hospital Publishing, Inc., Chicago, IL, 1991.

REFERENCES

1.   ASTM D1518-85, Standard Test Method for Thermal Transmittance of Textile Materials, American Society for Testing and Materials, Sept. 1985, Philadelphia, PA.

2. E. A. McCullough and L. K. Schoenberger, Liquid Barrier and Thermal Comfort Properties of Surgical Gowns, IER Report 90–07A, September 1990, Kansas State University, Manhattan, KS.

3. E. A. McCullough and M. K. Song, Liquid and Thermal Comfort Properties of Reusable and Disposable Gowns, Technical 92-13, July 1992, Kansas State University, Manhattan, KS.

4. T. L. Vigo, Antibacterial fiber treatments and disinfection, *Text. Res. J. July*: 454–465 (1981).

5. R. J. Good, Contact angle, wetting, and adhesion: A critical review, *J. Adhesion Sci. Technol.* 6(12):1269–1302 (1992).

6. G. M. Olderman, Liquid repellency and surgical fabric barrier properties, *Eng. Med* 13(1):35–43 (1984).

7. S. S. Block, *Disinfection, Sterilization, and Preservation*, 4th ed., Lea & Febiger, Philadelphia, PA, 1996.

8. R. C. Weast, M. J. Astle, and W. H. Beyer, eds., *CRC Handbook of Chemistry and Physics*, 69th ed., CRC Press, Boca Raton, FL, 1988.

9. C. Lentner, ed., *Geigy Scientific Tables*, Vol. 1, *Units of Measurement, Body Fluids, Composition of the Body, Nutrition*, Medical Education Division, Ciba-Geigy Corporation, West Caldwell, NJ, 1981.

10. C. Lentner, ed., *Geigy Scientific Tables*, Vol. 3, *Physical Chemistry, Composition of Blood, Hematology, Somatometric Data*, Medical Education Division, Ciba-Geigy Corporation, West Caldwell, NJ, 1984.

11. Internal Research, W. L. Gore & Associates, Inc., Elkton, MD.

12. D. H. Bangham and R. I. Razouk, *Trans. Faraday Soc.* 33:1459, 1463 (1937); *Proc. R. Soc. London Ser. A* 166:572 (1938).

13. R. J. Good and C. J. van Oss, The modern theory of contact angles and the hydrogen bond components of surface energies, *Modern Approaches to Wettability: Theory and Applications* (M. E. Schrader and G. I. Loeb, eds.), Plenum Press, New York, 1992, pp. 1–27.

14. O. M. Olderman, Liquid repellency of surgical barrier materials, *J. Ind. Fabrics* 3(2):30–41 (1984).

15. A. W. Adamson, *Physical Chemistry of Surfaces*, Interscience Publishers, New York, 1960, pp. 264–279.

16. E. W. Washburn, The dynamics of capillary flow, *Physical Review XVII*(3) (1921).

17. H. G. Heilweil, Measuring fabric air permeabilities over broad pressure drop ranges, *Notes on Research* 465, Textile Research Institute, Princeton, NJ, 1992, pp. 273–283.

18. P. W. Gibson, Factors influencing steady state heat and water vapor transfer measurements for clothing systems, *Text. Res. J.* 63(12):749–764.

19. W. C. Hinds, *Aerosol Technology: Properties, Behavior and Measurement of Airborne Particles*, John Wiley and Sons, New York, 1982.

20. T. Shikata, T. Karasawa, K. Abe, T. Uzawa, H. Suzuki, T. Oda, M. Imai, M. Mayumi, and Y. Moritsugu, Hepatitis B e Antigen and infectivity of hepatitis B virus, *J. Infect. Dis.* 136(4), (1977), pp. 571–576.

21. A. G. Wedum, W. E. Barkley, and A. Hellman, Handling of infectious agents, *J.A.V.M.A.* 161(11):1557–1567 (1972).

22. J. J. Lamb and T. E. Ferari, Review of Test Methods for Wet Penetration of Aseptic Barriers, Draft Report, DHHS, FDA, April 1981.
23. J. T. Schwartz and D. E. Saunders, Microbial penetration of surgical gown materials, *Surg. Gynecol. Obstet. 150* (1980), pp. 507–512.
24. P. L. Brown, Protective clothing for health care workers: Liquidproofness versus microbiological resistance, *Performance of Protective Clothing: Fourth Volume, ASTM STP 1133* (J. P. McBriarty and N. W. Henry, eds.), American Society for Testing and Materials, Philadelphia, PA, 1992, pp. 65–82.
25. J. Pournoor, The Effect of Pressure Profiles on Measuring the Barrier Effectiveness of Medical Textiles, Presentation at the Clemson Medical Textile Conference, Greenville, SC, September 1994.
26. E. A. McCullough and A. Fosha, A Comparison of Different Pressure Methods for Measuring the Barrier Properties of Surgical Gowns, IER Report 94-11, Kansas State University, Manhattan, KS, December 1994.
27. E. A. McCullough and L. K. Schoenberger, A comparison of methods for measuring the liquid barrier properties of surgical gowns, *Performance of Protective Clothing: Fourth Volume ASTM STP 1133* (J. P. McBriarty and N. W. Henry, eds.), American Society for Testing and Materials, Philadelphia, PA, 1992, pp. 83–98.
28. ASTM F23.40 Biological Resistance Subcommittee Meeting, San Antonio, TX, January 21, 1993.
29. U.S. Patent 4,382,990, Coating Composition for Fibrous Polyolefin Sheets, E. I. duPont de Nemours & Company, Wilmington, DE, May 10, 1983.
30. W. L. Gore & Associates, Inc., Elkton, MD.
31. K. W. Altman, et al., Transmural surgical gown pressure measurements in the operating theater, *Am. J. Infect. Control 19*(3):147–155 (1991).
32. B. Miller, The penetration of liquids into fiber networks, *J. Appl. Polym Sci. Appl. Polym. Symp. 47*:403–415 (1991).
33. W. C. Beck, M. H. Meeker, and G. M. Olderman, Demise of Aseptic Barrier Committee: Success and failure, *AORN J. 38(3)*:384–388 (1983).
34. J. A. Moylan, K. T. Fitzpatrick, and K. E. Davenport, Reducing wound infections, *Arch. Surg. 12*:152–157 (1987).
35. U.S. Department of Labor, Occupational Safety and Health Administration, Occupational Exposure to Bloodborne Pathogens: Final Rule, 29 CFR Part 1910.1030, *Fed. Reg.* December 6 (1991).
36. G. L. Telford and E. J. Quebbeman, Assessing the risk of blood exposure in the operating room, *Am. J. Infect. Control 21*(6):351–356.
37. M. C. White and P. Lynch, Blood contact and exposure among operating room personnel: A multicenter study, *Am. J. Infect. Control 21*(5):243–248 (1993).
38. S. Popejoy and D. Fry, Blood contact and exposure in the operating room, *Surg. Gynecol. Obstet. 172*(6):480–483 (1991).
39. D. Bell, Human immunodeficiency virus transmission in health care settings: Risks and risk reduction, *Am. J. Med. 91*(Suppl. 13B):249s–300s (1991).
40. G. Kelen, T. DiGiovanna, L. Bisson, D. Kalainov, K. T. Sivertson, and T. C. Quinn, Human immunodeficiency virus infection in emergency department patients: Epidemiology, clinical presentations and risk to health care workers—The Johns Hopkins Experience, *J. Am. Med. Assoc. 262*:516–522 (1989).

41. Staff Article, Study of surgical patients finds high rate of HIV, HBV, and HCV, *Hosp. Infect. Control December*:171–172.
42. C. O. Williams, S. Campbell, K. Henry, and P. Collier, Variables influencing worker compliance with universal precautions in the emergency department, *Am. J. Infect. Control* 22(3):138–148.
43. J. Dutkiewicz, L. Jablonski, and S. Olenchock, Occupational biohazards: A review, *Am. J. Ind. Med. 14*:605–623 (1988).
44. P. P. Shadduck, D. S. Tyler, H. K. Lyerly, M. W. Sebastian, C. Farnitano, K. T. Fitzpatrick, A. J. Langlois, and J. A. Moylan, Commercially available surgical gowns do not prevent penetration by HIV-1, *Surg. Forum 41*: 77–80 (1990).
45. H. R. Kotilainen, W. H. Cyr, W. Truscott, N. M. Gantz, L. B. Routson, and C. D. Lytle, Ability of 1000 ml water leak test for medical gloves to detect gloves with potential for virus penetration, *Performance of Protective Clothing: Fourth Volume, ASTM STP 1133* (James P. McBriarty and Norman W. Henry, eds.), American Society for Testing and Materials, Philadelphia, PA (1992).
46. C. D. Lytle, Personal communication, FDA, Rockville, MD (1990).
47. D. W. Bradley, Transmission, etiology, and pathogenesis of viral hepatitis non-A, non-B in non-human primates, *Advances in Hepatitis Research* (F. Chisari, ed.), Masson, New York 1984, pp. 268–280.
48. D. D. Ho, M. S. Moudgil, and M. Alam, Quantitation of human immunodeficiency virus type 1 in the blood of infected persons, *N. Eng. J. Med. 321*(24):1621–1625 (1989).
49. D. O. White and F. J. Fenner, *Medical Virology*, Academic Press, Orlando, Fl., 1986.
50. D. O. White and F. Fenner, *Medical Virology*, Academic Press, Orlando, FL, 1994.
51. S. L. Gorbach, J. G. Bartlett, and N. R. Blacklow, eds., *Infectious Diseases*, W. B. Saunders Company, Philadelphia, PA, 1992.
52. C. Lentner, ed., *Geigy Scientific Tables*, Volume 6, *Bacteria, Fungi, Protozoa, Helminths*, Medical Education Division, Ciba-Geigy Corporation, West Caldwell, NJ, 1992.
53. N. R. Krieg and J. G. Holt, *Bergey's Manual of Systematic Bacteriology*, Vol. 1, Williams & Wilkins, Baltimore, MD, 1984.
54. P. H. A. Sneath, N. S. Mair, M. E. Sharpe, and J. G. Holt, *Bergey's Manual of Systematic Bacteriology*, Vol. 2, Williams & Wilkins, Baltimore, MD, 1986.
55. Q. N. Myrvik, N. N. Pearsall, and R. S. Weiser, *Fundamentals of Medical Bacteriology and Mycology*, Lea & Febiger, Philadelphia, PA, 1978.
56. C. D. Lytle et al., Important factors for testing barrier materials with surrogate viruses, *Appl. Environ. Microbiol. September*:2549–2554 (1991).
57. C. D. Lytle et al., Filtration sizes of human immunodeficiency virus type 1 and surrogate viruses used to test barrier materials, *Appl. Environ. Microbiol. February*:747–749 (1992).
58. G. Spies, Presentation to ASTM F23.4 Subcommittee, Monsanto Company, St. Louis, MO, January 22, 1990.
59. J. R. Nelson, Personal Communication, Nelson Laboratories, Inc., Salt Lake City, UT (1987).
60. C. D. Lytle and W. H. Cyr, Surrogate viruses for testing barrier materials, Poster at the Fourth International Symposium on the Performance of Protective Clothing, June 20,

Montreal, Quebec, Canada. From Food and Drug Administration, Rockville, MD. and W. Truscott, Baxter Healthcare Corp., Irwindale, CA, 1991.

61. D. Korniewicz, Protective attire: How safe are we? Presentation at AORN Congress, April 11, Atlanta, GA, 1991. From Johns Hopkins University School of Medicine, Baltimore, MD.

62. M. L. Rapp, T. Thiel, and R. J. Arrowsmith, Model system using coliphage phi-X174 for testing virus removal by air filters, *Appl. Environ. Microbiol. March*:900–904 (1992).

63. G. E. R. Lamb, Use of ambient aerosols for measuring filter efficiencies, *Notes on Research*, November, TRI, Princeton, NJ, 1993.

64. Interfacial Dynamics Corporation, Ultraclean Uniform Latex Microspheres, 7/94 catalog.

65. P. L. Brown, Bacterial Challenge testing: Diffusion Technique, Research Report, March 24, W. L. Gore & Associates, Inc., Elkton, MD, 1982.

66. Shirley Institute, Tests of Effectiveness of GORE-TEX® Fabric as a Bacterial Battier, Didsbury, Manchester, February 10, Ref. No. 3081666, 1983.

67. B. Y. Litsky, Evaluation of Bacterial Filtration Properties of Surgical Gown Materials, University of Massachusetts, Amherst, MA, February 8, 1984.

68. A. Weber, K. Willeke, R. Marchioni, T. Myojo, R. McKay, J. Donnelly, and F. Liebhaber, Aerosol penetration and leakage characteristics of masks used in the health care industry, *Am. J. Infect. Control* 21(4):167–173 (1993).

69. Joklik and Willett, *Zinsser Microbiology*, 16th ed., Prentice Hall, Englewood Cliffs, NJ, 1976.

70. McGraw-Hill Encyclopedia of Science and Technology, Vol. 5, McGraw-Hill, New York, 1971, p. 117.

Index

551